A comprehensive and multi-faceted guide to ~~Hollick brilliantly guides the reader intellect varied terrains of the sciences, psychology, phi up a vibrant picture that amounts to a new vi.~~ ___ *21st century.* *A veritable tour de force.* **David Lorimer,** Programme Director, Scientific and Medical Network

This is a landmark book. It succeeds in carrying out a careful, grounded and thoroughly accessible critique of science, of an immense scope, while at the same time carrying its readers forward in their intuitive and spiritual knowing, through a series of reflective practices. Not stopping there, in its final chapters it builds a powerful vision of a future spirituality which can worthily complement the scientific way of knowing. By presenting the current ferment of spiritual and scientific thought, organising it into a coherent scheme and making it comprehensible to the general reader, Malcolm Hollick makes a unique and vital contribution to the current pivotal epoch in human development. **Chris Clarke,** Visiting Professor in Applied Mathematics, Southampton University, UK

An extremely impressive survey of recent scientific advances, clearly written, and accessible for the layperson. Informative, wide-ranging and very readable, and should appeal to those who have a scientific background, as well as those who are interested in spirituality, yet find modern scientific writing daunting. **George Chryssides,** Senior Lecturer in Religious Studies, University of Wolverhampton

Malcolm Hollick's The Science of Oneness *is a wonderful book. It is a comprehensive study of the nature and limitations of our science and how it can help us connect with the sacred reality of Spirit underlying all manifestation. Providing a sound theoretical basis, the author also gives practical steps for an inner and outer journey. Enjoyable reading. Highly recommended.* **Ravi Ravindra,** author of *Science and the Sacred,* Professor Emeritus at Dalhousie University in Halifax, Nova Scotia

This hits the nail on the head. We need a worldview that is integrated and shows that we are all parts of a greater whole – and that this view has its foundations in science, and is not mere speculation. I hope this will find the broad readership it deserves – and that we all deserve if enough people are to understand and embrace the holistic worldview to make a real difference in our crisis-racked and dangerous unsustainable world. **Ervin Lazslo,** Founder and President of the Club of Budapest, Visiting Professor at Stuttgart University

The Science of Oneness is an enjoyable journey through the challenges of frontier science as we confront the mysteries of a cosmos far more extensive, grand, mystical and conscious than previously realized. **Edgar Mitchell**, Sc.D, astronaut

This clearly written and accessible book is a breathtaking review of science from physics and cosmology to the functioning of the brain. Woven throughout is an equally broad appreciation of diverse forms of spiritual wisdom, both ancient and modern. This feast for the intellect is supplemented by carefully structured exercises, questions, and inspiring quotations that guide the reader into an inner realization of the unity that glows in every page of the book. I recommend it to those seeking wholeness and purpose in our seemly fragmented world. **Vic Mansfield**, Professor of physics and astronomy, Colgate University, and author of *Head and Heart: A Personal Exploration of Science and the Sacred.*

This is a fine and impressive book. Fine because Dr Hollick presents us with a cogent summary across a broad range of fields in science and probes a number of leading edges; he writes as one who is thoroughly familiar with his subject matter. Impressive because he directs our attention towards a viable science of oneness and shows us how a scientist can sustain the rigour of academic training and use it to good effect in examining spirituality, emotional literacy and experiential learning. As you might imagine from the broad canvas, this book will challenge you. But it is well written and Dr Hollick takes care to find ways to illustrate the areas he covers with analogies that help the lay reader's grasp of the topic. The Science of Oneness also offers us a fresh approach to combining rational analysis with relating the material to one's own subjective experience and understanding. I thoroughly recommend it. **Joycelin Dawes**, Co-Author and Editor of *The Quest*

With The Science of Oneness Malcolm Hollick takes us on an exciting and informative journey though the universe from moment of the Big Bang to the mysteries of human consciousness, from quantum entanglement to chaos and self-organization, from the nature of time to the evolution of life, and from ways of knowing to the mysteries of our own existence. A very rewarding journey. **David Peat**, Pari Center for New Learning

The Science of Oneness *is an invitation to a participatory, intimate relationship with all others and with the natural world itself. Unless a oneness with the Earth is realized, we may not be capable of mustering either the wisdom or the desire required to insure our future as a species. Malcolm Hollick's vision is the right medicine, at the right time, for the right reasons. This compelling book should be widely read.* **Larry Dossey, MD**, author of *The Extraordinary Healing Power of Ordinary Things*

The Science of Oneness

A Worldview for the Twenty-First Century

Malcolm Hollick

BOOKS

Winchester, U.K.
New York, U.S.A.

First published by O Books, 2006
An imprint of John Hunt Publishing Ltd., The Bothy,
Deershot Lodge, Park Lane, Ropley, Hants, SO24 0BE, UK
office@johnhunt-publishing.com
www.o-books.net

USA and Canada
NBN
custserv@nbnbooks.com
Tel: 1 800 462 6420
Fax: 1 800 338 4550

Singapore
STP
davidbuckland@tlp.com.sg
Tel: 65 6276
Fax: 65 6276 7119

Australia
Brumby Books
sales@brumbybooks.com
Tel: 61 3 9761 5535
Fax: 61 3 9761 7095

South Africa
Alternative Books
altbook@global.co.za
Tel: 27 011 792 7730
Fax: 27 011 972 7787

Text copyright Malcolm Hollick 2006

Design: Jim Weaver Design
Cover design: Book Design, London

ISBN-13: 978 1 905047 71 0
ISBN-10: 1 905047 71 1

All rights reserved. Except for brief quotations in critical articles or
reviews, no part of this book may be reproduced in any manner without
prior written permission from the publishers.

The rights of Malcolm Hollick as author have been asserted in accordance
with the Copyright, Designs and Patents Act 1988.

A CIP catalogue record for this book is available from the British Library.

Printed by Maple Vail Press, USA

To Christine

Contents

Acknowledgements

This book reflects the whole of my life's journey so far, and hence I am indebted to every person and every experience that has helped to shape me. My journey has involved far more people than I can recall, let alone thank individually, and I can mention only a few of the most significant here.

Without my parents, I would not exist, and I would not be who I am. And without my wife Christine's loving support, encouragement and patience I might never have finished the manuscript.

My successive Heads of Department at the University of Western Australia, particularly Professor Jörg Imberger, respected and supported my academic freedom. They gave me the space to gather much of the material and write the first version, even though I am sure they were often bemused by my intellectual wanderings.

Many people helped me discover my inner self and gave me the courage to move outside the academic straitjacket. In particular, thank you Loretta, Tarquam, Elizabeth and other teachers, friends and colleagues in The Australian Transpersonal Institute; Lizzie, my circle dance teacher; and all my friends and colleagues in the Findhorn Community and the University for Spirit Forum.

Especial thanks go to George Chryssides, David Lorimer, Ravi Ravindra, Edgar Mitchell, Ervin Laszlo, Vic Mansfield, Chris Clark, Joycelin Dawes, Larry Dossey, William Bloom and David Peat for spending precious hours reviewing the manuscript and writing endorsements.

Prologue

I embarked on the journey of this book because I wanted to help make the world a better place. Along the way my life was changed by what I discovered.

After decades studying the issues of environmental destruction, poverty and war, I concluded that the foundations for a better future are missing. We are at a time in the history of civilization when the certainties of the old worldview have disintegrated, and the concrete of a new worldview has not yet set. A key task, it seems to me, is to lay new philosophical and spiritual foundations on which we can build our future. New technologies, policies and laws are vital, but they will fail without strong foundations. Nothing less than radical change in our understanding of the universe, and of the meaning and purpose of life will do.

That's all very well, you may respond, but we can't simply go out and invent a new worldview. Unlike technologies, the myths, the *zeitgeist* of an age cannot be invented but emerge from the process of history. But history can be helped along by ideas. And our personal choice of worldview can contribute to the larger social movement towards a new mythology for our age. So this book is not just another work of popular science. It also reflects my own search for understanding and meaning, and is my contribution to the emergence of a new worldview.

Over the last few centuries the scientific revolution has transformed our beliefs, values and ways of life. It has brought a new awareness that we live in a global village, a new understanding of the universe and our place within it, and magical technologies that have made us richer than the greatest emperors of old. And, in its arrogance and lack of wisdom, science has brought our planet to her knees, and replaced traditional religions with what is often called Scientism. Scientists are the new priesthood, the oracles consulted on every issue who are listened to with a respect bordering at times on awe. Scientific ideas have penetrated deeply into every area of thought and life, but, as we shall see, the worldview from which they flow is often that of the old gods Newton and Descartes. The insights of more modern scientific heroes demand of us a different way of seeing the universe, and a different understanding of the meaning and purpose of our lives.

The entrenchment of science means that the emerging worldview will be based on scientific ideas and methods, or at least will make sense in scientific terms. So I set out to explore what was happening at the cutting edges of science in order to create a coherent picture of the universe and our relationship to it that could form the basis of a more just, sustainable and peaceful future. What I found exceeded my wildest expectations, leading me on a personal journey that transformed my beliefs and life. Rather than the mechanical, alienating, meaningless vision of reality bequeathed to us by classical science, I found a living, conscious, interconnected, meaningful and purposeful universe of which we are co-creators. And I found a vision which resonates strongly with the beliefs and wisdom of many ancient spiritual traditions throughout the world.

Mainstream science denies the validity of knowledge based on faith, and yet science itself, like all worldviews, rests on unprovable beliefs and assumptions about the nature of reality and truth. The science of oneness is no different. As we shall see, its basic tenets include a belief, supported by much evidence, that ultimate Reality is Consciousness or Spirit, not matter or energy as science currently holds.

Scientists often claim that scientific knowledge is neutral and value free. By contrast, the science of oneness is unashamedly rooted in spiritual values such as love, compassion and wisdom. And it sees critical subjectivity as an essential complement to objectivity in the elucidation of inner aspects of reality. But unlike religions, the science of oneness is non-dogmatic. It draws inspiration from all disciplines and faiths, is open to any alternative idea supported by evidence and reason, and lays no claims to *The Truth*. Thus, the science of oneness steers a middle course between the opposing dogmatisms of religion and science.

In exploring the emerging worldview, I learned a lot about the nature of knowledge. My intellectualism became balanced by an appreciation of experiential, intuitive, spiritual and other ways of knowing. I began to understand the limitations of human knowledge, including science, and to look for ways to test the reliability of what we think we know that complement the scientific method. So the science of oneness is more than a coherent scientific portrait of reality; it is also a holistic way of knowing that comes from personal experience, intuition and an inner sense of rightness as well as from reason and modern science.

As already stated, my aim is not just to satisfy intellectual curiosity, but to change lives and the world. Hence, this book goes beyond clear, simple explanations of the findings of science to explore ways of perceiving and interpreting reality that have the potential to affect personal, community and planetary futures. If this book is to fulfil my aims, it is important that you, my readers, engage critically with the ideas presented. Many of the concepts may be new to you, and you may feel ill-equipped to evaluate them. But you have your own life experience, imagination, and conceptual frameworks by which you make sense of the universe. I encourage you to bring all these, to bring your whole self, to bear on the material. Question and challenge it, and reinterpret its meaning where appropriate. Seek what makes coherent sense of the whole of reality, not just its separate parts; and seek what makes sense physically, emotionally and spiritually as well as intellectually. Minimize your subjective biases and prejudices by discussing your reactions with your friends and peers, but in the end accept what makes sense to you, and reject what doesn't. Trust your inner voice, not mine. Remodel your understanding of reality as necessary, and act out of what you learn.

The journey of this book began more than a decade ago. I am not the person I was when I set out. My ideas have evolved and developed in the process of writing, and are continuing to shift. No amount of editing and reworking can smoothly and seamlessly integrate this dynamic process into a fully consistent text. I am sure there are contradictions in places, and things that I would change now. This book is the story of an on-going journey, not the description of a final destination. It is not a set of ideas to be accepted or rejected, but an invitation to join me on this life-long voyage of discovery.

At times the terrain is dry, and the going hard and steep. We've a long way to go in a short time, and the pace may be fast. And often the beautiful woods for which we're headed may be concealed by the myriad trees. I've focused on key concepts and avoided technical jargon as far as possible, but it's still a rugged journey – one that I believe is necessary to reach the distant peaks of understanding for which we're headed. The view from the summit is stupendous, giving us new and glorious insights into the nature of the universe, our planet and ourselves; and bringing a new sense of identity and purpose, and a new hope for the future.

Before we set off, let me sketch a rough map of where we're going. Part I is preparation. It aims to improve our understanding of the nature and limitations of science, and the rich variety of other ways by which we gain knowledge. And arising from this discussion, it suggests some defining features of a more holistic science of oneness.

In Part II, attention focuses on the systems sciences that are revealing how seemingly independent objects are connected and interact to form coherent wholes that are creative, autonomous and evolving, and that display elementary purpose and will. Part III then takes us into the steep and rugged country of quantum and relativity physics, where we explore what is being revealed about the fundamental nature of physical reality. These two Parts represent opposite but complementary approaches to a unified reality: the One underlying the Many, and the Many emerging from the One.

Current theories of the origins of the physical universe are described in Part IV, revealing a stunning ignorance behind adherence to the big bang theory. This leads to an excursion into more radical alternatives. First, we examine the possibility that the most fundamental reality is Mind or Consciousness rather than particles of matter or quanta of energy; and then we look at the cases for and against cosmic design rather than chance evolution.

Part V moves on from the physical to the life sciences, addressing the questions: What is life? How did it start? How do complex organisms develop from a single cell? And how do organisms evolve? From here it is a natural step to explore the nature, origins and evolution of mind and consciousness in Part VI. Finally, in Part VII we face the ultimate Mystery – why does anything exist rather than nothing? We explore the case for the existence and nature of a Cosmic Consciousness, often called Spirit or God. We then go on to consider how the existence of this spiritual reality affects us, and our relationship to it. Our journey concludes with a guided meditation into the heart of the Universe in Part VIII.

How to use this book

Finally in this Prologue, I want to say a few words about how to use the book. My aim is to engage the intuitive, imaginative mind as well as the intellect. And so the presentation of concepts and ideas in the chapters is complemented by Review and Reflections sections at the end of each

Part. The Reviews draw out the key points made in the preceding Part, and place them in the larger context of the journey so far. It is an opportunity for me to allow my own inner knowing freer rein than in the main chapters which are devoted to presenting the scientific information. It is a chance for me to stand back and make more personal comments on what we've discovered. I recommend that the Review sections be read immediately following each Part, as concluding comments.

By contrast, the Reflections sections are designed to encourage inner knowing about the subject matter in you, my readers. They contain some entertaining activities, and evocative quotations from spiritual, literary, poetic and scientific sources. I hope these will entice you to pause and reflect on their deeper significance for your worldview and life. Some of the Reflections also include guided meditations designed to deepen your experience and inner knowledge of the subject. These are an opportunity to relax and let your subconscious take over after the hard mental travelling, and to enjoy wherever your thoughts take you.

If you are unfamiliar with such inner work, do not feel comfortable with it, or simply find it difficult to switch ways of thinking frequently, you may choose to leave the Reflections out on a first reading. To make it easier to do this, the sections are distinguished by gray box rules. However, if you do skip them, I encourage you to return later to tap the power of your subconscious and deepen your understanding.

I have included a range of materials in the Reflections sections in the hope of appealing to diverse interests and ways of learning. Hence, you are likely to find that only some of the materials appeal to you, and that you need to be selective. Whenever you do the Reflections, therefore, I suggest you read through each section fairly quickly, and then choose to spend time on the activities, quotations or meditations that draw you.

I recommend that you do the Reflections, particularly the meditations, when you will be undisturbed for at least 30 minutes and have the opportunity to relax. And I suggest that you have materials available before you start that you can use to express your experience afterwards: perhaps paper and pens, art materials, a musical instrument or anything else that appeals to you. For the longer guided meditations, I also recommend that you take the time and trouble either to have a friend read them slowly to you, or to record them with appropriate pauses for your meditative responses. Such meditations are much more powerful if you can relax and listen rather than having to read them at the same time as reflecting on them.

Whatever you choose to do, I hope you enjoy yourself!

If (the teacher) is indeed wise he does not bid you enter the
    house of his wisdom,
but rather leads you to the threshold of your own mind.
The astronomer may speak to you of his understanding of space,
but he cannot give you his understanding.
The musician may sing to you of the rhythm which is in all space,
but he cannot give you the ear which arrests the rhythm, nor the
    voice that echoes it.
And he who is versed in the science of numbers can tell of the
    regions of weight and measure, but he cannot conduct you
    thither.
For the vision of one man lends not its wings to another man.
And even as each one of you stands alone in God's knowledge,
so must each one of you be alone in his knowledge of God
and in his understanding of the earth.

Kahlil Gibran *The Prophet* **pp.67-68**

Part I:
Towards a Science of Oneness

Science is a key institution in our civilization. It is the source of much of our knowledge and many of the magical technologies on which our way of life depends. And the scientific method is widely regarded by scientists and non-scientists alike as the best path to reliable knowledge. Indeed, many scientists claim that it is *the* way to *Truth*, and deny the validity of alternative approaches. From the other side, critics argue that this arrogance is a sign that 'scientism' has assumed the mantle of the religions whose bigotry scientists fought so fiercely in the past.

Before exploring what modern science has to tell us about reality, it is important that we understand how the scientific method works, and something of its strengths, weaknesses and limits. It is also important to appreciate that it is just one amongst many paths to knowledge, and that it is helpful to integrate a variety of ways of knowing as we seek the truth about existence.

Chapter 1 outlines the nature, strengths and limits of science. Chapter 2 then explores the relationship between our perceptions of the world and outer reality. This is followed in Chapter 3 by brief descriptions of four other ways of knowing that complement the scientific approach, and which, together with science, form the basis of the science of oneness as presented in this book. Finally, Chapter 4 explores ways in which we can test the reliability of knowledge, and gain the wisdom to use it well.

Part I finishes with a short Review of what has been found so far, and Reflections to encourage us to pause, look inwards, and deepen our understanding.

1   Science and Knowledge

This Chapter outlines the scientific method, the claims made for it, and the criticisms leveled at it. It reveals that science is not an infallible route to reliable knowledge.

An Overview of the Scientific Approach to Knowledge

Scientists are almost as diverse as the human race. But they mostly share certain beliefs, values and methods that constitute the scientific approach to knowledge. Perhaps their most fundamental belief is that the material universe obeys fixed laws of nature. They claim that these laws can be discovered by standing back as objective observers who watch and measure what happens, conduct experiments under carefully controlled conditions, search out patterns and relationships in the data, and devise theories to explain the observations. The results of science are sufficient evidence for the power of this approach.

However, nature is so complex that scientists must simplify the real world in order to tease out her laws. They do this, both experimentally and theoretically, by studying ever-smaller parts in isolation from their surroundings until they understand how the basic components work – whether they be organic molecules in biology, or sub-atomic particles in physics. Armed with this knowledge, they then theoretically reconstruct the workings of whole organisms or stars. And once phenomena can be explained and predicted in this way, nature can be bent to our will, yielding the myriad technologies that have transformed the world.

However, accurate predictions are possible only when all factors can be measured, and theories can be expressed in the universal and precise language of mathematics. So important is this that qualities, events and processes that can't be quantified or expressed mathematically have limited scientific value. Hence, they tend to be ignored or quantified indirectly. For example, the feelings of people and animals are secondary to their observable behaviors; experiences such as music must be expressed in terms of frequencies and amplitudes of vibrations; and explanations focus on physical causes rather than purposes or motives. In some cases, scientists go so far as to deny the reality of these subjective factors despite their centrality to the quality of human life.

As far as possible, scientists follow rational, logical procedures, devising experiments, observations or theories that are not affected by their beliefs, feelings or physical presence. This is another reason why mathematics and measurement are so important, as they minimize the effects of culture and personal values. But scientists are human, and make mistakes like the rest of us. They guard against this fallibility by publishing their methods and results so that other scientists can repeat the experiments or observations, and check their interpretations and theories. The more often results are duplicated by different investigators, the more sure we can be that they are correct. A scientific theory, therefore, is a hypothesis that has been verified so often that it is accepted as true, eventually acquiring the status of a law of nature. But nothing in science is ever 100% proven. It often happens that new observations or reinterpretations of data reveal that past theories were mistaken or partial.

Technical Limits to Science

Despite bold claims made for the scientific method, there are technical limits to what science can discover. Perhaps the most important is between outer and inner knowledge. Science is a powerful tool for exploring the physical, material world. But it cannot come to grips with our inner worlds of personal experience, feelings, artistic creativity, consciousness and spirituality. These are intensely private worlds, not open to objective observation by others, and not readily communicated in words, let alone mathematics. Scientific explanations in terms of biochemistry and brain activity tell us how this inner reality works, but fail to capture the quality and significance of the experiences. Quantitative, objective science cannot tell us what it is like to be alive, to watch a glowing sunset, to listen to moving music, to savor delicious food, or to make love. And yet these are the things that are most important to us as living beings. These are the things that bring joy and meaning to life. We will return to this issue at various points in our journey, particularly in Part VI.

Even in the material realm, science cannot answer all questions. As an example, imagine measuring the length of a river or coastline. The more closely you follow all the little wiggles, the longer it gets without ever reaching a final answer. And, as we'll see in Part II, many processes are inherently unpredictable. Even Newton's law of gravity cannot tell us where the Earth will be in its orbit millions of years hence, and systems and processes often evolve in unexpected directions.

There are also physical limits to what we can observe, and theoretical limits to what we can explain. Cosmologists cannot see the outer limits of space because whatever is there is moving away from us faster than light can bring the information to us. And 'singularities' at which our mathematical models break down make it impossible to explore aspects of phenomena such as black holes or what came before the big bang. One day we may overcome these particular limits, but there will always be boundaries to the knowable.

Science and Belief

Despite its claims to objectivity and reason, scientific knowledge rests on a foundation of unprovable beliefs and assumptions. As the famous philosopher of science Karl Popper expressed it: "scientific discovery is impossible without faith in ideas which are of a speculative kind, and sometimes even quite hazy; a faith which is completely unwarranted from the point of view of science and which, to that extent, is 'metaphysical'."[1] And evolutionary biologist Conrad Waddington claimed that "a scientist's metaphysical beliefs have a definite and ascertainable influence on the work he produces."[2]

A good illustration of the influence of belief comes from the field of parapsychology. When scientists undertake identical parapsychological experiments, those who believe in paranormal phenomena such as extrasensory perception are more likely to get positive results than those who don't.[3] And Robert Matthews argues that the findings of parapsychology are rejected by mainstream scientists not because the research lacks quality, but because they do not believe the results are possible.[4] (This example is discussed further in Chapter 20)

The belief in the existence of fixed laws of nature has served science well, but its validity cannot be proved. Indeed, as we'll see in Part III, some scientists argue that natural laws evolve and are more like ingrained habits than fixed rules. In a similar way, scientists may discover a mathematical universe not because that is its nature but because that is what they are looking for. As Professor John Ziman put it: "Physics is the harvest of this (mathematical) fisherman, whose net only catches fish larger than the size of its mesh, and who proudly proclaims as a 'law of nature' that all fish are larger than this size."[5] Nature may reveal other faces to those, such as indigenous peoples, who approach her differently.

Mathematics itself is based on idealized concepts, such as the geometric point and line, and on axioms that seem self-evident but cannot be proved. As long ago as 1931 Kurt Gödel proved that it is impossible for a mathematical description of the world to be both complete and internally consistent. If it is consistent, it must have left something out; and if what is left out is included, it will become inconsistent.[6] Thus, as physicist and mathematician John Barrow expressed it: "if religion is defined to be a system of thought which requires belief in unprovable truths, then mathematics is the only religion that can prove it is a religion!"[7] And, in the words of Darryl Reanney, mathematics can never explain and encompass "the total richness of the world's truth. There will always be truth that lies beyond."[8]

So important is the role of assumptions in science that it is worth examining one of the most fundamental beliefs in more depth: the so-called 'law' of cause and effect. This concept is so embedded in modern common sense that it is hard to accept that it is not a universal law. But according to physicist Max Planck: "Causality is neither true nor false. It is a most valuable ... principle to guide science in the direction of promising returns in an ever progressive development."[9] How can this possibly be?

In the classical scientific view, everything happens as the result of an endless sequence of causes and effects. But a careful look at the way the world works shows things are not so clear cut. Imagine a spring being stretched. Does the tension in the spring cause the increase in length, or does the increase in length cause the tension? Suppose we hang a weight on one end of the spring and fasten the other. It's natural to think of the weight as the cause, and the change of length as the effect. But now suppose we stretch the spring in order to attach the ends to two fixed pegs. In this case, it's natural to think of the increase in length as causing the tension. In both cases, what we call the cause is the thing we can change and control, and what we call the effect is the change that results from this cause. So which is cause and which is effect often depends on our viewpoint.

Now consider a simple ecosystem. Once upon a time rabbits and foxes lived in a wood. The grass was plentiful, and so the rabbits were well fed and had lots of kittens. Plenty of rabbits meant plenty of food for the foxes too, and so most of their cubs survived. But as the number of rabbits increased, there came a time when they ate the grass faster than it grew, and they became hungry. And as the number of foxes grew, so the number of rabbits they ate increased until there weren't enough to go round. As a result, the number of foxes and rabbits fell, the grass recovered, and once again there was a time of plenty for all.

This story is illustrated in the following diagram. Looking at the outside loop of boxes and arrows, we can see that more grass causes an increase in rabbits, which causes a reduction in the amount of grass, which causes a fall in the rabbit population, which allows more grass to grow. And similarly, the inner loop suggests that more rabbits causes an increase in foxes, which causes the rabbit population to fall, thus causing a drop in fox numbers. But which is the first cause in these loops? Is it the availability of more grass, or the increased number of rabbits, or the reduction in the amount of grass, or the reduction in rabbit population? And which are causes and which are effects?

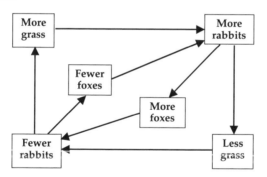

Figure 1   A simple ecosystem

A further difficulty with cause and effect concerns the role of will and intention. Suppose I want to go for a walk in the country tomorrow. This means getting up early, and, if I'm to do that, I want to go to bed early which means eating earlier than usual. Hence, I'll have to stop writing soon in order to prepare my meal. But there is nothing in the events of today to cause me to stop work early. The cause is an anticipated future event – what I want to do tomorrow. A scientist might argue that the 'real' cause is the state of my brain now, including the imagined future, and that it has nothing to do with actual future events. That brain state, in turn, might be 'caused' by happy memories of previous walks in the country. In a mechanistic sense, this may be true, but it doesn't adequately reflect my experience of the 'pull' of the future, and the causative role of my will.

This situation can be clarified by considering four types of cause identified by Aristotle. The *material cause* is the matter out of which something is made. The *efficient cause* is the external action that produces an effect, such as a force causing motion, or heat causing a rise in temperature. The *formal cause* is the inner activity which directs an object's growth and development towards a specific form such as a crystal or plant. But any formal cause must have a final form to which it tends, and hence formal cause implies *final cause* – the design or purpose of the object. In the case of a plant, its final cause is to grow, develop and reproduce itself.

Neither the 'existence' nor the 'non-existence' of (final causes) can be 'proved' in the scientific sense of the word.

A. Rapoport, p.173

Science focuses on material and efficient causes. It believes that form is simply the product of a sequence of efficient causes, and hence the idea of formal cause is unnecessary. The concept of final cause is rejected as false because the universe is mechanistic and the result of chance. Design and purpose have no place. However, as the above story illustrates, efficient cause is unable to provide a satisfactory explanation when we ask 'why?' something happens rather than 'how?'. In these cases we need answers that tell us about purposes and goals, which imply the 'pull' of the future rather than the 'push' of the past. Scientists themselves often make use of final causes, as when a biologist argues that an animal's fur evolved to keep it warm. But she would vehemently deny that this indicates purposeful behavior by nature.

The closer we look at the world around us, the more examples we find of final causes. Imagine a tree flowering in spring, then forming 'nuts' containing seeds which fall to the ground in autumn. The explanation by efficient cause might run something like this. The warmth of spring causes the tree to produce flowers, which are pollinated by insects. The fertilized ova develop into nuts and seeds under the control of chemicals produced by the tree, as determined by its genes. But this explanation fails to tell us why the tree produces flowers and seeds. It does so, of course, in order to reproduce itself. So it is the future survival of the species that causes the tree to produce seeds.

The idea of final cause brings meaning, value and purpose back into the mechanistic universe. These inner aspects of reality are what matter most to us, and are critical to the quality of human existence. And yet science denies their validity.

Science, Subjectivity and Culture

Beliefs and assumptions illustrate the way subjectivity intrudes into science. In practice, science cannot be purely objective because scientists, like all humans, are subjective and emotional. The choice of research field, the design of experiments, the theories pursued, and the interpretation of data are all influenced by personal beliefs and values, and the whole gamut of human emotions including ambition, pride, frustration, fear and joy. Scientists are also strongly swayed by aesthetic considerations, such as the simplicity and elegance of their theories. In the words of ecologist Edward Goldsmith: "science is in fact as emotional as it is subjective, for scientists are humans and as such have not been designed by their evolution to be unemotional any more than they have been designed to be objective."[10]

> The scientist never completely succeeds in making himself into a pure spectator of the world, for he cannot cease to live in the world as a human among other humans, or as a creature among other creatures, and his scientific concepts and theories necessarily borrow aspects of their character and texture from his untheorized, spontaneously lived experience.
>
> **David Abram (1996) p.33**

Story after story recounts how famous scientists wrestled with some knotty problem, before getting a flash of insight which they knew intuitively to be correct. Only later were the experiments done and the logical theories worked out that justified the insight. One of the best known examples is the discovery of the ring structure of the benzene molecule. After months of effort, Friedrich Kekule was dozing by the fire when he dreamed of a snake which curled round and took its tail in its mouth. That was all the hint he needed. And according to physicist Kip Thorne: "(Einstein) was motivated not by experiment, but by his deep, intuitive insight into how the laws of physics *ought* to behave."[11]

Writing about his life as a scientist, A. Cuthbertson claimed: "Scientists must ... call on their knowledge, their intuition, their imagination, to make sense of these fragmented glimpses of nature. ... (Their) description is an informed fantasy ..."[12] And a similar thought was expressed by Einstein when he wrote: "Physical concepts are free creations of the human mind, and are not ... uniquely determined by the external world."[13]

Imagination is more important than knowledge.

Albert Einstein
Quoted by Olshansky and Dossey (2004)

Pure objectivity is also impossible because the scientist is an interactive part of what is being studied. This is well-recognized in quantum physics and social science where the choices and behavior of the experimenter may affect the results. But it is also apparent in other fields, for example in work with experimental animals and even plants which have been shown to respond to the emotions of their carers. Ultimately, science is subjective because nature answers the questions asked, and it is humans that formulate the questions and interpret the responses.

Science is also a creature of its time and culture. In Copernicus' day, the natural motion for an object was believed to be a circle. But by Descartes' time, it was believed that undisturbed objects move in a straight line at constant speed. This change was not the result of a scientific discovery so much as a philosophical shift. In a similar way, Copernicus' idea that the Earth moves round the sun and Einstein's theory of relativity were both accepted before there was evidence of their validity because they complemented the spirit of the times. But other ideas have languished and been rediscovered only decades or even centuries later when the cultural context had shifted. In other words, an idea or observation becomes a scientific discovery only when it is recognized as such by the culture in which it is embedded.

The cultural context of science includes its funding and management. It has been estimated that as many as 40% of scientists work directly or indirectly for the military, and many others are funded by profit-seeking corporations. Consciously or unconsciously, the projects undertaken, the methods used, and the results obtained are strongly influenced by the source of funds. This is nowhere more clearly demonstrated than in the long-running dispute over the health effects of smoking. Research funded by tobacco companies generally shows lower effects than independent studies by government agencies and universities.

It is not only the source of funding that channels science, however. Panels of established scientists control what is published in scientific journals, and often reject not only sub-standard research but also work that challenges establishment views. One of the most sacred tenets of science is the freedom to think the unthinkable and offer it for testing. And yet on 24 September 1981 an editorial in *Nature,* one of the world's most prestigious scientific journals, was headlined "A Book for Burning?" We shall meet the ideas of Rupert Sheldrake, which sparked this extraordinary outburst, at various points in our journey.

Panels of established scientists similarly control the allocation of many research grants. In 2004, 33 scientists from 10 countries signed a statement arguing that the big bang theory of the origins of the universe is deeply flawed. They claimed that funding for research into alternative theories was inadequate because the funding panels that screen proposals are dominated by supporters of the big bang.[14]

Science led the attack on entrenched religion, opening the way to a culture of intellectual freedom. But nearly 50 years ago, R. G. H. Siu warned of the rise of 'scientism', claiming that "Many scientists are on the threshold of emulating theologians of the sixteenth century."[15] The situation has worsened in the intervening years. Scientists are now guardians of vast treasure houses of esoteric knowledge, written in arcane mathematical symbols and technical jargon that none but the initiated can comprehend. They have become a white-coated priesthood that interprets these scriptures for the masses. They have become arbiters of true knowledge, and of the paths by which it can be obtained. Some are also prone to pontificate on matters of belief rather than evidence, assuring us repeatedly, for example, that the universe and life are meaningless and purposeless, that we are here by chance, and that there is nothing beyond death – amongst other unprovable verities.

The Light and the Shadow of Science

Science is a great human achievement and has brought many blessings. In the view of its protagonists, it has banished old superstitions and religious mumbo jumbo, freeing humanity to see and relate to the universe as it really is. It has wrested control of knowledge from priests, and made it openly available to all. It has brought health, longevity, and unprecedented wealth to billions, opening up previously undreamt of opportunities for personal development and creativity. And it gives us the power to overcome the challenges facing humanity, and to seek our destiny amongst the stars.

By contrast, critics blame science for many of the world's ills. The science that brought us so many benefits also brought weapons of mass destruction, over-population, and the greatest environmental destruction the planet has faced for hundreds of millions of years. In a blink of the cosmic eye, humanity has become a virulent plague, destroying the world that gave us birth, and endangering our civilization and even our species.

Many scientists play responsible and invaluable roles in the alleviation of poverty, and in the pursuit of peace and sustainable development. But too many others continue to work on the means of violence, the whims of the rich, or technologies that destroy nature. And too many argue that science is morally neutral, casting the blame for the misuse of scientific knowledge on politicians, generals, terrorists, businessmen and other decision-makers. Science as an institution, and scientists as individuals, must shoulder their share of responsibility for the world's problems together with the civilization which gave science birth and in which it is embedded.

Sadly, the shadow of science runs far deeper than these practical outcomes. As a major historical movement, its worldview has permeated all areas of our culture and lives, and shares responsibility for the alienation, apathy and despair that haunt today's world. To nineteenth-century physicists, the natural world consisted of predictable, predetermined machines made of isolated atoms interacting through impersonal forces. The universe itself was seen as moving ineluctably towards a lifeless, uniform state of heat death – a view still shared by many physicists. From this perspective, we inhabit a cosmos that lacks freedom, meaning or purpose; that has no place for the human experience of life; and that is without love, beauty, passion or creativity.

Faced with this vision, metaphysician E. A. Burtt bewailed the loss of "a world rich with color and sound, redolent with fragrance, filled with gladness, love and beauty, speaking everywhere of purposive harmony and creative ideals;" a world that had been replaced by one "hard, cold, colorless, silent and dead; a world of quantity, a world of mathematically computable motions in mechanical regularity."[16] In similar vein, environmental philosopher Freya Mathews reflected that:[17] "We inhabit a meaningless and arbitrary world, and our own lives are ... imbued only with the value and significance that we attach to them. It is impossible, in such a context, not to suffer at some level from feelings of isolation, alienation and angst ... At the core of our identity, where we would expect to find the meaning of our existence, we discover only vacuity." And philosopher Bertrand Russell concluded: "Only within the scaffolding of these truths, only on the firm foundation of unyielding despair, can the soul's habitation henceforth be safely built."[18]

The mechanistic view of the physical world infiltrated biology, psychology and the social sciences from where it continues to affect every aspect of our lives. Over 300 years ago, Rene Descartes stated: "I do not recognize any difference between the machines made by craftsmen and the various bodies that nature alone composes." Biologists still use this machine metaphor, describing cells as *factories* under the *control* of DNA molecules which *organize assembly* of molecules into larger *structures* according to a *program* encoded in molecular *machinery*. And, despite changing attitudes, higher organisms are still frequently regarded as unconscious, unfeeling machines without interests, rights or value. Hence, factory farming and animal experiments are justified on the basis that animals don't suffer, their cries meaning "no more than the creaking of a wheel" as Descartes expressed it.

It is a small step from there to Descartes' more famous dictum that "I think therefore I am." Identifying with our minds, we regard our bodies as mere instruments for getting what we want in life. Encouraged by a medical profession dedicated to the repair of faulty machines, most of us take stimulants and tranquillizers (including alcohol, caffeine and nicotine), indigestion pills and pain killers so we can ignore our bodies' cries of distress; and we patch them up with surgery and pharmaceutical drugs when they break down.

This radical disconnection from other creatures and our own bodies is reinforced by the scientific method. The goal of objective observation encourages emotional withdrawal from what is being studied, and reduces nature and people to objects to be experimented upon, manipulated and controlled. Quantification requires us to perceive separate objects that can be counted and measured rather than interconnected wholes. And mathematics encourages us to inhabit a virtual reality of symbols rather than the real world of our senses and experience. The replacement of direct relationships with computers and communications is bringing a similar disconnectedness to the population at large.

In their efforts to emulate classical physics, behavioral psychologists claimed that our thoughts, feelings and actions are programmed responses to our experiences in life. A few decades later, many geneticists similarly claim that particular behaviors are controlled by specific genes; and many neuroscientists are intent on proving that consciousness and our spiritual impulses are no more than side effects of brain activity. According to these views, we are nothing but complex robots assembled from genetic plans and programmed by our genes and environment. Hence, personal and social problems can be solved by redesign and reprogramming, leaving us stripped of identity, consciousness and will, and at the mercy of faceless experts.

So powerful is the classical imagery of atomism and mechanism that it has come to dominate the social sciences too. Thus individual 'social atoms' respond to the forces of rational self-interest, whether as consumers, voters, or players on the stock market. And just as the myriad molecules of a gas obey statistical laws, so the behavior of mass society can be predicted by statistical models. We have become mere cogs in the machine, manipulated by advertisers and the media, and caught up in a competitive struggle for survival in which there is no room for compassion, cooperation, community or nature.

Charles Darwin was familiar with the theories of social evolution, and viewed nature in this distorting mirror. Thus, he saw a world of unbridled competition in which only the fittest survive, and nature as red in tooth and claw. Darwin's vision in turn fed back to reinforce the social theories, and was used to justify the extreme hardships of the industrial revolution. Right wing theorists still argue that human society is evolving according to natural laws which weed out those less fit to survive in the larger interests of the human race.

Positive changes in scientific beliefs, values and attitudes are happening, but they permeate our cultural worldview slowly. Despite huge advances in scientific knowledge in the last century, many aspects of our worldview are still dominated by the ideas of classical science. If we are to meet the challenges facing humanity, we urgently need a new worldview based on modern science, complemented by other beliefs, values and ways of knowing.

In Conclusion

Science has proved itself to be an extraordinarily fertile and reliable road to knowledge of the material universe. As a result, it has showered us with magical technologies that have transformed human civilization beyond all recognition. But science is far from infallible. There are limits to what it can reveal even in the outer material realm, and it is a poor instrument for investigating our inner worlds of consciousness, beliefs, values and feelings. Despite its claims to objectivity, science is riddled with subjectivity and is based on unprovable beliefs, as are all other human activities and knowledge.

Science is a major contributor to many of our problems. In particular, it has had a profound impact on our worldview, leading many to believe we inhabit a world of predetermined, lifeless machines at the mercy of blind forces; a world lacking meaning, purpose, consciousness, beauty or love; a world of separation and alienation of mind from body, person from person, and human from nature; and a world of unbridled, ruthless competition. Is it any wonder that such a worldview has given birth to a society of alienated, despairing, lonely, powerless people?

The good news is that many branches of modern science are contributing to a new, more positive worldview. Combined with other ways of knowing that are discussed in the following chapters, this is leading to a new science of oneness that is imbued with spirit and life, beauty and love; and a worldview that will enable us to live in harmony with each other and the planet.

2 Perception
and Reality

The first chapter examined the methods, strengths and limits of the scientific approach to knowledge. In the process, it became apparent that science is far from being an infallible way to unravel the secrets of existence and the nature of reality. In this chapter, we put science into the broader context of human knowledge by examining the relationship between our perceptions and outer reality. In the process, we will shed light on three questions: Are our perceptions true representations of what actually exists 'out there'? What does it mean to 'know' something? Are there limits to human knowledge?

These issues are vital to the emerging science of oneness. If we are aware of our own mental processes and the limits to what we can know, we will be more accepting of other ways of knowing, and more open to alternative perceptions and interpretations, thus enriching our knowledge and enhancing our wisdom.

Sensations, Models and Facts

Sensory data are the only means we have of finding out about the outside world. If I'd been born blind, I would have a very limited idea of the visual world; and if I were deaf, sound would consist of vibrations sensed in other ways. In the words of Deepak Chopra:[1]

> We believe that if we can touch a flower's petals, see its colors, or smell its fragrance, then we experience the intrinsic nature of the flower. Not so ... What we perceive of the flower is ... a reflection of the intrinsic nature of our nervous system ... A bee's eyes perceive the nectar, but not the flower. A snake perceives the flower as an infrared object; a bat perceives the

flower as an echo of ultrasound. ***Reality, then, depends on who is doing the looking and with what apparatus.*** (Emphasis added).

Sensory nerves all over my body continually send impulses to various parts of my brain. Apart from their strength, these impulses are all the same regardless of what causes them. Somehow my brain sorts sounds from sights and smells – a remarkable feat given that blind people can learn to 'see' with a camera linked to sensors on their tongue![2] Somehow my brain sifts important messages from ones which can be safely ignored. And somehow it extracts meaning from this jumble of raw data. Most scientists agree that the brain does this by creating a mental 'map', or 'model' of reality which it can compare with incoming data. This model develops throughout our lives, as our brains integrate our life experience and knowledge.

The power of our mental models is demonstrated well by Gregory Bateson's story of his encounter with an optical illusion.[3] Imagine a large, closed trapezoidal box with a peephole at the front, in which the left-hand end of the back wall is further from the observer than the right-hand end. Distorted outlines of windows are painted on this wall so that they appear rectangular when viewed through the peephole, thus giving the impression that both ends of the wall are the same distance away. Having carefully examined the apparatus, Bateson donned a pair of glasses that made it difficult for him to judge the distance to any object. He was then asked to reach into the box with a stick, touch a piece of paper on the left-hand side, and swing it across to hit another piece on the right-hand wall. This is what happened:

> The end of my stick moved about an inch and then hit the back of the (box) ... I tried perhaps fifty times, and my arm began to ache. I knew ... I had to pull in as I struck in order to avoid that back wall. But what I did was governed by my image. I was trying to pull against my own spontaneous movement. ... I never did succeed in hitting the second piece of paper, but, interestingly, my performance improved. I was finally able to move my stick several inches before it hit the back wall. And as I practiced and improved my action, my image changed to give me a more trapezoidal impression of the room's shape.

This story clearly demonstrates that there are no 'facts' waiting to be observed independently of our mental models. In practice, fact and model are inextricably interwoven, and interpretation of observations is inevitably grounded in our cultural and personal worldviews. Another illustration of this idea is provided by the pattern of black patches below. To most people, it appears random and meaningless at first. And then suddenly they may see the head of a giraffe. The marks on the paper are the same whether we see the giraffe or not. The giraffe is not an extra element in the pattern, but its meaning; a meaning that emerges in the process of seeing, and that we impose. We are not free to 'see' anything we choose in the pattern, but we could not see a giraffe unless we had an image of what a giraffe looks like in our memories. And unless we are told it is a giraffe, we may see other possibilities, such as the Pierrot face that my partner and I both saw.

Figure 2   An illustration of pattern interpretation

Our brains have evolved to seek meaning in the patterns of our environment, and we cannot separate what we see from the process of seeing. The idea of a giraffe arises from our efforts to make sense of the pattern, and once we have that idea we interpret the pattern in its light. We see what we expect to see. Hence, perception is not passive reception of given facts but an active process in which we participate in shaping the result. When Galileo first looked at the moon through his telescope he simply got a clearer view of the pattern of light and dark. But gradually his mind created meaning from it, and then he saw mountains and valleys.

Instruments provide us with objective, concept-free information, but it is our minds with their mental models that interpret it and give it meaning. Thus, scientific discoveries aren't inherent in the raw data, but emerge from the perception of meaning. The experiments we design are influenced by the theory we want to test, and color nature's response. As physicist Paul Davies put it: "time and time again (physicists) unwittingly massage their data to fit in with preconceived ideas. Sometimes, several different independent experimenters will carefully measure the same quantity and consistently get the wrong answer, because it is the answer they have come to expect."[4]

From this perspective, it is not surprising that new scientific theories often arise because of changes in worldview rather than new facts. Thus, during scientific revolutions scientists may see different things when they look with familiar instruments in places they have looked before. In medieval times, people saw cannon balls travel in a straight line angled upwards, and then suddenly fall vertically because that was what they believed happened. But what they saw changed when Galileo introduced the idea of a parabolic arc. In much the same way, some diseases are rare until they are described, and then suddenly they become common!

Scientific knowledge is not a purely rational, objective edifice built upon a foundation of solid facts as we are often led to believe. And the facts of nature are not fixed, objective bits of information waiting to be discovered, but patterns that will be revealed by the appropriate ideas.

The eye takes in 10 million bits of information per second and deals consciously with 40.

The ear takes in 100,000 bits of information per second and deals consciously with 30.

The skin takes in 100,000 bits of information per second and deals consciously with 5.

We can smell 100,000 bits of information per second and deal consciously with one.

We can taste 1,000 bits of information per second and deal consciously with one.

After Manfred Zimmermann (1989)

We don't experience reality in the raw as it actually is because our brains would be overwhelmed by the flood of data. We would be so busy processing it that we wouldn't have time to catch food, find a mate or avoid danger. So evolution has equipped us to filter information on a 'need to know' basis, and to rely on models rather than reality. This enables us to survive and reproduce, but not to comprehend the world as it actually is, or the nature of ultimate reality. But our mental models are far more than passive interpreters. They guide our actions, and thus determine how we interact with our environment. And modern technologies enable us to change the planet to match our model of reality, no matter how limited and inappropriate it may be. That is one reason why we need a new worldview.

Language as Mental Model

We absorb the core ideas and structures of our mental models from our families and societies, and it is only the decorative trimmings that are truly personal. Language is a key factor in this process, not only enabling us to share our thoughts, but also forming the basic framework of concepts and ways of knowing on which we build our worldviews.

We name every creation, discovery, and new idea. And new activities quickly acquire verbs, such as 'to surf' the web or 'to google' (ie to look for information on the internet using the Google search engine). But words are more than labels that facilitate efficient communication. Helen Keller, blind and deaf from birth, told how language brought her mind alive, and gave meaning to her sensations. And the way the language we learn as a child structures reality actually determines how we see the world as we grow up.

The Inuit people, for example, depend for their survival on being able to distinguish different types of snow. Is it wet, falling in large flakes, or dry and powdery? Is it hard-packed and frozen, or lying in soft, deep drifts? As a result, they have about 150 words for snow, and can perceive subtle differences that escape outsiders.[5] Similarly, the English word 'love' corresponds to three words in Greek and over 90 in Sanskrit, implying an impoverished understanding of loving relationships in the English-speaking world.

Another example is the perception of color. Humans can detect about 7.5 million shades, but most languages divide this spectrum into a few named colors. In English we commonly use about eight, including orange and yellow. But the Zuni people of North America give orange and yellow the same name. In consequence, they score badly on tests which distinguish between yellow and orange, not because they're color-blind but because they don't think of them as different.

As already noted, the language of science is mathematics. So extraordinarily powerful is it at modeling the natural world that many scientists believe it reflects a parallel mathematical universe whose theorems we discover rather than create.[6] Wherever we stand on this issue, however, mathematics has no greater claim to represent the truth of reality than any other model or language. As we saw in the last chapter, it is based on many unprovable assumptions, and can never be a complete and consistent description of the world.

Language, whether verbal or mathematical, not only provides the conceptual foundations for our mental models, but also has a deeper influence on our way of knowing. Indigenous peoples read the landscape like a book. Animals, plants and rivers speak to them as the printed word now speaks to us. Trees and mountains call out, the shape and color of a rock hold meaning. The hunter enters the mind of his prey, and reads the signs of its presence in the land. The legacy of this time lives on in our many onomatopoeic words: birds whistle, leaves rustle, the wind roars and a stream tinkles.[7]

This close relationship to the land is reflected in the structure of indigenous languages as well. They are replete with verbs representing the flow of activity rather than the myriad nouns for objects and categories that dominate European languages.[8] For Native Americans, language is "a living thing, an actual physical power within the universe. The vibrations of its words are energies that act within the transforming processes we call reality."[9] And when Australian Aborigines walk the Songlines of their Dreaming, the hills and creeks, rocks and waterholes prompt memory of the story, and they actively recreate and revitalize the land through its telling.

David Abram argues that abstract thought in isolation from the natural world was made possible by the development of the alphabet. Written language has a life of its own separate from the speaker and the land. It becomes possible to pause, reflect, question and debate without destroying the integrity of the story because the storyteller no longer has to hold everything in her mind. And when the written text speaks, the voices of the forest and river fade, and the association between language and land loosens. Reading draws us into profound connection with ink marks on the page, thus weakening our participation in the natural world. And our urban, technological way of life weakens those links still further.

Language is currently undergoing changes no less radical than the introduction of the alphabet. It is no longer confined to the spoken and printed word, but includes broadcast media, electronic documents of all kinds, digital images, audio and video recordings, films, the vast mass of multi-media materials on the internet, computer games, and more. The great benefit of this development is that information and ideas are increasingly available in forms that suit different ways of thinking and learning. The danger is that these technologies will isolate us still further from the real world, as we become increasingly immersed in virtual realities. We may be better informed about nature and the challenges facing us, but ever more alienated from direct experience of that reality.

The Way the Brain Works

We have seen in the last two sections how our perceptions are determined as much by our language and mental models as by external reality. The subjectivity of our knowledge is also shown by the structure of the brain and nervous system.

When I see something, my eyes send nerve impulses to a part of the brain called the LGN (short for lateral geniculate nucleus). But for each nerve coming from my eyes, the LGN receives inputs from over 80 fibers coming from other parts of my brain. This diluted visual information is then analyzed in yet other areas, which are themselves linked to many parts of my body. Thus, in forming visual images, our brains integrate information from the eyes with huge amounts of information from other sources, many of them internal.[10] The brain as a whole has about 100,000 times more internal connections than external sensors, and hence we are naturally more 'tuned in' to what is happening inside than to what is going on outside.

This inner life is most obvious in young children who spend many hours in imaginary worlds of their own. As we grow up, we learn to differentiate between this inner world and outer reality, but we never live completely in the outer world. We adults continue to have fantastic dreams, and love being transported into other worlds by art, fiction, drama, film, virtual reality, music and dance. If we're mentally healthy, we learn to balance the imaginative projection of our inner reality onto the outside world with inner acceptance of outer reality. But what we perceive is always a mixture of the two.

The way the brain works also differs from person to person, giving us a rich diversity of perceptions, experiences and expressions of outer reality. This is well demonstrated by the range of human intelligence. For many years, intelligence has been associated mainly with the IQ test, and educational systems have focused primarily on related mental skills and intellectual knowledge. But research has shown that intelligence takes a variety of forms.

Over 20 years ago, Howard Gardner challenged the dominance of IQ by suggesting that we each have a unique combination of different ways of solving problems or creating things.[11] He initially identified 7 *intelligences,* as he called them, associated with activity in different parts of the brain. He has since included another two, and still more have been suggested by other researchers. His nine intelligences are briefly described in the Table.

Verbal-linguistic intelligence Thinks in words. Skilled at reading, speaking, storytelling, listening, debating, writing and foreign languages.	*Musical intelligence* Thinks and learns in sounds, rhythms and patterns. Good at singing, playing instruments, remembering tunes, and composing music.
Mathematical-logical intelligence Performs well at mathematics, logical reasoning, problem-solving, classifying information, abstract concepts, and controlled experiments.	*Intrapersonal intelligence* Self-reflective, able to understand and work with one's strengths and weaknesses, dreams and desires, emotions and thoughts.
Visual-spatial intelligence Thinks in pictures and mental images. Performs well at reading, writing, visualization, visual arts, designing and making things.	*Interpersonal intelligence* Relates to and understands other people, sensing their thoughts and feelings. Good at verbal and non-verbal communication, conflict resolution, organization and cooperation.
Bodily-kinesthetic intelligence Learns through movement and physical skills. Well coordinated, with good balance, dexterity and motor skills. Communicates through movement, touch, body language, acting and mime.	*Existential intelligence* Able to contemplate the big philosophical questions such as 'Why are we here?' and 'What is the purpose of life?'.
Naturalist intelligence Keen observers of nature who relate well to animals, and readily understand natural processes. Good at sensory perception and pattern recognition, and at classifying organisms.	

Table 1   Howard Gardner's Multiple Intelligences

Amongst other proposals, Daniel Goleman's *emotional intelligence* has attracted a lot of interest, and is, in essence, an amalgam of intra- and inter-personal intelligences.[12] *Spiritual intelligence*, according to Danah Zohar and Ian Marshall, is about the meaning and purpose of life, creativity, discrimination, moral sense, good and evil, and envisioning a better future.[13] Its qualities include flexibility, self-awareness, going beyond convention, facing and transcending suffering and pain, inspiration by vision and values, reluctance to cause harm, seeing connections between diverse things, and asking 'Why?' and 'What if?'.

Only one of these, *mathematical-logical intelligence*, is central to science. Hence, an exclusive focus on the scientific approach to knowledge denigrates the contribution of most types of intelligence, disempowers most people, and impoverishes our artistic and cultural lives as well as our collective understanding of the world.

In Conclusion

In Chapter 1 we explored the nature of the scientific method, discovering that, despite its strengths, science is not an unfailing path to knowledge. This chapter has shown that this weakness is not confined to science. Regardless of the approach used, our perceptions are not 'reality' as it truly is, but are shaped by the way our brains work, by our languages and culture, and by the mental models we form from our learning and experience. Further, we all live to some extent in different 'realities' resulting not only from our varied knowledge, but also from our different intelligences.

Awareness of this gap between perception and reality makes us less confident of what we think we know, and more accepting of other ways of knowing. It reduces our tendency to dogmatism, and opens us to the enrichment that comes from diverse views. We realize that a more reliable picture of reality may be revealed by integration of different ways of knowing than by seeking the one Truth, whether through science or religion.

In the next chapter, we will discuss other ways of knowing that complement and enhance the scientific approach.

3 Other Ways of Knowing

Science is an incredibly powerful way of learning about the material world, but it is not an infallible route to all knowledge. There are other complementary ways of knowing with their own strengths and weaknesses. In seeking a more accurate understanding of reality, and in facing the challenges of the future, we need a way of knowing that integrates and balances these various approaches. We need a way of knowing that not only gives us power over the outer material world, but also helps us to understand the inner worlds of belief, perception, intuition, spirituality, purpose, emotion, art, music, creativity and so on.

This Chapter describes four ways of knowing that complement the scientific approach and help define a new science of oneness. First comes an alternative form of reason and logic based on the ancient Chinese system of yin and yang. This is followed by an exploration of intuition and what I have called spiritual knowing. Third is a way of knowing through direct experience of, or relationship with, what is known. The chapter closes with a discussion of two different approaches to the idea of oneness.

Reason and Logic

Science is founded upon the strict rules of formal logic and mathematical analysis, but there are other forms of rationality. One that is particularly relevant to the science of Oneness is described here.

In western logic, an object, action or idea either is or is not of a particular type. There's no middle ground. This object is a chair, and not a table. This idea is true, and hence cannot be false. But whenever we create a category, we automatically create its opposite as well: chair implies the existence of things which are not chairs, true implies ideas which are false.

Inside implies outside, up implies down, good implies evil, beauty implies ugliness, and so on. In other words, each of these polarities is not really two different categories, but two faces of the same aspect of reality. Many worldviews, including the Native American[1] and Taoist, recognize this duality, and accept that something can be both A and not-A.

Taoists believe that nature always seeks balance. Any movement towards an extreme will eventually swing back towards its opposite, just as going further and further East ultimately brings you to the West. The poles in this process are called yin and yang. When the yang reaches its peak, it retreats in favor of the yin, and vice versa. In the words of J C Cooper, "Life gives way to death, death gives rise to new life; strength over-leaps itself and becomes weakness, success reaches its zenith and comes down in failure."[2]

Yin is seen as complex, physical, emotional, intuitive, female, passive, receptive, yielding, negative; yang as simple, spiritual, rational, intelligent, male, active, energetic, creative, strong and positive. Yin is the quiet contemplative stillness of the sage; yang the strong, creative action of the king. "For Taoism the soul, the feminine yin, wisdom, is the means by which the yang intellect attains insight and understanding, together partaking of the yin-yang dualistic nature."[3] The whole of reality – physical, mental and emotional – is seen as infinitely divided in this way into what Taoists refer to as the 'Ten Thousand Things'. (See Table 2)

This cyclic view of nature implies a center: the pivot which guides the circle or spiral, the point of balance and stillness, the focus from which energy emanates and to which it returns, the point at which the tension of opposites is resolved and opposing forces rest in equilibrium, the point from which things can be seen whole. "All happiness, all wisdom, depends on the balance and harmony of the opposites."[4]

Yin and yang have vital need of each other, since they cannot exist in isolation. And their interaction is creative and transformative. Change is not a response to some cause, but is a spontaneous outcome of the dynamic interplay between yin and yang. Yet when we see them as poles of a single whole, the opposites cease to exist. As Chuang Tzu expressed it: "There is in reality neither truth nor error, neither yes nor no, nor any distinction whatsoever, since all – including the contraries – is One."[5] These ideas are captured by the ancient Chinese symbol of T'ai-chi T'u which shows the yin as dark and the yang as bright. Its symmetry is rotational, suggesting continuous cyclic movement. And the two dots express the idea that the extreme contains within itself the seed of its opposite. (See Figure 3)

Pleasure – pain	Individual – community
Happiness – sadness	Nature – culture
Fulfillment – emptiness	Consensus – authority
Thinking – feeling	Rights – responsibilities
Reason – intuition	Competition – cooperation
Thought – action	Wealth – poverty
Objectivity – subjectivity	Justice – injustice
Good – evil	Equality – inequality
Harmony – discord	Stable – unstable
Wisdom – folly	Flexible – rigid
Chaos – order	Dynamic – static
Quantity – quality	Out breath – in breath
True – false	Health – sickness
	Stress – ease

Table 2   *A few of the 'ten-thousand things'*

Figure 3   *Yin-Yang symbol*

Western culture is based on the ideal of progress. We seek continually to eliminate the negative and increase the positive, aiming for some elusive utopia. After the Second World War, we set out to eliminate poverty, sickness and hardship through economic growth, but that growth has now become an end in itself. And instead of a better society we are completing the circle with a new generation of jobless, homeless, addicted, suicidal people; of broken homes, crime and environmental degradation. Taoism shows why. We cannot eliminate the negative, because without it the positive has no meaning. We cannot 'progress' for ever in one direction without provoking a natural swing back towards balance. We need to stop looking for single Truths or The Good or The Best, and to become comfortable with the paradoxical nature of existence from the human perspective. The science of oneness needs to complement one-sided western logic with the concept of balance between polar opposites.

Intuition and Spiritual Knowing

Intuition and spiritual knowing arise from three sources: the inner world of our subconscious minds, our relationships with what is known (discussed in the next section), and our connection with spiritual reality including cosmic Consciousness.

The nature of intuition can be understood most readily by considering the way we make decisions. Few of us use rational analysis alone. Instead we rely strongly on 'gut feelings' and hunches. We do what we feel is right or beautiful, backed up by analysis, rather than the other way around. The strategies of great statesmen and generals are mostly the result of intuition, although analysis may be used to work out the operational details. Similarly, the inspiration of poets, musicians, painters and sculptors flows from intuition, not reason. And, as described in Chapter 1, the great discoveries of science usually start with a hunch that proves to be right, and may end with a conclusion based on aesthetics.

Intuition means gaining knowledge or insight by direct perception without reasoning. It involves looking deeply into something rather than standing back to get an objective overview. It leads to an understanding that is not accessible to the thinking mind, and that often changes our perception of the way things work or are related. And it often brings a new sense of the unity and wholeness of a phenomenon.

But intuitive flashes seldom happen by chance. They are usually the culmination of a long period of intense work and thought, followed by sleep or relaxation during which the insight comes, or we make a decision that we 'know' is right. At this point, our whole body relaxes, releasing the tensions of the unresolved problem.[6] In practice, it is possible to detect the right answer by being aware of this physical response – an approach to decision-making and research that has been formalized in kinesiology.[7]

Spiritual knowing goes beyond intuition, but its nature is widely misunderstood. Spirituality is not the same as faith in one of the established religions, although that can be an expression of it. Nor is it spiritualism, with its séances and communication with the dead. According to William Bloom:[8]

- It involves the instinctive sense of being in and connected to a universe filled with beauty, mystery and meaning.
- It includes the desire to explore and deepen that connection.
- It provokes general enquiry into those deepest of questions: Why are we here? Where did we come from? Where are we going?
- It involves the instinct to stretch beyond our ego boundaries and love others.

To Roberto Assagioli, the founder of Psychosynthesis, spirituality is "concerned with considering life's problems from a higher, enlightened, synthetic point of view, testing everything on the basis of true values, endeavoring to reach the essence of every fact, neither allowing oneself to stop at external appearances nor to be taken in by traditionally accepted views, by the way the world at large looks at things, or by our own inclinations, emotions and preconceived ideas."[9]

And for Alison Leonard "the spiritual journey is a search for meaning: for a deeper, broader awareness of my existence than the material or even the emotional or the psychological can give. It is a search for the truth: the truth for myself, of myself, and the truth for others, and then the search for the links between those differing glimpses."[10]

Spiritual knowing takes many forms including:[11]

- A new understanding about the nature of the universe;
- An experience of going beyond normal space and time;
- A sense of unity with all life;
- A clear recognition of what makes sense, or is true;
- A sense of profound mystery, wonder and awe;
- A clear perspective on one's life and of its meaning and purpose;
- A sense of being a channel for a larger, stronger force which guides one's life;
- Feelings of joy, ecstasy, peace, freedom, gratitude, compassion, beauty or similar.

Spiritual knowing may come in dreams; or through mental images, voices or words. There may be a sense of light, or normal reality may appear transformed for a moment, or an unbidden memory may arise. But for many, it is not so clear or dramatic. There may be a vague feeling about something, or an inner pressure to take some action; or a sequence of apparently chance events may take their lives in a new direction.

Spiritual knowing is often accompanied by a deep sense of truth that may not be rationally justifiable but carries unshakeable conviction. We see this in religious teachers like Jesus, the Buddha and Muhammad; in visionaries like Martin Luther King, and spiritual revolutionaries like Gandhi. More surprisingly, the founders of modern physics were also highly spiritual, including Einstein, Schrödinger, Heisenberg, Planck, Pauli, Bohr and others, most of whom won Nobel prizes for their work.[12]

It's important to be clear that spiritual experiences and spiritual knowing are not the preserve of especially spiritual or gifted people. They happen to millions, maybe billions, of ordinary people around the world, of diverse beliefs including agnostics and atheists. And we can encourage and develop this gift in various ways.

The old advice to sleep on a problem is still sound. But we can also cultivate mental and physical relaxation to facilitate the breakthrough of insights from the subconscious. Methods that work for me and many others include spending time in nature, vigorous physical exercise, gardening, craft work, yoga, tai chi and sitting meditation. These are a tiny sample of the range of practices available, and we each need to find the ones that suit us best so that we can bring the power of our subconscious to bear on our lives. If you would like to explore spiritual knowing further, I recommend *The Quest*.[13] The existence and nature of a spiritual dimension of reality is discussed in Parts IV, VI and VII.

Intuitive and spiritual knowing contrast strongly with the scientific method as it is normally described, but are actually important though unrecognized components of it. They are essential complements to 'normal' science as we move towards a more holistic science of oneness.

Knowing as Relationship

The scientific method encourages objectivity and disconnection from what is being studied. But the information gained in this way misses dimensions and richness that only a close relationship can bring, as this section describes.

I cannot truly know another person by collecting information about them, but must enter a relationship with them. I cannot fully understand nature by studying science, but must encounter it directly in my own being. I cannot become a musician without entering a relationship with my instrument and embodying the music. I cannot learn spirituality by listening to a Teacher, but must experience the reality of the spiritual realm within myself. I cannot fully understand an idea without entering a dialogue with it: questioning, testing, comparing, making it my own. And I can't understand love until I fall in love.

To know anything deeply, I must enter into relationship with it. And the knowing that comes in this way cannot be learned or taught, or stored in a database. It is a living response to that which is known that engages our bodies, emotions and spirit as well as our minds. It is a dynamic process that changes and develops with time.

Philosopher Martin Buber distinguished two forms of relationship: I-It and I-Thou. The scientific way of knowing typifies an I-It relationship with what is known. It is objective, impersonal, disengaged, unfeeling, and utilitarian – concerned with measurements and calculations, causes and effects, manipulation and control. Its primary aim is to gain information without giving anything in return. It is epitomized by Francis Bacon's approach of extracting information from nature by force, interrogation and torture.

Despite this, not only is the scientist changed by the knowledge received, but also the object of study is often affected by the interaction. Examples abound: parapsychology experiments are affected by the beliefs of the researchers; tests on laboratory animals can be changed by loving carers; plants respond to human emotions; instruments seem at times to respond to the 'vibes' of their users; and the state of quantum particles are determined by the physicists' objectives.

I-It relationships are not restricted to scientists. In our daily lives, we all tend to treat others as Its. Shop assistants, bank tellers, the postman and refuse collectors don't exist for us as people, but are simply means to various ends. We don't make eye contact, don't ask after their well-being, and are interested only in their efficiency. And as consumers, we clutter our lives with things that we don't really want and quickly discard.

By contrast, an I-Thou relationship is personal, engaged and feeling. It is a sensual, non-verbal, non-conceptual communication that often occurs without conscious awareness other than, perhaps, some inner sense or feeling of connection. It is an intimate encounter without concern for the outcome in which there is a mutual giving and receiving, touching and being touched, seeing and being seen. It is not manipulative or exploitive but a caring participation in the existence of the other, and the manifestation of the phenomenon being studied. It is a partnership in which we encounter the unity of self and other; a synergy through which each becomes more fully itself. It is a process in which nature reveals herself through our minds.

Most of us experience I-Thou relationships with our parents, families, partners, friends or pets. But we are less familiar today with such relationships with plants, rocks, machines or ideas. Let's explore what these are like with the help of a few examples.

I can relate to a tree as a beautiful picture or a graceful dancer; as an example of a particular species, or as an ecosystem of tree, lichens, epiphytes, insects, and birds. Or I can measure and quantify it, and simulate it on my computer. In all these, the tree remains an object, an It. But I can also relate to the tree as a Being that is aware of my presence and touch and love; that draws my attention and 'speaks' to me of its life story, offering me companionship, empathy and shelter. I am not so skilled as the shamans who read trees like a book, or as Dorothy Maclean, one of the founders of the Findhorn Community, who was instructed by the spirits of the plants on how best to grow them.[14] But nevertheless I can encounter the tree itself relatively free from conceptualizations about it, or abstractions of it. And in doing so, I don't need to deny the other ways of knowing, which deepen and broaden my connection with the tree as a Being, enriching my encounter.

The relationship of indigenous peoples to nature and their land exemplifies the I-Thou type of relationship, and is described well in these short quotes from David Peat:

> (W)ithin the indigenous world the ... knower and the known are indissolubly linked and changed in a fundamental way.[15]
>
> Coming-to-knowing means entering into relationship with the spirits of knowledge, with plants and animals, with beings that animate dreams and visions, and with the spirit of the people.[16]
>
> (K)nowledge for them is ... a living thing that has existence independent of human beings. A person comes to knowing by entering into a relationship with the living spirit of that knowledge.[17]
>
> Indigenous knowing is a vision of the world that encompasses both the heart and the head, the soul and the spirit. It ... is a science of harmony and compassion, of dream and vision, of earth and cosmos, of hunting and growing, of technology and spirit, of song and dance, of color and number, of cycle and balance, of death and renewal.[18]
>
> While Western thought grasps at the surface, the Indigenous heart, mind, and being seeks ... that inner voice and authenticity that lies within each experience and aspect of nature. In engaging (this) a person is not preoccupied with measuring, comparing, classifying, categorizing, or

fitting things into a logical scheme, but rather with seeking a relationship that involves the whole of one's being.[19]

The way this knowledge is learned is inseparable from the land and from the people who live on it. ... the heart of traditional knowledge cannot be translated, written down in a book, or transposed to an individual living thousands of miles away ... Knowledge belongs to a people, and the people belong to the landscape.[20]

Within a traditional society knowledge is a process that transforms and brings with it obligations and responsibilities.[21]

It is also possible to form an I-Thou relationship with modern materials and machines as illustrated by the following stories, which are examples of knowledge gained by direct, hands-on experience rather than through teaching or books. In the first, Mike Cooley describes a good tradesperson – a dying breed in our high-tech world:[22]

We have ordinary maintenance fitters who go to London Airport if a generator system is causing a problem. The whole aircraft might perhaps be grounded because of it. One of these fitters can listen to the generator, make a series of apparently simple tests – some of the older fitters will touch it in the way a doctor will touch a patient – and if it is running, will be able to tell you from the vibrations whether a bearing is worn and which one. The fitter will subsequently make decisions about the reliability of that piece of equipment, and upon that decision, the lives of 400 people may directly depend. ... if you asked those "ordinary people" to describe how they reached that decision, they could not do so in the usually accepted academic sense ... yet that conclusion will be right, because they will have spent a lifetime accumulating the skill and knowledge and ability which helps them to arrive at it.

The second story is about a traditional African blacksmith:[23]

The (Tanzanian) blacksmith's knowledge is entirely based upon accumulated empirical experience, which has been 'inherited' from one generation of blacksmiths by the next. The knowledge is part of the blacksmith's senses and cannot be separated from him. If you give a blacksmith a piece of scrap, he will first test it. He weighs it a couple of times in each hand, heats it in the furnace and observes how long it takes to reach a certain color. Finally, he beats it and looks at the sort of sparks it gives, listens and feels how the iron 'responds' and perhaps he will also smell the sparks. Meanwhile he mumbles, as if he is talking to the steel. After testing it, he knows what sort of steel he is working with, furthermore, he even knows for which purpose this steel is best suited. He cannot convert his knowledge into the percentage of carbon content, but he 'knows'.

Sadly, I-Thou relationships are easily displaced by our emphasis on objectivity and intellectual knowing, and by the utilitarian I-It values of our industrial, consumer society. But they are an essential aspect of the new science of oneness.

Approaches to Oneness

The science of oneness is based on the belief that everything that exists is one whole, and that the apparent independence of objects, events and processes is illusory. Broadly speaking, there are two approaches to understanding this oneness that I have called "the one from the many" and "the many from the one." The first is the approach of mainstream science and modern western culture. And the second is more akin to the approaches of mystics and indigenous peoples. The science of oneness seeks to harness the strengths of each in a synthesis of the two.

The conceptual separation of objects, events and processes is deeply rooted in western culture as well as in modern science. From this perspective, the idea of oneness arises from the perception of connections between independent phenomena that ultimately link the many into one whole. Two examples of this process are classification and systems science.

Classification is a basic scientific activity. Biologists group individual plants and animals into species, similar species into a genus, related genera into families, and so on all the way up to the three basic divisions of life. In a similar way, geologists classify rocks, agricultural scientists classify soils, and physicists classify atoms. And in everyday life we classify objects as chairs or tables, which belong to the more general category of furniture, and so on.

This process not only makes life easier by simplifying nature's complexity, but also reveals a kind of oneness in which everything is related in a hierarchy that extends all the way from sub-atomic particles. But this simplification comes at a cost. Whenever we group things, we tend to lose sight of the richness and diversity of the individuals, and to see any object simply as an example of a category once we've identified it. That yellow flower is just a dandelion, and with that judgment I fail to notice its magical beauty and symmetry, its rosette of dentate leaves, and the uniqueness of that particular flower. I tune out the particular object, seeing simply an example of a type of weed. This is a perception of unity, not of wholeness, because it reduces individual richness to a flat, uniformity.

In the last few decades, the one within the many has been revealed by other fields of science too. As we'll see in Part II, systems sciences have shown how objects and processes are connected to form interactive wholes with different properties from those of their isolated parts; and how small systems are connected in ever-larger wholes up to the scale of ecosystems, the planet Earth and beyond. Also, quantum and relativity physics, which we'll discuss in Part III, have both shown that the universe is an indivisible whole beneath its apparent fragmentation.

These deeply holistic visions have inspired many people, but seem to have made little impact on the way most scientists go about their work. Their approach is still mostly analytical, and relatively few scientists study systems as wholes, entities or beings in their own rights. Thus, systems scientists still tend to focus on individual parts and processes, and then use mathematics to simulate the functioning of the whole. Similarly, medicine still tends to lose sight of the organism amongst the organs and biochemical processes, whilst biologists often fail to see the ecological wood for the trees. And physicists are still intent on discovering the fundamental particles that they believe will explain life, the universe and everything.

The alternative approach that sees the many emerging from the one is more difficult to describe briefly because it is a different type of perception to that of modern science. It is most easily approached through the ideas and methods of Johann Wolfgang von Goethe who was a prolific scientist as well as a famous poet.

Goethe developed an intuitive and holistic approach to science which involved two steps.[24] First, he looked actively at what he was studying in order to see it clearly and deeply, free, as far as possible, from the blinkers of established theories, classifications and mental models. He then deepened his understanding by reliving this sensory experience over and over in his imagination, seeking to comprehend its unity without analytical or abstract thought.

Underlying Goethe's method was a belief that every phenomenon is a manifestation of the potential of a deeper reality. He sought to understand that underlying whole through the particular example he was observing. Hence, instead of using the intellectual mind, which sees what our mental models lead it to expect, Goethe sought an encounter with the phenomenon as it really is through sense perception and a change of consciousness. Rather than fragmenting the phenomenon by analysis, he focused on the unique whole with its inner diversity and richness. Rather than standing back as an objective observer, he entered fully as a participant into the sensory experience. Rather than quantifying what he was studying through measurement and mathematics, he sought to make it visible to the imaginative mind as it really was. Rather than replace the phenomenon with an equation, he sought insight into what he had seen. Rather than seek causes, he looked for circumstances in which the phenomenon occurred.

These ideas can be illustrated by analogy with reading. When we see the word 'cat' we don't spell out the letters, but immediately see the whole word and apprehend that it means a particular kind of animal. In this way, we encounter the deeper reality behind the printed text. Goethe sought to read nature in a similar way. By contrast, normal science is like seeing the separate letters c, a, t and seeking to comprehend their meaning by measuring their shapes and spatial relationships, and looking for an equation that will reproduce the pattern.

Another illustration is Goethe's way of seeing plants. He believed that each part of a plant is a different expression of one organic whole. Hence, he saw every part as a metamorphosed leaf, with a gradual transition from stem to leaf to petal to stamen. The truth of this perception is clearly apparent in water-lily flowers in which the transition from petals to stamens occurs in stages so that one form flows into the other. This way of seeing uncovers no new facts, but reveals the unity and flow of growth, development and transformation through which plants create their form.

At a deeper level, Goethe believed each particular plant is an expression of an archetypal Plant. This archetype is not an abstract generalization of a 'typical' plant, but a dynamic whole with the potential to take on the form of any plant. Thus, the plant kingdom consists of diverse expressions of this single archetype. And individual plants are not produced according to an inner blueprint, but emerge from the Plant by a self-determining process of manifestation.

Seeing in this way reveals that the many *are* one; that the many emerge from the One rather than the other way round. Each particular example is a unique expression of the whole, and hence cannot be considered independently of the whole. The interaction of the parts enables this whole to create itself; and the whole is the context that gives meaning to the parts. But this wholeness is not a separate attribute. It is a collective expression of Oneness in which all the parts participate. It is the innermost identity, soul or spirit of the person; the character or 'treeness' of the tree. In this One we encounter an autonomous Being with its own goals, direction and purpose expressed as a drive to actualize its potential.

For those of us who have absorbed the scientific worldview of mechanical cause and effect, genetic determinism and all the rest, Goethe's approach to Oneness is hard to understand and harder to accept. It requires us to enter an I-Thou relationship with the phenomenon being studied; to allow nature to reveal herself rather than imposing our own theories; and to open ourselves as intimate parts of the phenomenon rather than being detached onlookers. We are asked to encounter the One as a lover, not measuring and analyzing every part, but responding to its whole Being – body, mind, emotions and spirit – and allowing ourselves to be visible without barriers or defenses.

This is the spirit in which John Muir, founder of the Sierra Club in the USA, lay on the rocks 'thinking like a mountain' in order to understand the formation of the Sierra Nevada mountains. And this is the spirit in which Dorothy Maclean, of the Findhorn Community in Scotland, learned to communicate with plants and receive instructions on how to grow them.[25] It is the opposite of the classical scientific approach enunciated by Francis Bacon in which information must be extracted from nature by force, interrogation and torture.

Goethe's way is not simply a different technique, it is a different approach to science. Its philosophy and values are based on critical subjectivity rather than objectivity. It emphasizes intuition and spiritual knowing over analysis and mathematics. It focuses on the uniqueness, diversity and richness of phenomena, rather than the similarities between them. And it aims for an intimate relationship between nature and the scientist. But Goethe's way is complementary to mainstream science rather than a competitive alternative. The science of oneness needs scientists who can work in both ways; who can simultaneously see the separateness of the parts and their organic unity.

In Conclusion

The science of oneness is a new, holistic approach to knowledge based on a belief in the unity of all existence. It includes what can be known through the scientific method, but goes beyond this to encompass the humanities and creative arts; the contributions of intuition, spirituality and relationship as well as the intellect; and the wealth of perspectives brought by diverse cultures, languages, religions and types of intelligence. The science of oneness uses other forms of rationality as well as mathematics and logic; and understands the emergence of the many from the One, as well as the connection of the many to form the One.

4 Reliable Knowledge and Wisdom

One of the greatest challenges facing any way of knowing is to discriminate trustworthy, reliable knowledge from nonsense. The great strength of the scientific approach is its ability to test knowledge through objectivity, quantification and replication. But these methods are not applicable to other potentially valid ways of knowing such as intuition, spiritual knowing, and relationship. So how can we make sure that the science of oneness produces reliable knowledge and not delusion, fantasy, or mass hallucination? This is the subject of the first two sections of this chapter. The third section goes on to explore how we may apply the science of oneness wisely.

Here is such a storm of superstition
And humbug and curious passions, where will you start
To look for the truth?

Christopher Fry

Testing the Reliability of Knowledge

We can never be sure that any knowledge, scientific or otherwise, is true. But we can do our best to test its validity. In the case of the science of oneness, with its diverse ways of knowing, this can best be done by balancing the polarities of intellectual and intuitive, objective and subjective knowledge. On the one hand, we need to cultivate subjective insights that come through intuition, spiritual knowing, and I-Thou relationship; and trust the knowledge and wisdom of the heart. On the other hand, we need constantly to challenge it with our heads; maintaining a vigilant, questioning and critical attitude. We must be willing to live with the uncertainty that what we currently believe to be true may turn out to be mistaken. And we need to complement our personal inner knowing with knowledge that arises from collective processes.

The Reflections sections at the end of each Part of the book will help to develop skills in intuitive, spiritual and relationship knowing, and encourage confidence in this heart knowledge. The remainder of this section suggests some key questions for assessing the reliability of any knowledge, and the next section goes on to describe an appropriate method of harnessing the collective insights of a group.

How trustworthy and credible is the source of this knowledge?

We take much of what we 'know' on faith in an authority such as a parent or teacher; doctor, scientist, priest or guru; newspaper, TV, book or sacred text. This is inevitable because we cannot test everything for ourselves. So an important part of assessing the reliability of knowledge is to judge the trustworthiness of our sources. We can do this by asking questions such as: "Has this person studied the subject, or have direct experience of it?" "Is this person or organization known for their truthfulness and integrity?" "Is there a hidden agenda or motive here?" "Is this paper (or magazine, TV channel, author, etc) known for the accuracy of their information?" "What have other experts said about this idea?" "What is my gut feeling about this?" "Does it make sense to me intuitively as well as intellectually?"

For desired conclusions ... it is as if we ask ourselves, "Can I believe this?", but for unpalatable conclusions we ask, "Must I believe this?

T. Gilovich (1991) p.83

Does this idea work when applied in practice?

Jesus advocated testing spiritual knowledge by its practical results when he said "you will recognize them by their fruits."[1] But this leaves us with the challenge of distinguishing good fruits from bad. In other words, before we can assess the pragmatic outcome of any knowledge, we must first answer questions such as: "What do we want?" "What is of value to us?" "What is right, just or good?" "How should we balance competing goals?"

Without wise answers to these questions, pragmatism can lead us dangerously astray. For example, scientific knowledge is often judged by its ability to predict the behavior of the material world, and hence by the power it gives to create technologies. On this criterion, the natural sciences are spectacularly successful. But if what we actually want is to lead happy, fulfilled lives within harmonious societies and flourishing ecosystems, we might reach a less positive conclusion.

Is this idea or information self-consistent, and consistent with validated knowledge obtained in other ways, and from related fields of study?
The 'bootstrap' theory of physics proposes that the most fundamental law of nature is that all other laws must form a consistent whole. If they did not, the natural world would disintegrate due to its internal inconsistencies. Hence, in principle at least, we can test any knowledge by its consistency with all the rest. In practice, however, inconsistencies abound because our knowledge is fragmented and incomplete.

For non-scientific knowledge, we can ask not only "Is this idea internally coherent?" but also "Is this idea similar, different but complementary, or contradictory to accepted 'truths' of our culture, and of other cultures?" "Does this idea add a new dimension or depth to existing knowledge?" "Is this idea ethical and just within our culture?" and "Does this feel right in the context of our society?"

Does this make sense according to an established system of reasoning or logic?
'Reason' is taken here to mean the ability to solve problems or reach the 'truth' of some issue through thought processes. Reason must be coherent and consistent, but need not conform to modern western logic. It could, for example, be the Taoist yin-yang or the reasoning of an indigenous people. Sometimes it may be helpful to investigate how knowledge stands up to different types of logic.

It is important to recognize that, when there are competing ideas, we don't necessarily have to choose one as right and reject the others. It is possible simultaneously to hold different ideas, each of which expresses a facet of the truth, and that collectively give a deeper insight into reality than any one alone. The 'truth', so far as it exists, may lie in the creative tension. We can see this clearly in the world of quantum physics where the basic components of matter reveal themselves as either particles or waves depending on the situation (See Chapter 9). And a similar tension exists between visions of Spirit as transcendent (eg a father God) or immanent (eg our deepest selves).

Finally, come questions about the extent to which knowledge is shared. Is it supported by other people with appropriate experience and/or

qualifications? Are the methods used and the results obtained publicly available and open to scrutiny? How many times has the process leading to this knowledge been repeated and found to be valid? Does this knowledge arise from only one culture or sub-culture, or has it been found to be valid across cultures?

The most reliable knowledge results when many people with appropriate expertise from different cultures independently undertake the same activity, have the same experience or observe the same thing, and draw similar conclusions. However, caution is still required if the people have been trained in the same discipline, whether it be science or mysticism, and are influenced by the same teachers, leaders or peers.

Further challenges arise in applying this test to inner knowing. My experience, intuition and inspiration are highly personal, and I am the greatest expert on them. There is no way I can communicate exactly what I feel and think, or what an experience means to me. And no matter how clearly I communicate, and no matter how carefully you try to duplicate the conditions, you cannot repeat what happened to me exactly because your inner world is not my inner world. And yet comparison of our experiences with those of others is the only means we have of ensuring that we don't succumb to personal delusions. This dilemma is discussed in the next section.

The power of systematic questioning can be illustrated by considering the validity of the Australian Aborigines' worldview. This makes little sense to modern western minds, and is of scant use to us in interpreting our experiences, or in deciding how to live our lives. European colonists saw it as a failure because Aborigines had never 'progressed' beyond hunting and gathering. But western scientific theories are of similarly little use to traditional Aborigines. Their Dreaming myths provided shared, coherent, logical images of the material world which were consistent with experience, and formed a reliable basis for action within their culture and environment.[2] These myths enabled them to survive for 100,000 years in a very harsh and changing environment, and to develop a way of life with more time for relaxation, art and ritual than most modern westerners enjoy. By contrast, many European explorers died in country regularly inhabited by Aborigines, despite their scientific knowledge and modern equipment. And European travelers continue to die in the outback when their technology fails. So which is the more reliable view of reality? Given the longevity of their way of life, it can be argued that the Aboriginal worldview works better. After all, our much-vaunted scientific culture is in danger of destroying human civilization after a mere few hundred years.

Cooperative Inquiry

Replication of research is one of the keys to the success of science. It weeds out errors, and builds confidence through repeated verification. But how can intuitive and spiritual knowing, and the experiential knowledge that comes from relationships be replicated and verified?

Ken Wilber argues that mysticism is a scientific inquiry into the spiritual realms.[3] Having been trained in practices such as meditation and yoga, seekers are able to explore the spiritual realms for themselves. The validity of their discoveries is then tested by comparison with the experiences of other practitioners. In this way, false claims and wrong turnings are eliminated. However, John Heron and Jorge Ferrer disagree with this interpretation, arguing that the Teacher controls the experience through his or her guidance.[4] Imagine, Ferrer suggests, that a student of Buddhism reports an experience in meditation of the loving, personal Christian God. His teacher would not regard this as evidence that Buddhist beliefs are mistaken, but as a delusion or wrong view. The student would be told to return to his meditation until he saw things 'as they really are'.

Another approach to collective inner knowledge is cooperative inquiry in which a group carries out 'do-it-ourselves' research into their own subjective experiences. It uses a process developed by John Heron and others which relies on critical group reflection rather than external authority. The method has been applied to a wide range of subjects including exploration of professional practices, organizational structures, alternative lifestyles, and different states of consciousness. It could equally well be used by scientists to tap into their collective experience and creative inner knowing; or by any group to probe their worldview or understanding of some aspect of science.

Heron notes that, in traditional social research, the researcher contributes the thinking behind the project, and the subjects contribute the action to be studied. In cooperative inquiry, by contrast, these separate roles are replaced by a relationship in which all participants are co-researchers and co-subjects. Everybody is involved in design and management of the project and the interpretation of the results. And everybody experiences the activity being explored. "This is not research on people, but research with people."[5]

At their first meeting, the group agrees the topic of inquiry, develops a set of questions to investigate, plans how they will carry out their investigation, and agrees how to record what happens. In the second stage, the planned activities are carried out and recorded, with each participant engaging experientially with the research. In the process, understandings may be deepened and clarified, creative insights may emerge, or unexpected events may happen. In the words of one participant: "It demands ... a willingness and capacity to release our grip on favorite conceptual maps, or at least hold them very loose, and to develop subtle discernment in order to avoid self deception ..."[6]

In the final stage, the team reassembles to share their results. They seek to grasp the whole pattern intuitively as well as to understand the results conceptually. As a result, they may develop their original ideas further, reframe them, or reject them and pose new questions. They may decide to pursue the inquiry further, choose a new aspect of their topic to investigate, or change their inquiry methods.

This cycle is repeated several times to check and develop ideas and discoveries, investigate different aspects of the topic, or develop new skills. At the end, all the results are collated, the outcomes are clarified, and either a written report is prepared, or a collaborative presentation is given using dance, drama, drawing or other expressive art. The process naturally leads to a sense of responsibility and a willingness to apply the lessons in action.

As described, the process is deceptively simple. In practice it requires considerable skill to form a cohesive, cooperative and self-critical group in which the involvement of each participant and the group process are balanced. The group dynamic needs to include a degree of hierarchy as this enables guidance to be given by those with greater vision, skill and experience. However, collaboration is essential to connect the individual with a supportive community of peers and enable the input of different views. And autonomy allows expression of the creative potential of each individual.

If the science of oneness is to produce reliable inner knowledge, it is essential that group processes, such as Cooperative Inquiry, be adopted that facilitate open and replicable inquiry into these ways of knowing.

Wisdom

Our world is flooded with information, scientific knowledge and technical 'know-how', but this is not enough. In order to make good decisions, we need wisdom. And lack of wisdom has contributed to the morass of problems we face as a civilization and planet.

Where is the life we have lost in living?
Where is the wisdom we have lost in knowledge?
Where is the knowledge we have lost in information?

T S Eliot

Wisdom is a complex quality. At its heart is the idea of pursuing the best ends by the best means, but our society often seeks the best means without worrying about ends. For example, if something is technically possible, we do it regardless of the consequences. And we pursue economic efficiency, productivity and growth without ever asking ourselves why these are important, if they are right, or whether they help us attain what we really want.

Rational knowledge contributes to wisdom, but is not sufficient. Siu argued that it needs to be integrated with intuitive and spiritual knowledge[7], whilst Ralph Waldo Emerson claimed that it must be tempered by the moral sense of what is right,[8] and Nicholas Maxwell emphasized the importance of seeing what is of value, and striving for it.[9] Often, wisdom emerges through asking the right questions, rather than from finding the right answers; and through remaining open to new ideas and information. Wise decisions do not necessarily conform to norms of right or wrong, but may redefine what is acceptable.[10]

The philosophers of ancient Greece believed that to be wise is to love harmony, beauty, and truth, to know yourself, and to strive for moral perfection.[11] Thus wisdom entails extraordinary personal development. The wise person is balanced, integrated, and whole; self-aware, and in harmony with him- or her-self; empathic, understanding, patient, compassionate and humorous. The wise know how to learn from others' mistakes as well as their own, and are not afraid to change their minds as experience dictates. They weigh evidence and advice, and make clear, sensible and fair judgments that take both long- and short-term consequences into account.[12]

Whatever you want to weaken
Must first be convinced of its strength
What you want to overcome
You must first of all submit to ...

Lao Tzu, Tao te Ching

The association of wisdom in western thought with balance, wholeness and harmony parallels the Taoist search for balance and harmony between yin and yang. We can never know for certain when we are in what Taoists call 'the Way', but imbalance is revealed when our actions have unexpected results, and is more likely as we move towards one end of any spectrum.

Harmony does not mean uniformity, however, which is like a song with one note, or a dance with one pose. Rather, harmony is a dance in which wide-ranging notes and movements blend into a pleasing whole. It is not only compatible with diversity, but is impossible without it. A harmonious community emerges when people with different talents, tastes and tempers all contribute to its creation.[13]

Where can we find such wisdom today? In our complex modern world, no single person, no expert group, no government, can possibly be the fount of all wisdom. But the community at large is replete with ordinary people who collectively have great wisdom. We need to tap this wisdom, and learn to trust it.

The best way to run the world is to let it take its course
– and to get yourself out of the way of it!

Lao Tzu, Tao te Ching

This is not to say that wise leaders are not needed. But wise leaders listen, and draw on the wisdom of their communities. They meld varied opinions into a coherent shared vision, and express that vision in ways that people understand, and that inspires them to work for its achievement. And having inspired them, wise leaders fade into the background, letting their followers believe they did it themselves.

This is quite different to the approach espoused by social scientists who seek to understand human behavior and social systems in order to predict the effects of different policies. At root, this amounts to developing better means of enabling unaccountable experts, unscrupulous politicians and business managers to manipulate people. Rather, the path of wisdom is one of self-reliance, and faith in the collective wisdom of communities that work together to identify and resolve their own problems.

Wisdom seeks to integrate scientific knowledge, human experience and inner knowing into a living whole, and to elucidate the meaning of facts and ideas. It is concerned with the psychological, emotional and social impacts of our knowledge, and in ensuring its good use. In all these areas, scientists have no more authority than anyone else, and the direct experience and inner knowing of each one of us is equally relevant. The search for wisdom requires that we individually and collectively interrogate our knowledge, asking questions such as:

- Does this idea or action reflect an extreme view, or is it somewhere in the middle ground? How does it affect the balance amongst 'the 10,000 things'? Would it bring my model of reality, and my consequent actions, towards the central point of balance and harmony, or carry me towards an extreme?
- If there are conflicting facts and theories, is one more appropriate to a positive worldview? Are there deeper insights to be gained from holding them in tension? Is there a way to integrate them into a coherent whole?
- How well does this insight compare with the teachings of the wise figures of history such as the Buddha, Jesus, Gandhi or the Dalai Lama? Does it spring from love, compassion and understanding for all beings? Or does it come from my emotional needs, and the desire for recognition and power?
- How can we make wise use of this knowledge? How does it affect the way I live my life? If I acted on this knowledge, what effect would it have on the physical, emotional, psychological and spiritual well-being of myself, other people and other beings? Would following this insight reduce environmental degradation and resource depletion, violence and war, and/or hunger and poverty?

The science of oneness cannot guarantee wisdom, but it is more likely to lead towards it than methods based on the intellect and mainstream science alone. The science of oneness seeks to integrate all kinds of knowledge and all ways of knowing, and in doing so naturally has to confront questions about goals and ends, and the rightness and justness of means. Its focus on I-Thou relationships between knower and known requires development of personal attributes such as inner stillness, the ability to listen, and empathy. And its emphasis on personal interpretation of the significance and meaning of scientific 'facts' through cooperative inquiry is consistent with a hands-off style of leadership and trust in the wisdom of the community.

In Conclusion

The science of oneness welcomes and values perspectives and insights from all sources of knowledge. But it takes nothing on trust, continually challenging and questioning what we think we know, testing its validity in the fires of skepticism, and sharpening individual insights on the whetstone of collective knowledge. And beyond reliable knowledge, it seeks the wisdom to use that knowledge well, not only for the benefit of humanity but also for all beings, and the Earth.

Review of Part I

Science is meaningless because it gives no answer to our
question, the only one important for us: 'What shall we do and
how shall we live?'

Leo Tolstoy
Quoted by Siu (1957) p.139

Like me, you probably grew up in a culture dominated by the worldview of science. And, like me, you probably absorbed its tenets into the depths of your subconscious as part of the way things are. But, as we have seen, science has its limitations like all human knowledge. Despite its success at unravelling the mysteries of nature, science is based on unprovable beliefs, is riddled with subjectivity, cannot illuminate our inner experience, and has a dark side.

Science must shoulder its share of responsibility for the problems we face as a civilization, as well as accepting accolades for the benefits of technology. Perhaps the most deep-rooted harm comes from the bleak classical vision of a lifeless, uncaring, mechanistic universe at the mercy of blind forces; of a world of separation and alienation of mind from body, person from person, and human from nature; and of a world of unbridled, ruthless competition that lacks consciousness, spirit or love. Is it any wonder that it has given birth to a society of alienated, despairing, lonely, powerless people?

If the universe means a vast machine to us, our whole being will
unfold that meaning in the individual, in human relationships,
and in society as a whole.

David Bohm
Cited by Birch (1990) p.56

In criticizing science so strongly, I do not want to encourage rejection. Science is still the best way ever discovered of finding out how the material universe works, and of harnessing the forces and resources of nature for human purposes. Indeed, the findings of science have been, and remain, an inspiration to me, helping me find the meaning and purpose of my life. But it is important that we see science for what it is, and do not hold unrealistic expectations of it. My aim in this Part has been to counter the tendency to elevate science to the status of holy writ, and to put it into the larger context of human knowledge. I want to encourage a constructively critical attitude, and inspire the confidence to interpret the significance and meaning of science's findings for ourselves, drawing on our own knowledge and inner wisdom.

Science opens a big window onto reality, but we do well to remember that the glass is tinted by the structure of our brains and nervous systems, our personal experiences, the language in which we think, and the culture in which we grew up. Science cannot reveal reality as it actually is, nor tell us why things are the way they are, nor advise us how to use its power wisely. Nor is science an appropriate vehicle for exploring our inner realms. For all these aspects of knowing, we need to open other windows too; to balance the weaknesses of science with complementary ways of knowing in order to gain a more complete, holistic picture.

And so the vision of a science of oneness is of a temple of human knowledge with many windows opening onto reality through which stream beams of light, fusing to form a multi-faceted, rainbow-hued image of the whole. An image which remains incomplete and fuzzy, but nevertheless reflects the Oneness of all existence. And a temple to which we all may bring our small contributions of experience, insight and wisdom, and from which we may go out inspired to act wisely in the world.

Like a bee gathering honey from different flowers, the wise man accepts the essence of different scriptures and sees only the good in all religions.

The Srimad Bhagavatam

Scientific, religious and other authorities always paint their particular brand of knowledge as Truth, and denigrate others. But once we realize that all knowledge is at least partly subjective, our view of authority begins to change. We no longer accept uncritically the superiority of expert knowledge, nor push our own beliefs, values and perceptions onto others.

We learn instead to respect and value different ways of understanding and living in the world. We begin to see that the marvelously diverse images from science, the humanities and the arts, and from the religions and cultures of the world are all partial representations of the true Reality. We begin to see that each reflects a different facet of the wondrous, jeweled whole. And we learn to accept and rely on our own inner authority.

The conceit … is remarkable and monumental in supposing man, with a finite brain, to have all the answers.

J. C. Cooper (1981) p.79

When I decide to believe a scientific theory or follow a particular spiritual teacher, I choose what seems right for me. No-one can make that choice but me. Unavoidably, therefore, I am responsible for my beliefs and actions. As John Heron put it: "No authority resides in anything external unless you first decide to confer that authority on it."[1] But to rely on my own authority alone would be to risk delusion. Hence, I need to share what I think I know with others who will give me honest and critical feedback through a structured process such as cooperative inquiry. And I must be willing to change my ideas in response. In this way I can dare to trust what I know from my own subjective experience, intuition and spiritual awareness, and to build new insights from science into my evolving worldview.

In the West when we want to emphasize a point we may quote an important authority. In traditional societies, personal experience, what a person has actually seen or done or heard in the bush or during a dream or vision, is considered to be the most valuable form of knowledge.

David Peat (1994) p.72

We are free to co-create our own worldviews; ones that, unlike classical science, are positive and life-affirming. This does not mean we can reject well-established scientific facts and 'laws of nature'. But it does mean freedom to choose between competing theories and alternative interpretations of them; freedom to make fresh syntheses of scientific and other knowledge; freedom to interpret their meaning for our lives in our own ways; and freedom to bring wisdom to bear in how we apply the findings of science. And so I encourage you to allow your life experience to speak, and your intuition to guide. Open as many windows onto reality as you can, and gaze fearlessly through them. Tune in to your inner self and whatever outer spirit you may believe in. Trust your inner resonance to tell you what rings true. Choose what brings life, freedom, creativity, love, harmony, beauty and wisdom into the world. But be cautious and hold what you know lightly, remembering that no knowledge is absolute or forever.

The highest wisdom has but one science – the science of all, the science explaining all creation and man's place in it.

Leo Tolstoy, War and Peace

Throughout the rest of the book, I have sought to apply the methods of the science of oneness to uncovering the emerging worldview, and to revealing the deep sense of unity that underlies all things. I have tried to choose wisely amongst competing theories and interpretations, and to distil out the knowledge and insights that we can trust, at least for the time being. I have endeavored to weave science, spirituality and other sources of knowledge into a coherent whole, and to paint a new vision of the universe and our place within it. In the process, I have discovered a Reality that is conscious, purposeful and meaningful, full of spirit and life, power and love.

Reflections on Part I

Before embarking on these Reflections, let me remind you of their purpose and how to use them. They are intended to deepen your understanding of the material presented in the preceding chapters by helping you tap into your intuitive, subconscious knowing. And I hope they will encourage you to trust your own judgment in interpreting and choosing between alternative theories and points of view rather than relying on external authority. If any of the sub-sections do not draw you, do not speak to your heart, then please omit them. And if you don't feel ready to engage in this activity, then please skip the whole section. But if you do so, I encourage you to return to it later.

These Reflections are divided into three sections. First is a meditative review and questioning of our journey so far to help you become more aware of your inner responses to it. This is followed by a brief reflection on the power of final causes in our lives. And the final section contains several thought-provoking quotations on knowledge and wisdom.

A Meditative Review of Our Journey So Far

I invite you to choose a time and place where you will not be disturbed for half an hour, and to make yourself comfortable. Perhaps sit in your armchair or other favorite place, with a notebook by your side in which to record any insights or feelings that arise.

Once settled, begin to focus your attention inwards. Notice your breath as it moves in and out through your nose or mouth, and down into your lungs. Notice the movement of your chest and abdomen, drawing the air in and gently releasing it again. For several breaths, just observe it without trying to change anything.

Now, if you wish, let your attention move to other parts of your body. Is any part tense? Are there any aches or pains? Any discomfort? If necessary, change your position gently to get more comfortable.

Is your mind busy with everyday concerns? Gently try to let these distractions go and return to watching your breath. And as you breathe, say to yourself:

> Breathing in peace
> Breathing out anxieties
> Breathing in relaxation
> Breathing out tensions

After a few minutes, when you feel peaceful and relaxed, you might like to read the questions below one at a time, allowing yourself to reflect on each one in turn. You also may wish to write down any reactions or insights that you want to remember.

How did you feel when the limitations and subjectivity of the scientific method were described? Perhaps you were surprised? Relieved? Disappointed? Angry? Or what?

Does my description of the dark side of science, particularly its negative worldview, seem accurate to you? If not, in what ways does it seem wrong?

Do you think the scientific worldview has an emotional or psychological effect on you personally? If so, in what way?

Does the idea feel right to you that we can never know the absolute truth about reality, and that human knowledge is fluid and uncertain? If it does feel right, does this make you feel anxious, insecure or threatened? Or liberated? Or indifferent? Or what?

How does the idea of the tension of opposites represented by yin and yang affect you? Does it make sense, or is it confusing? Does it reflect your experience of reality, or is it new and strange?

How do you feel about integrating the scientific method with subjective experience and inner knowing as a way to understand reality?

Are you aware of any past experience or relationship in your life that helped form an important part of what you know about the world now?

Are there any experiences you would like to have that would deepen your understanding of life and reality? How might you go about getting them?

Can you remember a time when you had a sudden intuition about something that turned out to be right? If so, do you feel that this was a chance event, or an opening into another way of knowing?

Are you aware of an inner 'voice', 'authority' or sense of knowing that you use to judge the truth? Or do you rely on external authority for your beliefs and actions?

How does the idea of being personally responsible for your own 'truth' make you feel? Strong and confident? Vulnerable? Not good enough? Or what?

Do you believe spiritual practices can lead to valid knowledge? You might like to explore the reasons for your belief.

What activities or places help you most to relax and reflect on life? Are there any others that you feel drawn to learning or visiting?

Do you have any friends with whom you feel it is safe to share your deep thoughts and feelings, who would give you honest feedback, and with whom you might form a cooperative inquiry group to investigate issues raised by the science of oneness?

When you have finished reflecting on these qualities, gently come back to the room and the present. You might like to spend some time recording your reactions in words, pictures or some other way before returning to your normal activities.

The Power of Final Causes

In Chapter 1 we explored the concept of cause and effect, including the goals and purposes that Aristotle called final causes. Science denies the efficacy of final causes, and yet spiritual traditions, personal development facilitators, business management consultants, planners, politicians, visionaries and others all harness their power, and are convinced that it works. Whether we call it prayer, visioning, positive thinking, manifestation, strategic planning, or something else, the pull of the future can change us, our families and communities, and the world if we have faith in our visions and dreams.

You might like to recall an occasion on which a vision or dream of yours, no matter how small, came true. How did it happen? Why do you think it happened?

If you would like to discover more about the process of transforming your life in this way, I recommend Lucia Capacchione's book on Visioning.

Quotations about Knowledge and Wisdom

Do not believe in anything simply because you have heard it. Do not believe in traditions because they have been handed down for many generations. Do not believe in anything because it is spoken and rumored by many. Do not believe in anything because it is found written in your religious books. Do not believe in anything merely on the authority of your teachers and elders. But after observation and analysis, when you find that anything agrees with reason and is conducive to the good and benefit of one and all, then accept it and live up to it.

The Buddha

Knowledge ... is a loving union, and what is loved cannot just be the information gained; it has to be the real thing which that information tells us about. ... first comes the initial gazing, the vision which conveys the point of the whole. This vision is in no way just a means to practical involvement, but is itself an essential aspect of the goal. On it the seeker's spirit feeds, and without it that spirit would starve.

Mary Midgely[1]

(The Native American) People's relationship with plants, animals, rocks, and trees serves them as a sort of electron microscope. By entering into direct relationship with the animals and the Keepers of the animals, The People were able to gain access to the knowledge they have about the world. Native people ... not only have knowledge that comes from direct experience, but access to the knowledge of the birds, insects, animals, rocks, and trees. This sort of process of knowing allows one to enter directly into the perception of nature at many scales and levels.

David Peat[2]

Mystics understand the roots of the Tao but not its branches; scientists understand its branches but not its roots.

Old Chinese saying

The wisest man is he who does not believe that he is.

Nicolas Boileau

Man is wise only in search of wisdom; when he imagines he has attained it, he is a fool.

Solomon Ibn Gabirol

The wisest man preaches no doctrines; he has no scheme; he sees no rafter, not even a cobweb against the heavens. It is clear sky.

Henry Thoreau

To know how little can be known.

Alexander Pope

It is in the stillness that all will be clarified;
it is in the stillness that you become at peace
and can find Me.

Eileen Caddy[3]

At the still point of the turning world. Neither flesh nor fleshless;
Neither from nor towards; at the still point, there the dance is,
But neither arrest nor movement. And do not call it fixity,
Where past and future are gathered. Neither movement from nor
    towards,
Neither ascent nor decline. Except for the point, the still point,
There would be no dance, and there is only the dance.

T S Eliot[4]

Part II:
The Science of Systems

Having outlined the science of oneness as an approach to knowledge, it is now time to explore the territory it reveals. The next few Parts explain and discuss the findings of modern science, and reinterpret them in the light of other sources of knowledge and the science of oneness.

We could start our journey at the beginning of time: with the big bang and the evolution of the cosmos, life and consciousness. Or with quantum physics and the fundamental particles of matter from which everything is made. But I have chosen instead to start with the science of systems which shows how the myriad separate objects of classical science interact and connect with each other to form coherent wholes. In the process, we will discover how the one emerges from the many, and how even non-living systems form creative, autonomous, evolving entities which display elementary purpose and will.

Chapter 5 begins with an introduction to simple systems, and moves on to describe more complex ones that behave chaotically and organize themselves. Chapter 6 explores the identity of individual systems and the way they connect with each other to form larger wholes. And Chapter 7 discusses the processes by which even non-living systems evolve. Finally, the Part ends with Review and Reflections sections.

5 The Nature of Systems

Self-regulating Systems

Imagine we take all the parts of a refrigerator and put them in a heap. The heap contains all the properties of all the parts, but it certainly isn't a refrigerator! Now imagine that we put the parts together. Like magic, we have an insulated container that keeps our food cool. The difference between the first and the second situations is that we have arranged the parts in a particular way, and the parts are interacting with each other. The motor, compressor, refrigerant and heat exchangers combine to remove heat from inside the cabinet, and the thermostat controls their action to keep the temperature constant.

The refrigerator has two key properties which are important in most systems. It is *open*, and its interactions include *negative feedback*. An open system is one that exchanges energy, materials, or both, with its environment. In the case of the refrigerator, it uses electricity to drive the motor, and it disperses the heat from inside the cabinet to the surrounding air. Without these two exchanges it couldn't work. In a similar way, an internal combustion engine takes in fuel and air, and discharges waste heat and gases; and I take in food, water and oxygen, and discharge urine, feces and carbon dioxide.

Interactions between the parts of a system often form loops. Part A may be linked to part B, which is linked to part C, which in turn is linked back to part A as shown in Figure 4. Thus a force or substance, energy or information from A is 'fed back' to A via B and C, and can influence the subsequent behavior of all three components. When driving a car, I (A) push the accelerator pedal (B), and my car (C) speeds up. Information about the car's speed is 'fed back' to me through my senses, and I adjust how hard I push the accelerator. The temperature inside the refrigerator is kept constant in a similar way, as shown in Figure 5. A sensor (thermostat) inside the cabinet sends information about the temperature to a switch. When the cabinet gets too warm, the power to the motor is switched on, and heat flows from the cabinet to the surrounding atmosphere. Once the temperature is low enough, the thermostat causes the power to be switched off again.

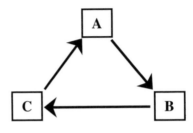

Figure 4   Feedback loop

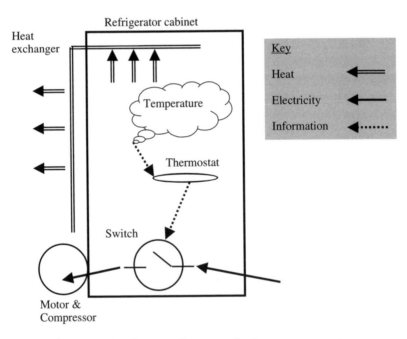

Figure 5   Schematic diagram of refrigerator operation

This feedback control does not keep the temperature truly constant, however, because the temperature at which the power is switched on cannot be exactly the same as the temperature at which it is switched off again. Self-regulating systems are always dynamic in this way, and never achieve a perfectly steady state.

The thermostat is an example of *negative feedback*, where the information is used to maintain a more or less stable condition oscillating about some average. The opposite, *positive feedback*, uses information about a change to increase that change, like a thermostat that makes the refrigerator colder and colder. Such runaway processes are common in real systems, as we'll see later, and would be disastrous if not tamed by complementary negative feedbacks.

This discussion reveals some other important characteristics of self-regulating feedback systems. One is *flexibility* to adapt to changing circumstances, as the refrigerator does when the outside temperature changes. The flexibility of systems varies widely, but there is always a range of conditions within which they can operate successfully. Outside this range, the system will not remain stable. In the case of the refrigerator, the weather may get so hot that heat leaks into the cabinet faster than the cooling system can remove it. And if the outside temperature falls below freezing, the refrigerator will be unable to prevent the inside temperature doing so too.

Systems cope best when all operating conditions are near the middle of their ranges. They become less able to adapt as one or more conditions move towards the extreme end of their operating range. For instance, a refrigerator can tolerate quite large changes in outside temperature if it is normally about 25° C. But if it is close to freezing outside, quite a small drop in temperature may make the food too cold.

Most domestic refrigerators are designed to work on AC electricity of a particular voltage and frequency, and cannot adapt to other energy supplies. But leisure refrigerators are more flexible, and can run on mains power, a car battery, bottled gas, or even solar energy. Such flexibility in a system is particularly valuable when the future is uncertain. After all, adventurous travelers cannot be sure what energy sources will be available at their next stop!

All parts of a system must interact smoothly, and be matched in size and characteristics. The motor of a refrigerator must be large enough to drive the compressor, but not too large or it will waste energy; the heat exchangers must be matched to the flows of coolant and heat; and the whole cooling system must be matched to the rate at which heat enters the cabinet. Any mismatch, any lack of *harmony* and *balance*, within a system will make it less efficient, and may prevent it working at all.

Another characteristic of systems is that they are *wholes*. By putting the parts of the refrigerator together in a particular way, a new property emerges – that of a working refrigerator – which is not inherent in any of the parts in isolation. Remove any part, interaction or process, and the refrigerator stops working. Also, the identity, purpose and functioning of each part depends on its relationship to the whole. For example, the cooling system of the refrigerator could be a room heating system if it was arranged to extract heat from outside the building and dissipate it inside.

Finally, self-regulation implies *purpose*. Once the thermostat has been set, the refrigerator works by itself to achieve its purpose of keeping the temperature at the set level regardless of changes in the weather. However, such a simple system has very little *autonomy*, and is unable to set its own goals. Also, it cannot repair itself when it breaks down, nor replace itself when it wears out. More complex systems can do all these things, as we shall see.

So far I've concentrated on mechanical, human-made systems such as the refrigerator because they are relatively simple and are familiar to everyone. But the concept of a system can be applied to many other things including living organisms.

Chaos

When my son was young he had a cuckoo clock on his bedroom wall, complete with ornate carvings, and a gentleman who came out when it rained. The pendulum of this clock was in two parts. The upper part was moved to and fro by the clockwork mechanism, and the lower, heavy part simply hooked onto it. Most of the time this arrangement worked fine as long as I was careful setting the pendulum going. But sometimes, due to a knock or for no apparent reason, the pendulum would go crazy, swinging wildly around with no discernible pattern. In classical physics, the pendulum and the clockwork which keeps it swinging epitomize the mechanical regularity of the universe. Yet here was one which seemed intent on disobeying all the neat theories. I didn't realize it at the time, but this was a perfect example of what has come to be known as chaos.

Classical science claims that we can predict the behavior of any system if we know enough about it. Yet in the last few decades scientists have been discovering that predictability is the exception rather than the rule, and that most systems behave erratically, or chaotically, in many situations.[1] Chaos took part in the creation of the universe. It can be seen in the motion of stars in their galaxies, planets in the solar system, and atoms in chemical reactions. It appears in the flow of a stream, the dripping of a tap, and the beating of a heart. Chaos helps explain the unpredictability of the weather, rogue waves that sink ships, the variability of bird and insect populations, the spread of epidemics, the metabolism of cells, and the workings of our nerves.[2]

If you watch from a bridge as a leaf floats down a stream, you may see it trapped by a small whirlpool, circulate a few times, and escape, only to be trapped again further down the stream. Trying to guess what will happen to a leaf as it comes into view from under the bridge is an idle pursuit in more senses than one: the tiniest shift in the leaf's position can completely change its future course.

Ian Percival (1989)

So were we mistaken in believing that nature obeys strict laws? The answer is no. Chaotic behavior is not lawless. All the examples I have mentioned obey the laws of nature and, in principle, can be predicted by applying those laws. In practice, however, we can only predict the *typical* behavior of a chaotic system, not its *actual* behavior. For example, we can calculate a sequence of positions for a chaotic pendulum, all of which are perfectly valid according to the laws of physics. But this is only one of an infinite number of *possible* sequences, and it is very unlikely that any actual pendulum will follow this exact one.

An error as small as 15 meters in measuring the position of the Earth today would make it impossible to predict where the Earth would be in its orbit in just over 100 million year's time.

C. Murray (1989)

We cannot predict what chaotic systems will actually do because they are incredibly sensitive to tiny disturbances and errors. Even when we measure the properties of a system as accurately as we can, there are always tiny differences between our measured values and the actual conditions. Also, life is too short to measure *all* the properties of a system, and we normally ignore those which have a negligible effect on the behavior we want to predict. Similar small errors arise because we cannot carry out our calculations with absolute precision, and have to be content with a limited number of significant figures. Thus, there are always tiny discrepancies between the true answers to our calculations, and the numbers that we take as the answers. Even if we understood a chaotic system completely, and our mathematical model of it was perfect, these tiny errors in measurements and calculations would cause our long-term predictions to differ wildly from the actual behavior of the real system.

Imagine a container of gas molecules, billions of billions of them. If the positions and velocities of every molecule were known at one time, Newtonian physics claims it is possible to predict their future movements. But the lifting of a pool ball on the star Alpha Centauri would destroy the prediction within a fraction of a second.

After David Peat (1987) p.51

One of the most popular illustrations of this sensitivity is the so-called *butterfly effect*, named after the butterfly which flapped its wings in Beijing one day, thus causing a storm in Washington a week later. Far fetched? Not really. One of the founders of chaos theory, Edward Lorenz, stumbled on the phenomenon one day when he was using a computer to simulate the weather. He repeated some of his predictions, using data generated by the first simulation as the initial weather conditions. He naturally expected that if he fed the same meteorological conditions into the computer, he would get the same weather predictions out. But, to his amazement, the results of the second simulation diverged rapidly from those of the first one.

Lorenz discovered that this happened because the initial data were not precisely the same in the two cases. In the first simulation, the computer was using data stored internally with an accuracy of six significant figures. In the second case, the data were rounded to three significant figures. These tiny differences of less than one part in a thousand were amplified over time by the simulated system until the two results were quite different.

This sensitivity to initial conditions creates huge difficulties for long-range weather forecasting. Even with an extremely dense network of meteorological stations and super-supercomputers, we still wouldn't be able to forecast the weather more than a short time ahead because small, undetected variations in conditions between the data points would grow and spread until the actual weather was quite different to that predicted. Weather forecasters are trying to overcome this problem in various ways, but useful weather forecasts for longer than a week or two ahead seem likely to remain a dream.[3]

The important point about chaos is this. It is not just that we have too few data with too little accuracy, and that our computers are too slow. Rather, we can never have good enough data or fast enough computers to be able to predict what chaotic systems will do for more than a short time ahead.

The universe itself cannot 'know' its own workings with absolute precision, and therefore cannot 'predict' what will happen next, in every detail. ... the Universe is incapable of computing the future behavior of even a small part of itself, let alone all of itself. ... the Universe is its own fastest simulator.

Paul Davies and John Gribbin, pp.41-42

But if long-range prediction is impossible, how is it that science and technology are so spectacularly successful? The answer is that not all systems behave chaotically. Many are what is called *linear*, or are more or less linear over a wide range of conditions. In linear systems, effects are always proportional to their causes. Push an object and it will accelerate. Push twice as hard, and it will accelerate twice as fast. Hang a weight on a spring, and it will stretch. Double the weight and the amount of stretch will double provided the weight remains below a critical value. Many other natural phenomena are also linear in most situations of interest – for instance wave motions of all kinds, electric and magnetic fields, and heat flow.

The most important consequence of linearity is that the properties of a system can be found by combining the properties of its components – the whole is equal to the sum of its parts as classical physics maintains. This makes it easy to predict the behavior of complex systems because it is possible to take them apart, study the bits, and then simply combine their properties. Also, errors due to imprecise data or calculations grow slowly in linear systems, and there is generally time to correct predictions in the light of actual measurements. For instance, we can predict the motions of the planets with great accuracy for years ahead. And every now and then we can up-date our calculations as the observed positions deviate from those predicted.

Until recently, scientists and engineers concentrated almost exclusively on linear systems, or systems that are almost linear. This has generally made it possible to avoid the effects of non-linearity in practical technologies, but chaos occasionally rears its head in disasters such as the unexpected capsize of a ship or the collapse of a building.[4] It wasn't stupidity or ignorance that led to this narrow focus, but the absence of any way to solve the equations describing non-linear systems – something that only became possible with the advent of high-speed computers.

In non-linear systems, effects are not proportional to their causes, and the behavior of the whole cannot be predicted from that of its parts. Hence, they must be studied as a whole, either in the real world or through computer simulations. Non-linear systems may behave quite predictably under some conditions, but unpredictably under others. In some cases, the zones of predictability are clear cut, but in others their boundaries are infinitely convoluted and detailed, making it very hard to tell whether a system is definitely stable or not in particular conditions.

Although chaotic systems are unpredictable in detail, there are similarities between them which suggest underlying regularities and laws that are yet to be understood fully. Also, the fact that they behave unpredictably does not necessarily mean that they cannot be controlled. If we understand a system well enough, it may be possible to monitor what it does, and then adjust it frequently to keep its behavior regular, much as a thermostat controls the temperature in a refrigerator.[5]

Chaos is even useful in some situations. The sensitivity of chaotic systems means that it may be possible to control them by making a tiny adjustment, observing what happens, and then making further adjustments. One example of this technique was to guide a spacecraft out of a small orbit around a point in space where the gravitational forces of the sun, moon and earth exactly balanced.[6]

> In this orbit (the spacecraft) was extremely sensitive to small changes of direction. The competing pulls of the three gravitational forces meant it was capable of reaching almost anywhere in the Sun-Earth-Moon system with minimal effort. The spacecraft's ... orbit was an island of stability in a chaotic sea of potential trajectories. ... NASA gently nudged the spacecraft out of its orbit and into a sequence of five lunar flybys. Nudging over the next year kept it on track ... This was a chaotic system where a small expenditure of fuel could get the spacecraft from A to B very efficiently. ... The chaos in the system gave NASA a degree of control ... that it could not otherwise have had.

More recently, scientists have worked out how to send spaceships to the moon with a fraction of the fuel used now by making use of chaos. The only drawback is that the journey would take a couple of years![7] Such control may not be possible, however, for complex natural systems like the weather. John von Neumann suggested making small local changes to the weather in order to trigger desirable changes to large weather systems. Edward Lorenz agreed that we could change the weather in this way. But if we did, he argued, we could never know what the weather would have been like without the intervention, and so we could never know whether we'd made it better or worse![8] Nevertheless research on ways to do this continues.[9]

Self-organization

Another key property of many systems is *self-organization*. An example is the emergence of regular shapes as liquids cool and solidify, such as those seen in snowflakes and crystals. If you look at snowflakes under a microscope, you will see that each flake is different to all the others, but that its six arms are identical. The molecular forces which hold the flakes together extend over extremely small distances, and it is still a mystery how the billions of molecules in each arm 'know' what the other arms are doing. Snowflakes and crystals form as the system moves towards *equilibrium,* or a state of uniform temperature in balance with its surroundings. Even more remarkable kinds of self-organization occur in *non-equilibrium* systems that are pushed away from equilibrium by energy flowing in from their environment.

If a layer of oil is poured into a pan and heated gently from below, the surface remains smooth at first. But as the heat is gradually turned up there comes a point when a pattern of hexagons appears. Why? When the heat input is low it is carried from the bottom of the pan to the surface of the oil by *conduction*, or the random collisions of molecules. But as the heat increases, a faster means of transfer is needed. The hexagons are *convection* cells formed by trillions of molecules flowing in coordinated streams upwards at the edge of the cells to the surface and then down again in the middle. Once again, it is a mystery how this coordination is achieved.

Now imagine that you fill your washing machine with small blue and red balls, and watch through the window as it turns. You might expect to see a purplish color, perhaps with an occasional flash of blue or red. You certainly wouldn't expect to see the window switching regularly back and forth between blue and red. And yet this is effectively what happens in 'chemical clocks' in which a mixture of reagents oscillates between two compositions. In other reactions, complex spatial patterns of concentration develop, or waves of high and low concentration pass through the reagents. If the container is large enough, some reactions even create their own boundaries, choosing where the concentration patterns occur. Many systems, particularly biological ones, also grow, develop, evolve, maintain and reproduce themselves, as we'll see later.[12] And recently, self-organizing reactions have been harnessed to create novel computers that can quickly solve mazes or guide robots.[10] In all these examples, unimaginably huge numbers of molecules coordinate their activity at distances tens of millions of times greater than those over which molecular forces are effective. It remains a mystery how they do it.

... if chemical clocks had not been observed, no one would believe that such a process is possible.

Ilya Prigogine and Isabelle Stengers (1984), p.147

At first sight, non-equilibrium self-organizing systems appear to contradict one of the most fundamental laws of nature. The *second law of thermodynamics* states that the universe is running down; that the order and structure we see around us are temporary, and will inevitably decay into disorder. In a process known as increasing *entropy*, hot things cool down, steel rusts, ancient buildings crumble, biological materials decay, and the sun one day will burn out. By contrast, self-organizing systems create order out of disorder, converting randomness into structure and pattern. But, like the simpler systems discussed earlier, self-organizing systems must be *open*, drawing energy from their surroundings to drive their processes. And this flow of energy increases entropy and disorder elsewhere. When the decrease of entropy within the system and the increase in entropy in its environment are both considered, it turns out that the second law of thermodynamics remains in force.

Self-organizing systems, like chaotic ones, are non-linear, and must include both positive and negative feedback. Positive feedback provides the creative force that amplifies tiny internal differences until they transform the system. And negative feedback provides the balancing restraint that prevents the system from destroying itself in an explosion of change. In self-organizing chemical systems, for example, one reagent stimulates production of more of itself in what is known as *autocatalysis*, whilst other reactions *inhibit* production. If the right chemicals are present, all that is needed to spark self-organization is a tiny *non-uniformity* in the concentration of reagents.

Another example of this balance between change and stability is the evolution of the early universe. A very slight non-uniformity in the distribution of matter enabled gravity to draw more matter towards the denser patches. This increased the gravitational imbalance and speeded up the concentration of matter. Without some constraint, however, this positive feedback would have continued until all matter disappeared into black holes (see Part IV). But as particles of matter move closer together they experience stronger and stronger electromagnetic forces pushing them apart again. And so a balance was eventually reached between gravity and electromagnetism, without which the cosmos would not exist.

Network Simulations

Network computer simulations have revealed a lot about the extraordinary power of complex, self-organizing systems. Imagine a fishing net with cords joined together by knots to form a grid. Each knot, or node, represents a component of the system such as a molecule, species or person, and can take on certain values. The cords, or links, between the nodes represent the way the components interact with each other. Hence, the value of each node is set in response to the values of the other nodes to which it is connected according to predetermined rules. In some cases, the connections may be regular, the links forming a normal net with a rectangular pattern of openings. But in other cases the connections may be more like a tangled web as shown in Figure 6. To start a simulation, a value is chosen for each node, the pattern of connections is set, and the 'laws' governing the interactions between nodes are determined. The values of all the nodes are then changed according to the established rules. This process is repeated over and over again to see what happens.

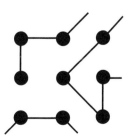

Figure 6   Part of a randomly connected network with 2 links per node

Of particular interest here are Stuart Kauffman's explorations of how networks behave if the connections between the nodes and the rules of interaction are set randomly, and changed after each cycle.[12] Not surprisingly, the behavior was chaotic if each node was connected to several other nodes. However, when each node interacted with only two others, stable patterns of behavior emerged despite the fact that the rules of the game changed continuously!

In one simulation, Kauffman used a network of 10,000 nodes, each of which could have the value 0 or 1 and was connected to two others. Which nodes were connected, and the rules for changing node values in response to these connections, were randomly selected and changed with each calculation cycle. To all intents and purposes, this network has an infinite number of possible patterns of values: 10^{3000} (ie 1 followed by 3000 zeros) compared with 'only' 10^{72} elementary particles in the universe! Despite this vast choice, the system settled into a repetitive cycle of about 100 different patterns. If any of these patterns was changed slightly, the network returned to the original cycle. But large changes, or a different initial pattern, could lead to a different cycle. However, in all there were only about 100 different cycles that the system adopted. In other words, the system 'visited' no more than 10^4 (100 x 100) patterns out of the potential 10^{3000}.

To illustrate just how extraordinary this is, imagine that each possible pattern is represented by a point on a large board, and that the points are spaced 1mm apart. If this board stretched right across the known universe it would hold only 10^{60} points, so 10^{2940} universes would be needed to represent all possible patterns. And out of this unimaginable array, the system always chooses to circle around one out of 100 sets of just 100 points!

In Conclusion

According to classical science, the material world is made of isolated, inanimate objects pushed around by impersonal forces, and at the mercy of the law of cause and effect. By contrast, this introduction to the nature of systems has revealed a world in which objects connect and communicate to form larger wholes, or systems, that regulate and organize themselves; that are flexible, adaptable, and creative; and that display primitive forms of autonomy and purpose. But beneath the surface of this unpredictable, chaotic, self-organizing world, we catch glimpses of a deeper order in which the behavior of trillions of molecules is coordinated in physical and chemical processes, and complex networks regularly choose a few preferred patterns from the their near-infinity of potential states.

In the next two chapters we will also see how all systems are connected into a single whole, and non-living systems evolve creatively towards ever-greater diversity and complexity.

6 Of Parts, Wholes and Holons

In this chapter, we explore what makes a system a system. First, we look at defining the spatial boundary, and then at the internal composition, structure, processes and functions. Finally, we investigate how systems connect with each other into ever-larger wholes, like nested Russian dolls from the smallest to the largest scales.

The Boundaries of Systems

Whenever we want to study or simulate a system we must first draw a boundary to define what is part of it and what is outside, in its *environment*. In some systems, the boundary seems obvious. You buy a refrigerator in a box, and plug it in. It is a complete machine that is designed and made as a whole, and that can be moved from place to place without changing the way it works or what it does. It only needs energy from its environment in order to work.

However, the boundary becomes less obvious if we change viewpoint. The refrigerator can be viewed as a collection of *sub-systems* including the cooling system, the control system, and the storage system. A mechanical engineer would be most interested in the cooling system, and might consider the rest of the refrigerator to be part of its environment. Similarly, the cabinet would be of most interest to an industrial designer.

Each of these sub-systems can be divided in turn into sub-sub-systems. Thus, the cooling system consists of an electric motor, compressor and heat exchangers. A company supplying components to the refrigerator manufacturer might be interested in only one of these, and would treat it as a whole system. This process of subdivision into smaller and smaller systems can be continued right down to the level of atoms and fundamental particles, the systems of interest to physicists.

In the opposite direction, the refrigerator is part of larger technological, social and economic systems. Depending on the reason for defining the system, it might be regarded as part of:

- A family household system;
- A food production, distribution and storage system;
- A manufacturing system in which raw materials, energy and labor are combined to produce refrigerators;
- An economic system in which refrigerators are made, transported, sold, used, repaired and eventually discarded or recycled;
- An energy system in which fossil fuels are burned to generate electricity which is distributed to users and dissipated as waste heat;
- And so on.

In all the above examples, the boundary seems clear once the viewpoint has been chosen. But in many cases it is more obscure. At what point, for instance, does a drop of sweat oozing out of a pore in my skin cease to be part of me and become part of my environment? At what point does an oxygen molecule absorbed by my lungs become part of me and cease to be part of the air? Are the myriad bacteria which live in my guts, and on which I depend for my nutrition, part of me or not? And so I could go on.

The boundaries between systems and their environments are largely arbitrary. We create them in our minds, and impose them on the world for our own purposes. In reality, systems interact seamlessly with each other from the smallest to the largest scales. Nevertheless, it makes sense to choose boundaries that minimize the number and strength of interactions between what is inside and what is outside. It would not be useful, for example, to choose a boundary that arbitrarily chopped the refrigerator in two, or that excluded some atoms of a molecule.

The division of the perceived universe into parts and wholes is convenient and may be necessary, but no necessity determines how it shall be done.

Gregory Bateson (1985) p.47

System Identity

Apart from its spatial boundary, what is it that makes a system a system? Is it the atoms and molecules of which it is made, its physical arrangement of parts, the functions it performs, or the processes that happen inside it? Let's explore some possibilities.

We may transpose a melody a few octaves higher or lower and it still remains essentially the same melody, although this transposition may be such that the two variations of the melody have no single individual tone in common.

A. Angyal (1981) p.37

We each developed from a fertilized ovum, to a blob of cells, a fetus, an infant, a child, an adolescent, and an adult. And we go on developing and changing into old age. Day by day, cells in my body die and are replaced. Molecules which yesterday were carrots and cows, are parts of me today. Yet despite this constant change, I am a unique person who others recognize as an identifiable individual. Even more dramatically, a butterfly starts as an egg, becomes a caterpillar, then a pupa, and finally an adult insect. Each of these stages is very different in form and behavior to the others, and yet we recognize a common identity underlying them all.

From these examples, it is clear that the identity of living systems must depend on something other than the quicksand of their changing bodies. And the same applies to many physical systems, too. There are old Dakotas still flying in which every component has been replaced over the last 60 years, and yet they are still regarded as the same aircraft. So what is the unique essence that makes any system recognizably itself as time passes and its material structure changes?

Richard Dawkins points out that the atoms in a gene change every few months. So genes that have survived for millions, or even billions, of years are not fixed physical structures but patterns of information which are passed from generation to generation like sacred scriptures.[1] Organizations are similar. A large corporation remains essentially unchanged as personnel come and go, its offices and factories move, and it is periodically restructured. What makes it an identifiable system are the rules and information which define who belongs, how they relate to each other, and what they do.

Even mechanical systems can be defined in terms of processes and information rather than structure. In the last chapter, I described the refrigerator in terms of physical components such as the cabinet and compressor. But a refrigerator may also be defined by its function to keep food cool; the processes of cooling and temperature control that it embodies; and exchanges of electricity and heat with its environment. When we think of it this way, the structure becomes less clearly defined, more fluid. For instance, a refrigerator does not need a motor and compressor in order to keep food cool, but a process that removes heat from the cabinet. This usually is powered by electricity, but it can be heat (such as a gas flame), magnetism, or even sound waves.[2]

We naturally tend to perceive systems as material structures because this is what we can see, hear and touch. Hence, we think of a system's functions and processes as being carried out within, or by, the structure. But the above examples suggest that the information and processes embodied by a system are more fundamental than its physical form. Structure is determined by information and process, not the other way around.

This distinction between the structural and process views of systems, is itself largely arbitrary. Many systems change very slowly compared to the human timeframe of years and decades. The sun and moon, rocks and non-radioactive atoms all appear eternal, and it is often appropriate to treat such systems as constant structures. But if we change our timeframe, our perceptions also change. In infinite time, everything changes and nothing is a fixed structure. But in an infinitesimal time span, nothing changes and everything is static![3]

The best image of process is perhaps that of the flowing stream, whose substance is never the same. On this stream, one may see an ever-changing pattern of vortices, ripples, waves, splashes, etc, which evidently have no independent existence as such. Rather, they are abstracted from the flowing movement, arising and vanishing in the total process of the flow.

David Bohm (1980) p.48-49

Although the embodied processes are fundamental, systems do have their own structures or physical identities which are maintained or reproduced unchanged over time with great persistence. This is illustrated well by the shape of chemical crystals, the reproduction of living things, and the network simulations described in the last chapter. The origin of such forms is one of the most enduring mysteries of science. Why this is so was explained by Rupert Sheldrake using the example of a foxglove plant. The plant has a definite position, mass, energy, and temperature; and is composed of a variety of chemicals in particular proportions. It has electrical phenomena going on within it; absorbs a certain percentage of the light falling on it; and transpires a certain amount of water per hour. But it is more than all these measurable quantities and rates; it remains irreducibly a foxglove. In Sheldrake's words:[4]

> As the plant grows, it incorporates into itself matter and energy taken from its surroundings; when it dies this matter and energy are released, and the form of the plant breaks down and disappears. As the material form of the foxglove comes into being and as it vanishes, there is no change in the total amount of matter and energy in the world, but there is a change in the way in which the matter and energy are organized. ...
>
> It is generally the case that the matter and energy of which things are composed have the potential to be present in many different forms, and so these forms cannot be fully explained just in terms of their material constituents and the energy within them. The form seems to be something over and above the material components that make it up, but at the same time it can be expressed only through the organization of matter and energy. So what is it?

The discovery of self-organization has gone a long way to explaining the mechanics of how particular forms come into existence, but has not completely resolved the mystery, as we shall see in Part V.

Holons and Holarchies

Imagine dividing the universe into systems and sub-systems. We might start by putting a boundary around each subatomic particle, and then draw a second set of boundaries around groups of particles that form atoms. We could repeat this process for molecules, cells, multi-cellular organisms, and so on. What we'd end up with is a foam-like structure with bubbles within bubbles within bubbles as shown in the diagram. Each system integrates the subsystems within it into a new whole, while preserving their independent identity.[5] At each level, "The many become one and are increased by one," as philosopher A N Whitehead expressed it.[6]

Each 'bubble' is both a whole and a part – an entity that Arthur Koestler dubbed a *holon*.[7] It is a complete system in its own right, and is simultaneously a subsystem within a larger whole. Reality is composed of holons, ranging from the tiniest subatomic particles all the way up to the universe, nested like Russian dolls one within the other within the other.

So if we are to understand the universe, we must understand the nature of holons, and the way they interact.

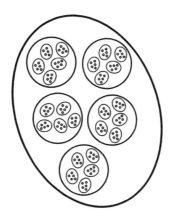

Figure 7   Schematic representation of a holarchy

Arthur Koestler also coined the word *holarchy* to describe the relationship between holons at different levels of organization, such as between atoms and molecules or cells and organisms.[8] Following Ken Wilber, I will refer to a holon's position in the holarchy as its *depth*. The more levels a holon integrates, the *deeper* it is; and the fewer levels it integrates, the *shallower* it is. Hence, a sub-atomic particle is shallow, and an animal is relatively deep.[9]

Every holon constrains the freedom of its constituent sub-holons. Why it does so is illustrated by the way our bodies work. If the heart doesn't function properly, we die. Similarly, the heart cannot do its job unless its myriad cells perform their functions. And so on down to the shallowest levels of the genes and other organic molecules. Hence, in order to survive, our bodies must ensure that their constituent organs and systems work properly.

Necessary though they are, these constraints create a tension between the desire of each holon to assert its individuality and identity, and the need for it to fit harmoniously into the larger whole. Each of us is an autonomous individual, but our freedom is limited by the society in which we live. And the distinct form of the hydrogen atom is modified when it joins with other atoms in a molecule.

Deeper holons normally nurture and protect their shallower holons because they need them to play their parts fully and harmoniously. Repression or destruction is a sign of disease, as when an animal's immune system turns against its own cells. The other side of this coin is that shallower holons have a responsibility to play their parts in maintaining the whole, and a right to have their needs met so that they can fulfill their potential. This is the tension that we all experience in our lives between self-assertion and obedience, independence and community, separation and participation, rights and responsibilities, yang and yin.

Control is also normally flexible. Each holon provides opportunities and sets limits within which its shallower holons are free to fulfill their functions as they choose. Thus, the body as a whole determines the heart's form and function, but the heart sets its own beat. Similarly, each cell is free to feed, respire, excrete, defend and reproduce itself provided it performs its functions within the body. And I am free to live my life how I choose within the legal and moral structure of my society.

This discussion may give the impression that deep holons have unlimited freedom. But in practice they cannot become whatever they choose. They are constrained by the resources made available by the shallower holons which they integrate, and the natural laws which they obey. Hence, the shallower holons define the potential, the possibilities and the limits of what the deeper holons can become. If carbon atoms had not been available, organic molecules could not have emerged, and, without organic molecules, life as we know it is impossible.

A holon has most influence over other holons that are closest to it in the holarchy. Thus, my body controls the nature, arrangement and function of its cells, but has relatively little influence over the molecules within the cells, or the cells of other people. In the opposite direction, the way my cells work affects me, but has no direct effect on other people; and I can influence what happens in my community, but have little effect on communities in other countries. But sometimes a tiny change in a shallow holon is amplified by positive feedback, thus triggering an avalanche of reorganization at deeper levels. For instance, mutation of a DNA molecule in one of my cells may transform my life, or kill me, or introduce a new trait into my offspring. Or a charismatic reformer may spark revolutionary change in society.

There are also less dramatic ways in which a shallower holon can assert its independence. One is to change the length of its cycle of activities so that changes in the deeper holon affect it less. For instance, a redwood tree may live for a few thousand years, and hence must be able to survive significant changes in climate if it is to live to a ripe old age. Few organisms are that tough. By contrast, plants that adopt an annual life cycle can avoid the stress of climate change by moving to a new location through seed dispersal, or by evolving new characteristics. For them, cold spells or dry periods of hundreds or thousands of years are of relatively little concern. Similarly, some organisms increase their independence by creating a barrier between themselves and their environment, such as a cell membrane that prevents desiccation during drought.

The Holarchy of Nature

Shallow holons have very little freedom or power. They are constrained by the demands of level upon level of deeper holons, and have few shallower holons which they can control. However, shallow holons are of fundamental importance because all deeper holons depend on them. Destroy all cells, and we would extinguish life. Destroy all atoms and we would reduce the cosmos to a sea of subatomic particles. The shallowest holons also set limits to what self-organization and evolution may produce by establishing the basic laws of physics and chemistry. By contrast, deep holons are very powerful, but are not fundamental. We humans have great power over the world around us, but we are vulnerable in our power: change the biosphere in any significant way and humanity may become extinct, but destroy humanity and the living planet would continue. Reflecting on such relationships, Ken Wilber claimed, somewhat tongue in cheek, that physics is the most fundamental but the least significant science![10]

We can work out whether one holon is deeper or shallower than another by imagining what would happen if one of them was destroyed. Another indicator is the number of holons, or *span,* at each level. As the depth increases, so the number of holons falls because each successive level integrates many shallower holons. Applying these tests to a few obvious examples, we can conclude that atoms are deeper than subatomic particles, multi-cellular organisms are deeper than cells, and the human mind is deeper than the global ecosystem, or *biosphere.*

We get some surprises, however, when we apply these tests to decipher the holarchic structure of the universe. We normally think of the biosphere as forming a higher level of organization than individual animals. But the biosphere would continue to exist if all higher organisms were destroyed, whereas the death of the planet would exterminate all life. This suggests that the biosphere is more fundamental than the plants and animals within it. Similarly, the extinction of life would not destroy molecules, but life cannot exist without molecules. Hence the biosphere is deeper than molecules. These conclusions imply that the biosphere came into existence at the same time as the first primitive organisms.

Now let's consider the relationship between individual plants and the population of plants of the same species. If we killed all the individuals, clearly the population would also disappear; and destroying the population would exterminate all individuals. So neither seems to be deeper than the other. Thus, contrary to some ecological thinking, they are different aspects of the same level of organization, with the population forming the society of which the individual is a member. Such societies of holons at the same level and within the same deeper holon exist at every level of the holarchy.

A final example of the holarchy of nature is the relationship between human civilization and the biosphere. Environmentalists often claim that human civilization is part of the biosphere. But holarchical analysis reveals that civilization is actually deeper than the biosphere because destroying the biosphere would destroy civilization, but not vice versa. Similarly, mind and consciousness are shallower than civilization, but deeper than the biosphere. However, this does not mean that we can mistreat the biosphere with impunity. The very existence of culture, mind and consciousness depends on the biosphere.

In Conclusion

The boundaries we perceive between systems and the way we define their nature are largely arbitrary creations of our minds as we seek to understand the world by dividing it into appropriate parts for particular purposes. In reality, the cosmos is a giant whole, with holon nested within holon all the way from the tiniest sub-atomic particles to the physical universe, life and consciousness. And within this holarchy lies a universal tension between freedom and control, creativity and determinism. Deeper holons constrain their constituents for their own ends, but within those constraints shallower holons 'do their own thing' and may sometimes spark a 'grass-roots' revolution from below. And deep holons are not free to become whatever they choose because their potential is defined by the resources and possibilities provided by their constituent holons. Everything is connected to, and dependent on, everything else.

7 Evolution

Evolution is not confined to life; it is characteristic of all self-organizing systems, and the whole holarchy of nature. And evolution has not finished its work. Change is often too slow to see, but the world we experience is a snapshot from a continuing drama. Life, consciousness and the cosmos continue to evolve, and will go on evolving for unimaginable eons to come. Who knows what further wonders the process will spawn? This chapter describes the broad principles of the way evolution works. Later chapters provide detailed discussions of the evolution of the cosmos (Part IV), life (Part V), and consciousness (Part VI).

> Whatever this process of evolution was, it seems to have been incredibly driven – from matter to life to mind.
>
> **Ken Wilber (1995) p.3**

Differentiation, Integration and Emergence

According to current theory, the universe began as a fireball of pure energy. As it cooled, particles condensed and then coalesced into simple atoms. Gravity drew this matter together into stars which forged the diversity of atoms we know today. And on at least one little planet, life was born. Life heralded the emergence of consciousness, and with the appearance of humanity came abstract thought, and the evolution of technology, cities, culture, and the arts.

It is no accident that these stages of evolution parallel the levels of the holarchy of nature as described in the last chapter. Indeed, it is possible to define evolution as a process which produces ever-greater holarchic depth. Subatomic particles are shallow but fundamental. Atoms, molecules, cells, higher organisms and mind form progressively deeper, more complex, and more significant levels.

There are two stages to evolution at any level of the holarchy. First comes *differentiation* which creates an increasing variety of holons of the same depth – the zoo of subatomic particles, the periodic table of atoms, myriad molecules, the diversity of life, and so on. This process produces an ever-richer 'society' at a particular level of organization without any change in holarchic depth. It is the transformation of one kind of atom to another, one molecule to another, or one species to another. The nature of the holons produced by this process cannot be predicted, but nevertheless they bear the marks of their common heritage. All atoms are recognizably atoms, all cells are clearly cells, and so on at every level of the holarchy. It is this process of differentiation at the level of living organisms that Darwin's theory of evolution seeks to explain.

In the second stage of evolution, cooperative societies of holons at the same level combine to form a new and deeper holon. This is a process of integration and emergence; of transformation from atoms to molecules, from molecules to living cells, from life to mind. It is a process of going beyond, transcending what exists now, and integrating the previously separate holons into a new whole; a process of bringing new and unpredictable qualities into being, and bringing more of the universe within the bounds of a single holon.

Once this integration has occurred, the first stage of multiplication and differentiation is repeated at this new level. But this does not mean that differentiation stops at the lower levels. The emergence of cells did not prevent the evolution of new molecules, and the emergence of mind did not stop the creation of more non-sentient organisms.

Both stages of evolution involve change, but not all changes are evolutionary. Many are simply reversible alterations in the state of the system, such as a rise in temperature, with no change in structure, function or process. A second type of change is reversible adaptation, as when the body adjusts to the thin air at high altitude. Again, there is no change to the nature of the system. Neither of these is evolutionary change.

Evolution fundamentally and irrevocably changes the form of the system, the way it works, or what it does, creating objects, processes and ideas that have never existed before. The outcome of evolutionary change is not predictable in detail, but evolution moves in a consistent direction towards increasing diversity, complexity, holarchic depth and autonomy.

The Role of Disturbances

The process of evolution depends on a balance between the creative urge of positive feedback and the conservatism of negative feedback. Systems are continually tested by the challenges of existence, and their survival depends on resisting change. Hence holons suppress disturbances whenever possible, and change only out of necessity. Most disturbances are eliminated quickly, but occasionally one is of a type and size that can grow and spread until it changes the nature of the holon itself. Dozens of mutant organisms may be conceived and die without reproducing, but then the first of a new evolutionary line is born. Would-be reformers come and go, leaving hardly a ripple on the surface of society, but then comes a Buddha, Christ, Marx, or Gandhi.

Disturbances may come from within the holon or from its environment. Thus, some genetic mutations occur spontaneously, and others are due to external factors such as radiation or toxic chemicals. Some political revolutions are sparked by international events, but most are the result of internal tensions. Some technological changes spring from changing patterns of demand, but others are simply the product of on-going research and development.

Chance and Necessity

When a new type of holon appears, its environment includes unmodified 'relatives' of the kind from which it sprang. Thus, a mutant organism coexists with normal organisms of its species, and a new technology sits alongside the old one. Evolutionary change occurs when a modified holon survives, either together with the original type or by replacing it. From this perspective, the old and new compete for survival, with *natural selection* weeding out the variants that are less well adapted to their environment. In essence, this is the Darwinian theory of the evolution of life which we will discuss in detail in Chapter 15.

However, the environment of holons also changes and evolves. A new species may be absorbed by its ecosystem without obvious change, or it may spark evolution of the ecosystem itself. And any change to an ecosystem, whether due to a new species or an external factor such as fire, drought, flood or climate change, acts as a disturbance that may severely challenge species which up till then have been well-adapted to their environment.

According to this theory, evolution is a process in which *chance* variations face the *necessity* of survival. Variants are regarded as chance events for three reasons. First, scientific theories are only approximations to reality. They may reliably predict a system's typical behavior, but cannot predict small deviations from normal which thus appear to be chance fluctuations. Second, even the best theories cannot predict the behavior of chaotic and self-organizing systems, as we saw in Chapter 5. And third, systems have limited control over the disturbances that arise. Indeed, until recently it was believed they had no influence, but it is now clear that self-organizing systems themselves generate many of the alternatives on which natural selection works.[1] For example, bacteria can guide their mutations to meet particular environmental challenges (See Chapter 15).

Not only are disturbances unpredictable, but so are their effects. There are times when a holon faces two or more possible futures. Close to such a *bifurcation point*, the outcome may be so sensitive to small disturbances that it is impossible to predict which way it will go. As an example, imagine a ball rolling down a pinball machine, and heading straight for a pin. There are two possibilities – deflection to the left or to the right. Which way the ball goes may be determined by roughness of the pin, a very slight deviation of the ball, a slight tilt of the board, or a small vibration of the machine. The more perfect the machine, the more difficult it is to predict what will happen.

As a result of these uncertainties, we cannot predict when the next genetic mutation will occur, what it will be, or what effect it may have. We cannot predict the next scientific breakthrough, or the impact it will have on society. And we cannot predict with confidence where the next political upheaval will occur, or what the outcome will be.

Evolution and time

Evolution often appears to happen in fits and starts – what biologists refer to as *punctuated equilibrium*. In the fossil record, periods of little change are interspersed with bursts of activity, and similar effects are seen if random 'mutations' are fed into computer models of evolution.[2] Times of rapid change between periods of stability also are common in history. However, rapid evolution may be the result of many tiny changes in quick succession rather than sudden jumps. For instance, a mouse could evolve to the size of an elephant in steps too small to see in just 60,000 years – a period too short to detect in the geological record. The transition would thus give the impression of a sudden big jump.[3]

Not only can rapid evolution occur in small steps, but also this turns out to be the better strategy. A large change is far more likely to make an organism less viable than more viable; or, as Richard Dawkins put it, there are many more ways to be dead than alive. Thus, a strategy of careful probing by small steps in possible directions is more likely to work than dramatic leaps into the unknown.

A similar process occurs in science and technology. Most innovations are just a small step beyond what has been done before, but we fail to see the shoulders the inventors were standing on. Early internal combustion engines were big and crude, unable to compete at first with better-developed steam engines. But the technology survived, and bit by bit it was refined until the unreliable, inefficient engines of the turn of the century became the sophisticated engines of today.

Although technology usually evolves in small steps, there are times when a leap to a new form occurs that is equivalent to the emergence of a new species. The transition from steam to the internal combustion engine is an example. Even more dramatic is the emergence of technologies that were conceptually impossible in an earlier era, such as heavier-than-air flight, nuclear power and electronics. Yet even in these cases, behind the creative leap there was a gradual evolution of theory and practice that eventually made the new form possible.

As evolution proceeds, the holons which emerge become ever more efficient at using resources and performing their functions. Hence, the number of plants and animals in an ecosystem increases with time, and we see a similar effect in the development of technology. When steamships replaced sail, they reduced the cost and increased the speed of transport, thus raising demand for shipping, and hence for ships as well. The jumbo jet had a similar effect on air travel.

In theory, evolution could work backwards, and return a holon to a previous form. But in practice this doesn't happen because there is only a minute chance that exactly the right disturbances will occur, and that the environment will select exactly the right variants. Hence, evolution introduces an arrow of time towards ever-greater organization and complexity that is opposite to that of entropy. Having said that, however, major evolutionary steps between holarchic levels can be obliterated by catastrophes such as the extinction of the dinosaurs 65 million years ago. Even more dramatically, if all life on earth was wiped out, the planetary holon would continue to exist at the shallower holarchic level of crystals and molecules.

Stability, Change and Revolution

Unexpected changes in environmental and social systems are disruptive to our way of life. We naturally prefer stability. Thus it is tempting to prevent evolutionary change by minimizing fluctuations and keeping systems away from bifurcation points. Unfortunately, in practice it is impossible to be sure where the bifurcation points are, and preventing change may lead to the build-up of tensions which must be released eventually. In other words, frequent, small evolutionary changes may be replaced by rarer but larger revolutionary ones. This is analogous to the build-up of stresses in geological faults. If they are released regularly, only minor earth tremors are felt, but when they accumulate for many years the result may be a catastrophic earthquake.

Healthy, resilient ecosystems typically experience quite large fluctuations, and attempts to stabilize them may be disastrous in the long term. An example is fire management in the redwood forests of the USA. Young redwoods grow on the clean forest floor created by fire. So fire prevention, which allows dead leaves and branches to accumulate, inhibits regeneration. Also, when a fire eventually does occur, the plentiful fuel ensures that it is fierce and destructive. A political illustration is the collapse of the communist bloc. Repressive governments and central planning prevented change for decades while internal pressures mounted. When a crack finally appeared, the whole edifice was swept away almost overnight.

A holon that is tested often by large disturbances experiences frequent but small evolutionary changes. This process helps it to remain well-adapted to current conditions, and thus relatively well able to cope with further change. By contrast, a holon that is tested by small disturbances evolves less frequently and in larger jumps. Reducing the size of fluctuations also may reduce the resilience of the system so that it is more easily disrupted. This may be one reason why many fisheries have collapsed despite being managed scientifically to get the maximum sustainable yield each year.[4]

Internal communication is another vital factor. The faster news of a disturbance spreads, the less likely the system is to be able to isolate and eliminate it without changing. This is obvious in dictatorships where technologies such as mobile phones, faxes, photocopiers, email and the internet help dissidents communicate without detection, thus increasing the size of the 'disturbances' they can create, and decreasing the effectiveness of surveillance. This was a significant factor in the collapse of the communist bloc, and is why China is trying so desperately to control access to the internet.

In Conclusion

Evolution is not confined to life, but occurs in many physical and social systems as well. Three conditions are essential for evolution to occur. There must be a source of variation which disturbs the system, and provides the seeds of change. There must be positive feedback to amplify these disturbances, enabling the seeds to grow and develop. And there must be counterbalancing negative feedback loops which prevent a runaway explosion of change.

Evolution occurs in two stages. First, is differentiation at a particular level of the holarchy, as represented by the proliferation of the species of life. Then there is the occasional leap to a deeper holon, a new level of organization, as represented by the emergence of life from a soup of molecules.

Given infinite time, or infinite opportunities, anything is possible.

Richard Dawkins (1986), p.139

The significance of these ideas from the perspective of the science of oneness is discussed in the following review section. And the processes of evolution are illustrated and elaborated in Parts IV – VI in relation to the cosmos, life and consciousness.

Review of Part II

Let's pause a while in our climb to rest and look back over the country we've crossed. What can we see?

Spread out before us is the holarchy of nature, with holons nested within holons all the way from the tiniest sub-atomic particles right up to the whole cosmos and perhaps beyond. We can see how each holon fits seamlessly into the whole, and yet is an independent system in its own right. It is a whole-part. And we can see how the emergent holons at each level of organization both create new wholes, and also embrace the parts without destroying their individuality. As Ken Wilber expresses it, they transcend and include, unfold and enfold, bring forth and embrace, create and love.[1]

Shallower holons create possibilities for deeper ones, both empowering and limiting them. And deeper holons nurture their constituents whilst constraining their freedom in the interests of the whole. Thus healthy holarchies enable each holon to maximize its potential both as an individual and as a part.

> Nature does not premeditate, she does not use mathematics; she does not deliberately produce whole patterns; she lets whole patterns produce themselves. Nature does what nature demands; she is beyond blame and responsibility.
>
> **Peter Stevens**
> **Quoted by Gleick and Porter (1991)p.40**

Continuing to observe the holarchy, we see what appears to be a boundary around each holon, within which it acts as an independent system. But these boundaries shift as our perspective changes, and we realize that they are not real – we have imagined them for our own convenience. We also notice that the components, even the form of each holon is fluid, dynamic, no more than a fleeting reflection of the embodied processes. The material entity that we see and touch is solidified information; a pattern frozen briefly in time.

Zooming in on the activity within each holon, it becomes clear that the way the parts interact is as important as the nature of the parts themselves. And so in order to understand them, we must study systems as wholes. Watching closely, we find that we can predict the behavior of many of the systems within our view. But others do unpredictable things, or re-organize themselves into unexpected forms. This emergent newness is immediately tested, eliminating the unfit, and welcoming the fit. And as we watch, quiet periods of stability are punctuated by times of revolutionary change. But always there is movement towards greater diversity, complexity, depth, and integration.

So much for the observable facts. But what is their significance? What do they mean? To the mechanistic mind, the remarkable behavior of systems is simply what inanimate matter does when it is connected. It arises by chance from the blind outworking of the same laws of nature that control every other phenomenon. Looking through the lens of classical science, many scientists see further evidence of the objective, material, deterministic nature of reality. Our understanding of systems enables us to probe deeper into how things work; to explain the mystery of form in terms of cause and effect; and to push even further into irrelevance the question of final causes, or why things are as they are. The behavior of systems is used as evidence against the existence of meaning and purpose; and to support the case that existence is a fierce competitive struggle.

But there is another, equally valid, interpretation which will come increasingly to the fore as our journey continues. Looking through the eyes of the science of oneness, systems grow, develop, adapt, maintain and actualize themselves. They evolve according to their own dynamic purposes within the broad constraints of their environments, producing ever greater diversity, complexity and integration. Reality is mysterious, purposeful, free, autonomous, creative and cooperative. And whilst we can no longer hope to control the future, we have the power to influence it by acting as small disturbances within the chaotic, self-organizing systems of which we are parts.

These discoveries raise deeper questions about the nature of the universe and the meaning and purpose of our lives. Why is matter self-organizing? Why does it assume certain forms and not others? Are creative emergence and the apparent direction of evolution really no more than the outcome of blind chance and necessity, or is the universe based on meaning, value and purpose? Is there a law of nature that drives matter towards ever greater complexity, balancing and complementing the destruction wrought by the second law of thermodynamics? Might the cosmos itself be conscious and purposeful, striving to actualize its unimaginable potential? Are we responsible for using our power as system disturbances wisely and for the good of the whole? We shall return to all these questions in due course.

A human being is a part of the whole, called by us "Universe" – a part limited in time and space. He experiences himself, his thoughts and feelings as something separated from the rest – a kind of optical delusion of consciousness.

Albert Einstein
Cited by Grof (1990) p.90

It's not that either of these interpretations of the holarchic nature of reality is 'right' and the other 'wrong'. Both are valid ways of looking at reality, and both provide useful insights. The systems view includes and potentially integrates the polarity between objective, materialist science and a spiritual view of reality. And if we can hold this polarity in creative tension, we can gain deeper insights from the synthesis. From this vantage point, purpose takes its place alongside blind chance, freedom flexes its muscles in opposition to determinism, creativity enters the fray against the destructive force of the second law of thermodynamics, chaos balances predictability, community asserts itself against individuality, cooperation infiltrates competition, responsibilities balance rights, and love and service displace impersonal relationships. One end of these polarities explains how things work, whilst the other enables us to explore why it is so.

We are free to choose for ourselves whichever interpretation, or whatever balance point between interpretations, resonates most strongly with our inner selves. If we accept that we are intimately connected parts of the cosmic whole, we surrender some of our treasured individual freedom. And if we recognize the creativity of nature, we lose the reassuring certainty of classical science's predictability. But that freedom and predictability were illusory. In return for these losses, we are freed from the shackles of determinism, and can be confident that what we do, the way we live our lives, does matter and may change the whole future of humanity. We shed our alienation and loneliness, joining the great brotherhood and sisterhood of creation; and we regain the mystery, adventure and joy of life. We become co-creators of the destiny of the universe; holons working in harmony with all the other holons of the cosmos.

Reflections on Part II

In this section, I've included several activities, reflective readings and guided meditations designed to help you deepen your understanding of systems, and connect with your intuitive, inner knowledge of them. I suggest you take your time, equip yourself with a notebook to record your thoughts, and enjoy yourself.

Remember, please feel free to skip any of the reflections that don't draw you, and to move on now to Part III if you wish. But if you do so, I hope you will return later to deepen your intuitive encounter with the ideas. And if you do any of the meditations, they will be more effective if you take the time to record them with appropriate pauses so that you can relax and listen.

Illustrations of Connection, Wholeness and Fragmentation

The properties of any system depend as much on the way its parts are connected as on the nature of the parts themselves. This gives nature almost endless possibilities to play with. You might like to explore this idea by seeing how many ways you can connect 5 toothpicks with BluTack.

All the toothpicks are the same in these arrangements, so it makes no difference if you interchange two parts. But the creative possibilities are much greater if the parts are different. How many different ways do you think it would be possible to connect the 5 toothpicks if they were painted different colors? The answer is in the notes at the back of the book.[1]

*   *   *

Scientists usually describe systems by their measurable properties. The following two activities illustrate the incompleteness of this approach.

Consider your home. Imagine measuring and analyzing the properties of all the parts of your house or apartment, including your furniture and personal possessions, and then putting them all together in drawings and specifications. Would that adequately represent your home? What would be missing?

Now consider your partner or best friend. If you knew everything about their anatomy and physiology, biochemistry and genetics, nervous system and brain, would you know them? What would be missing?

*   *   *

Zen Buddhist Master Thich Nhat Hanh wrote:[2]

> If you are a poet, you will see clearly that there is a cloud floating in this sheet of paper. Without a cloud there will be no water; without water, the trees cannot grow; and without trees you cannot make paper. So the cloud is in here. The existence of this page is dependent on the existence of a cloud. Paper and cloud are so close. Let us think of other things ... the forest cannot grow without sunshine .... Therefore you can see sunshine in this sheet of paper. And if you look more deeply ... you see not only the cloud and the sunshine in it, but that everything is here: the wheat that became the bread for the logger to eat, the logger's father ... the paper is full of everything ... The presence of this tiny sheet of paper proves the presence of the whole cosmos.

You might like to draw a mind map or system diagram of connections in your life. Start with something simple, such as the shoes you're wearing, or the breakfast cereal you ate this morning and work outwards from there.

Thich Nhat Hanh invites us to give thanks to the universe before we eat: "In this food I see clearly the presence of the entire universe supporting my existence."

A Guided Meditation on the Effect of Fragmented Thinking

I invite you to relax. Perhaps close your eyes. Take a few deep breaths.

And now allow your mind to picture a natural river flowing through woods and grasslands. Preferably one that you know from experience.

Imagine its channel with deep still pools and rushing rapids. Its banks perhaps lined with trees and undercut by the flow on the bends. The estuary where it meets the sea. The land around it with its vegetation and buildings. The little streams and trickles that feed it. The fish that live in it, and the birds that live on it.

Imagine looking down at the river from a satellite. It is a dynamic whole which integrates and connects the landscape. It branches like a tree, and like a tree it grows in geological time, reaching ever higher into its headwaters. How many tributaries does it have? And what is their total length? The answers approach infinity. And the smallest channel approaches the infinitesimal.

The microscopic rivulets that feed the river draw nourishment from every square meter of the land, ebbing and flowing with the day's rainfall and the year's seasons. The river is a symphony of rhythms from the deep bass of its great branches to the treble fluting of its rills; resonating at every timescale from the seconds of a brief shower, to the centuries of its growth and development.

Now watch what happens as we treat the river as a collection of independent parts. Farms replace forests on the watershed. As a result, storm runoff increases, and erosion speeds the growth of channels uphill as soil and chemicals are carried away downstream. Watch as towns spring up, their roofs and roads shedding even more water, and preventing the rain from percolating through the soil to the groundwater.

And downstream, see the eroded soil settle out in the river pools, burying the aquatic life and reducing the channel size. Watch as the runoff from a storm cannot drain away fast enough, and floods out over fields and villages, only to be followed by huge machines dredging and straightening the channel and confining the river with levee banks. And see how the river, treated as a drain, becomes a drain – a lifeless, artificial, ugly channel replacing the living, evolving beauty of nature.

And continue to watch over the years and decades, as the rhythms of the river speed up, and the harmony and balance of its processes decay.

> Everything that occurs, no matter how minute and local, is the outcome of all that has occurred before and is the ground for all that will occur thereafter.
>
> **Ervin Laszlo (1993) p.37**

When you are ready, gently release the river from your imagination, and return to the present time and place. You might like to record your experience in some way: a journal entry, poem, drawing, or other medium.

A Guided Meditation to Explore Chaos

We've grown so used to the power of science to predict events that the unpredictability of chaos can come as a shock. And yet our daily lives are chaotic, and very sensitive to small disturbances like burning my mouth at breakfast, missing my bus, or losing my keys. The following meditation explores the potential impact of chaos on our lives, and our emotional reaction to it. I suggest you do it when you will not be disturbed for half an hour; and have paper, pens and crayons or other materials handy for making notes or a drawing afterwards.

I invite you to sit in a comfortable chair, or lie on your back on the floor. Breathe slowly and deeply for a few moments. And relax as you breathe. Let go the tensions in your feet and ankles; in your calves, knees and thighs; in your groin and buttocks; in your belly and back. Relax your hands and arms; your shoulders and neck; the muscles in your face and scalp. Continue to breathe gently for a few moments.

Now recall an event in your life when things could so easily have been different – a narrow escape from injury, or a train you just missed, or a job you didn't get, or a person you didn't say hello to, or a similar near miss. You might like to spend about 15 minutes imagining what your life would have been like if you hadn't missed.

And now let your mind expand to imagine how the lives of others might have been different; how events in the world around you might have been different if you hadn't missed.

And reflect on the power that a deliberate small action may have in changing the course of history.

When you've finished imagining your alternative life, you may like to spend a few minutes writing or drawing or otherwise expressing this alternative biography and your reactions to it.

Now I invite you to sit back and relax again. Close your eyes and breathe slowly and evenly for a few moments. And reflect on these questions:

How does the uncertainty of your life make you feel? What emotions has it stirred? Do you feel threatened, cheated, anxious, exhilarated, or what?

How does the idea that you can change not only your own life but also the world around you make you feel? Do you believe it? Does it worry you that you can never know what effect your actions have?

You may want to spend some time pondering how this awareness and your feelings about it might change your attitudes to life, and to the people and things around you. Do you think this awareness will change you? If so, in what ways? How might this awareness change your life?

Record your responses if you wish before quietly returning to your normal activities.

Exploring Self-organization and Form

The extraordinary power of self-organization can be demonstrated by simple computer models. One of the best known is the game of 'Life', invented by John Conway in 1970. So fascinating is it that *Time* magazine complained in 1974 that "millions of dollars in valuable computer time may have already been wasted by the game's growing horde of fanatics."[3]

> The Life screen ... is a world unto itself. It has its own objects, phenomena, and physical laws. It is a window into an alternate universe.
>
> Shimmering forms pulsate, grow, or flicker out of existence. "Gliders" slide across the screen. ... Much of the intrigue of Life is the suspicion that there are "living" objects in the Life universe.
>
> **William Poundstone, p.24**

Amongst many other things, this very simple simulation shows that stable, structured patterns can evolve from initially random situations. I would love to describe 'Life' here, but sadly there isn't space to do it justice. If you would like to play the game online, and explore its behavior, go to:

http://www.bitstorm.org/gameoflife/ or
http://www.math.com/students/wonders/life/life.html.

To read about it, see Poundstone (1985).

*   *   *

Self-organizing systems often consistently create the same forms time after time. We encountered the mystery of how this happens in Chapter 5, and here we explore it further. I suggest you read this section slowly and meditatively.

Imagine a cup. What is it that makes it a cup?

It is not the matter of which it is formed as this might be clay or plastic, metal or wood. When the cup is broken or melted or burnt, the energy and atoms of which it was formed remain, regardless of the existence or non-existence of cups. And this same energy and matter could be formed into many different objects.

Imagine burning a plastic cup. Its carbon atoms combine with oxygen to form carbon dioxide molecules which are dispersed in the air. One of these molecules might be absorbed by the leaf of a nearby tree, where the carbon atom is incorporated into a sugar molecule and then becomes part of a protein within a leaf cell. This cell may be eaten by a caterpillar, and the carbon atom might end up in a butterfly that is eaten by a bird. And so on through endless cycles and transformations.

The matter and energy of which things are made can be present in many forms. Each carbon atom has the potential to be diamond or coal, a healing drug or gene, a computer or airplane. As Rupert Sheldrake concluded: "The form seems to be something over and above the material components that make it up, but at the same time it can be expressed only through the organization of matter and energy. So what is it?"[4]

A Meditation on Self-Organization

We tend to take self-organization for granted, or attribute it to an all-powerful Creator. In this meditation, we will encounter the miracle in ourselves.

I invite you to make yourself comfortable in your favorite chair, or lie on your back on the floor.

Now focus on your breathing ... Drawing in the life-giving air that sustains the self–organizing processes of your body. Breathing out waste gases from those same processes. And as you breathe let your mind examine each part of your body in turn.

How is it feeling? Relaxed or tense? Strained or peacefully at ease? Let go of any discomforts and tensions you find. Let your chair or the floor support you without any effort on your part. Your body is completely at rest.

And now let your mind gently contemplate the wonder of the self–organizing system that is you.

Consider how you began as a sperm cell in your father and an ovum in your mother. How the fertilized egg divided, and divided again. How this cluster of cells began to differentiate into different types of tissue and organ. How the developing fetus took on various forms before becoming recognizably human. And remember how your infant body and mind grew and developed through all the stages of childhood, creating the person that you recognize as yourself today. Allow yourself to be filled with wonder and awe at the miracle that is you.

Now watch your body as it rests here. Watch how it draws energy and nutrients from your last meal, and the air you are breathing. How it uses those nutrients to repair damaged or worn out cells and tissues. How the energy enables you to move and breathe. How your body rids itself of wastes.

See how your body harmonizes and balances myriad complex biochemical processes. How it maintains a constant temperature, and chemical composition. How it hunts down and exterminates invading organisms.

Watch the flow of thoughts through your mind, far exceeding the performance of the most advanced computers. See how your nervous system carries messages from your senses, and instructions from your brain.

Notice that what you have been watching are processes, or activities. It was the process of life itself that formed you from the fusion of two cells. And it is the process of life that maintains the integrity and identity of your body, just as your body contains and integrates the living processes within it. And so you might see yourself as the harmony of process and form.

Look again, and you may see that these processes, this form, are the embodiment of information. The information gleaned over countless generations of evolution and encapsulated in your genes; the information absorbed during your lifetime from your family, your community, your environment; information from the invisible realms of energy, consciousness and spirit.

And you may recognize the common root of the words 'form', 'inform' (to shape) and 'information' (that which gives shape).

Now I suggest you spend a few minutes quietly contemplating these and other ways in which you were created and are sustained.

And now I invite you to ponder these questions:

Is the being that is you the result of complex self-organizing biochemical and social processes, and nothing more?

Is your existence a meaningless, purposeless, chance event?

If you were to experience yourself as more than self-organized information and matter, what might this extra dimension be?

If your existence has meaning and purpose for you, what are they? Where do they come from?

When you're ready, and in your own time, return to the room and the present.

What did you experience? What questions arose in your mind? What answers did you find? Has the experience changed you in any way?

You might like to write down what you experienced, or share it with another person, or express it in some other way.

A Reflection on Polarities in Reality

In Part I we met the concept of yin and yang; the idea that there is no single truth, no right and wrong, but rather a balance, a tension between opposites such as good and bad, beautiful and ugly. And we saw how we can gain deeper understanding from the synthesis of these opposites. Despite its ancient Chinese lineage, this is a radical idea to modern western minds used to seeking out single, unambiguous facts. And so you might like to reflect on your reactions to some of the paradoxes we've met so far.

I invite you, once again, to sit in a comfortable chair, and take a few slow, deep breaths to relax your body and mind. And then gently reflect on polarities in your beliefs and daily life.

There may be tension between the scientific and spiritual worldviews.

On the one hand are scientists who tell us that reality is objective, material and deterministic; that the whole of reality can be discovered by the scientific method; and that efficient causes are sufficient to explain everything without the complication of formal or final causes.

On the other hand are those who believe there is an inner, spiritual dimension in addition to, or underlying, material reality; a dimension that must be experienced subjectively, and that brings meaning and purpose, love and beauty into existence.

Do you feel that you are at one end of this polarity? Can you see that there is, or could be, truth on both ends? Do you think there may be benefits in holding both views and aiming for a deeper synthesis?

I invite you to let go of your rational mind, and allow your intuitive, inner self to answer.

And now consider some of the other polarities we have encountered on our journey so far:

Whole and part	Individuality and community
Order and chaos	Competition and cooperation
Freedom and determinism	Rights and responsibilities.
Creation and destruction	

Do you experience these tensions in your life? In what ways? Can you see that there may be truth on both sides? Can you hold the tension, and gain deeper insights into your life? Tune in to your inner self and see how it responds.

Now, gently return to your conscious mind and the room around you. And take the time to write down any insights or thoughts that arose.

A Reflection on Evolution

Evolution cannot happen without changes for the forces of selection to work on. And so human progress depends, perhaps, on the maverick, the protestor, the eccentric. As Charles Handy put it:[5]

> George Bernard Shaw once observed that all progress depends on the unreasonable man. His argument was that the reasonable man adapts himself to the world while the unreasonable persists in trying to adapt the world to himself. ...
>
> In that sense we are entering an Age of Unreason, when the future, in so many areas, is there to be shaped, by us and for us; a time when the only prediction that will hold true is that no predictions will hold true; a time, therefore, for bold imaginings in private life as well as public, for thinking the unlikely and doing the unreasonable.

Are you 'unreasonable'? Do you feel you should be? How would your life change if you became a disturbance in the evolving river of life of which you are part? What effect do you think this would have on your family, friends and colleagues? On your community?

Part III:
Relativity and Quantum Physics

The last Part revealed how a single Reality emerges from the universe's myriad independent objects when we see how they are connected in the great holarchy of nature. It is now time to reverse the lens and explore the nature of the One out of which the many evolved. In Chapter 8, we peer with Einstein into the vastness of cosmic space and time. We encounter a universe in which time is elastic and laid out before us like space; in which distances stretch and shrink, and the mass of objects changes; a world in which space and time curve and ripple in dynamic unity. Welcome to the world of relativity, the physics of the very large.

Chapter 9 switches focus, zooming in on the mystery at the heart of matter. To classical physicists, the answer was simple – matter consisted of inert, indivisible atoms. But in the strange sub-atomic world of the quantum physicist, particles flicker in and out of existence, can be in two places at once, can move backwards in time, and can connect instantly with each other across the vastness of the universe. And at the limits, where quantum and relativity theory meet, physicists are striving to develop a unifying Theory of Everything that will reveal the mind of God.

Finally, we will pause for another Review of our progress and a well-earned time of "R & R."

In this Part, we grapple with some of the most difficult concepts of modern science. Often they seem more like the wild fantasies of science fiction than sober fact. And often they are impossible to explain in a way that makes sense in our everyday experience. At these points we must either abandon our quest, or simply accept what the physicists tell us they have found in the depths of their equations and equipment. The intellectual going is tough, so take a deep breath and here we go ...

8   The Physics of
the Very Large

3-D Space and 4-D Spacetime

We are all familiar with three-dimensional space. In order to describe the position of a light bulb in a room we must specify three dimensions: the distance along the length of the room from one end, the distance from one side, and the height above the floor. In a similar way, we can pin-point the position of an aircraft by specifying three coordinates: longitude, latitude and altitude. These coordinates enable us to calculate the distance between any two points in 3-D space using elementary geometry. This distance never changes, and it is possible to move freely in any direction, including back the way we came.

By contrast, relativity tells us that we must extend our thinking to the four dimensions of space and time – or what is called spacetime. Time in this coordinate system is just like another space dimension with one vital difference – we can only move in one direction, towards the future. Unlike movement in space, we cannot travel back the way we've come through time. In this 4-D coordinate system, an *event* can be specified by its 3-D location in space, plus its position in time.

As we dig deeper beneath this superficially simple idea, we will find our everyday perceptions of space and time shaken to the core.

The Perception and Measurement of Time

The sense of time passing is deeply embedded in human experience, culture and language. David Abram speculates that it stems from our spatial relationship with the landscape, thus uniting space and time eons before the advent of relativity physics.[1] He suggests that the concept of the future arose from our spatial awareness of the unknown that lies beyond the horizon, or, more immediately, of whatever is hidden behind a tree or rock. Thus, the unseen future unfolds as we move through the landscape. Similarly, the past is embodied in the trees and animals around us whose forms encapsulate their lives; and in the very ground beneath us in the layers of soil and rock which encode their history.

To us who are committed physicists, the past, present and future are only illusions, however persistent.

Albert Einstein
Quoted by Davies and Gribbin p.82

For those who live close to nature, embedded in her rhythms and cycles, it is natural to perceive time as circular. Night follows day, the moon waxes and wanes, and summer alternates with winter. Flood and drought, heat and cold, wind and calm, birth and death, activity and rest, come and go and come again. In our own bodies we feel the rhythms of breath, heartbeat and menstruation; of sleeping and waking. These cycles are intimately linked to the orbit of the moon, the movements of the sun and stars, and the migrations of animals. And so time and space are less distinct than they are for us. Within this cyclical time, people are alert to newness, and yet conscious that everything is a part of the eternal order of things. Each dawn is both new and yet the same. For such societies, there is no history, no progress. What happens now repeats what has happened before, as reflected in creation myths. Everything is a manifestation of cycles of greater or lesser duration.

This sense of cyclical time is highly developed in Hindu philosophy, which extends the earthly cycles to cosmic scales, with four *yugas* making up the great cycle, or *mahayuga,* of 4.32 million years, and a *kalpa* extending to 4.32 billion years. A day and a night of Brahma is two kalpas, and the life cycle of Brahma is 311 trillion years.[2]

We have largely lost the sense of cycles in our linear, time-driven culture. We experience time as flowing from the unpredictable future, through this fleeting present moment, to the unredeemable past. And yet science also reveals a deeply rhythmic reality of waves, vibrations and cycles. As we'll see, the fundamental essence of the universe is a vibrating energy field; gravity and electromagnetism travel through space as waves; subatomic particles are probability waves; molecules vibrate; planets orbit suns which orbit galactic centers; stars are born, live and die; and the very universe itself may go through phases of expansion and contraction, or cycles of birth and death.

But cyclic time is not the same as circular time. Whilst patterns and events recur, they are never repeated precisely. Today is different in lots of ways to yesterday, this spring is not the same as last spring, and I am older than I was a year ago. So time spirals rather than cycles. Indeed, relativity theory suggests that when the Earth returns to its starting point in space as it orbits around the sun, it actually has moved by one light-year along the time dimension in a curving 4-D spacetime spiral.

Whether we perceive time as flowing in a straight line or as spiraling cycles, we experience it as having a definite direction. We cannot retrace our steps and redeem the past. It's a bit like being able to move only north and never south again. Why is this? Why is time so different to the space dimensions? Why is there an irreversible arrow of time pointing like a signpost to the future as the only way to go?

Physicists have identified several processes that create arrows of time in nature. One is the second law of thermodynamics according to which order and structure in the universe inevitably decay towards disorder as steel rusts, hot things cool down, organisms die and decay, and machines wear out (see Chapter 5). Self-organization and evolution appear to negate this arrow by creating order and complexity out of nothing, but their creativity is always at the expense of decreased order, or increased entropy, elsewhere. So here are two distinct arrows of time, the one pointing towards chaos and the ultimate heat death of the cosmos, and the other towards increasing complexity as evolution of the universe proceeds. Electromagnetic radiation such as light and radio waves provides a third arrow. Whether its source is a star or a TV transmitter, radiation always spreads out as it travels. Never does diffuse radiation from the edges of space bring itself to a focus at a particular place.

We are now in a position to ask how we can locate an event on the time dimension. Neither our perception of past, present and future, nor our experience of cyclic time is of much help. What we need is an absolute measure from a fixed datum. For everyday events, we record the date and time. We calculate the date simply by counting the number of sunrises since an arbitrary starting point, as reflected in the many different calendars still in use. We record finer divisions of time by counting the swings of a clock pendulum, or the vibrations of a crystal. And for longer periods up to the age of the earth, we estimate the 'time before present' of geological strata and fossils by analyzing radiation and the concentrations of different chemical isotopes. But none of these provides an absolute datum that is independent of our planet, culture and technology. For that we need a 'cosmic clock'.

As we'll see in Part IV, cosmic time is thought to have begun at the moment of creation in the big bang, and the universe has been expanding ever since – coincidentally providing a fourth arrow of time. And so we can locate any event in time by correlating it either with the size of the universe as measured by the distance to the most remote galaxies, or with the density of matter which becomes more and more dispersed as the universe expands. This is a very coarse measure on human time scales, but it does provide us with an absolute yardstick to which every event can be related. And we can use geological time, calendars and clocks to provide more precise local measures as needed.

The Relativity of Space and Time

We get even further from the normal idea of time with Einstein's theory of relativity which shows that time varies between observers depending on where they are and how they are moving. This can be explained most clearly by starting with the idea of the relativity of space, dating back to Newton.

Imagine a bouncing ball in a moving railway carriage. If it hits the floor in the same spot twice, passengers will see it moving vertically. But someone standing on the ground outside and watching through the window will see it move horizontally as well because it travels with the train. Thus the positions at which events take place, and the distance between them, depend on the location of the observer. To illustrate this more precisely, suppose the ball bounces up 1m from the floor. Passengers see it move a total distance, up and down, of 2m. But an observer outside the train, would see it leave the floor when it is, say, 2m to his right and land again when it is 2m to his left. In this case, the total distance moved diagonally is 4.5m. Note that the observations from one position are just as valid as those from any other, and so there is no way to define the 'true' distance traveled.

Although space is relative in Newton's theory, time is absolute. In other words, the timing of events is the same no matter who observes them, and the interval between two successive bounces of the ball is the same for both observers. Hence, the observed speeds at which the ball moves are different because it travels different distances. If we now imagine a flash of light reflecting from a mirror instead of a bouncing ball, this means that the speed of light would vary between observers.

But Einstein claimed that the speed of light is absolute – the same for all observers. In other words, if one spaceship was traveling towards a star at almost the speed of light, and another was moving away from it at a similar velocity, the light from the star would appear to have the same speed to the crew of both ships. This is quite contrary to our normal experience. It is as if the rate at which my car approaches another one coming in the opposite direction depends only on how fast the other car is moving and is unaffected by my own speed; or as if the car behind always catches me up at the same rate no matter how fast I go.

This situation can be clarified by considering the railway carriage again. Suppose two balls are thrown with the same force at the same instant from the middle of the carriage towards opposite ends. Before they are thrown, both balls are traveling at the same speed as the carriage, and hence observers inside and outside the carriage will see them hit the ends simultaneously. If, however, two flashes of light are sent towards the ends of the carriage, the times at which they arrive will appear different. From the viewpoint of a passenger, the flashes will reach the ends at the same instant because they travel the same distance at the same speed. However, an outside observer will see the light reach the rear end first. This is because the rear end has moved closer to her and the front end has moved further away in the time the flash of light takes to reach its destination. Hence, the light traveling towards the rear has less distance to travel than that traveling towards the front, but the speed of both is the same.

In practice, these effects are far too small to see without special instruments, or until we learn to travel close to the speed of light. Nevertheless, they have been proved experimentally time after time, and have practical applications in technologies such as satellite navigation, and the synchrotrons used to probe the structure of matter. They also have amazing implications for our image of reality.

One consequence of relativity is that time does not pass at the same rate for everyone. Instead, we must refer to 'proper' time when the clock is at rest relative to the observer, and 'relative' time experienced by an observer moving in relation to the clock. For instance, if I were to travel in a spaceship away from the Earth at 90% of the speed of light, my on-board clocks would run at less than half the speed of those on Earth. Time wouldn't seem any different to me, but when I got back after a voyage of

2 years (ship time), 4 years would have passed at home. So, in the words of John Barrow, "The length of my life depends on how fast I am moving relative to the people who want to talk about such a concept."[3]

What is time? We physicists work with it every day, but don't ask me what it is. It's just too difficult to think about.

Richard Feynman
Quoted by Olshansky and Dossey (2004)

Gravity produces a similar, but smaller effect. Time runs slower close to a massive body like the Earth, and so a person living on top of a mountain ages slightly faster than one at sea level.[4] Indeed, very accurate clocks can detect the difference between the top and bottom of a tower!

Not only is the speed of light absolute, but also nothing can travel faster than light. This means we cannot access our own past or future, even though everything that has happened or will happen is spread out along the time dimension of 4-D spacetime. To understand why this is so, imagine light from some event on Earth traveling away through space. In order to catch it up and see it again, we would have to chase it at more than the speed of light, which relativity theory says is impossible. And in a similar way, electromagnetic events in our future are traveling towards us at the speed of light, which is faster than any messenger can come to warn us. The sun might have exploded a minute ago, but we won't hear the news for several minutes yet!

As we gaze at the night sky, the light we see began its journey anywhere between a few minutes ago if it came from a planet, and 15 billion years ago if it came from deepest space. Laid out before our eyes, therefore, are events spanning the whole history of the universe. But an alien in a distant galaxy would get the news in a different order to Earthlings because he is at different distances from the celestial objects being observed. Cosmic events might appear in the opposite sequence, or what appear to us to be simultaneous events might be spread out in time. Thus the pattern of cosmic events that we see depends on where we are, and how we are moving. An infinity of 'presents' exist, some of which lie in our past and some in our future.

Nevertheless, time is not anarchic, and there is an absolute sequence of events relative to the expansion of the universe that preserves the law of cause and effect. In other words, effects always lie in the future of their causes. If two people are close to each other in space, whatever they see will be in the same sequence even if they are moving fast relative to one another. But if they are far apart, the sequence will differ because of the different lengths of time it takes the news of events to reach them.

The effects of relativity extend beyond time to length and mass as well. Imagine looking at a broom handle 2m long from different angles in 3-D space. Side on, it appears its full length of 2m, but when viewed from one end it shrinks to no more than the diameter of the handle. And so the handle's length depends on the position of the observer. A similar thing happens in 4-D spacetime, although the effect is tiny until close to the speed of light. The handle's greatest length is when it is at rest relative to the observer, and it shrinks as the relative velocity increases until it is about half its rest length at 90% of the speed of light.

Now consider two broomsticks ridden by witches Agatha and Bertha. When they ride at the same speed alongside each other, their broomsticks appear to be the same length. But when they ride fast towards each other, they each see their own broomstick at its normal rest length, but the other one as shorter. So Agatha sees Bertha's stick as shorter than hers, and Bertha sees that Agatha's stick has shrunk. Surely they can't both be right? Which is really the shorter? The answer is neither. There are no absolute distances independent of the observer. And this applies to distances in space, such as between the Earth and the Sun as well as to witches' broomsticks!

Finally, we come to mass and an understanding of why the speed of light is an absolute limit. As an object accelerates, so it gains kinetic energy (= $0.5mv^2$ where m = mass, v = velocity). And Einstein's equation $E = mc^2$ (where E = energy, c = speed of light) equates energy and mass. So the faster an object moves, the greater its energy, and hence the greater its mass. At 10% of the speed of light this adds only 0.5% to the mass it has at rest. But at 90% of the speed of light the mass doubles, and it increases rapidly beyond that, becoming infinite at the speed of light. The higher the mass, the more energy is required to accelerate it further, and hence infinite energy would be required to reach the speed of light. This means that only objects without mass, such as photons of light, can actually attain that velocity.

To conclude this section, let's try to imagine the universe as seen by a particle of light, or, more imaginatively, a being made of light. At the speed of light, time stands still and distances shrink to nothing. Hence a photon or light-being visits all points along its path at the same moment, and crossing the universe is no distance at all. So it is everywhere and everywhen.

Spacetime

It has been known since Newton's day that gravity pulls all objects towards each other. But relativity reveals that the gravitational field affects the geometry of space itself. Unlike a magnetic field, which we can imagine as spreading out and filling space (See the next Chapter), gravity actually creates the shape of space. Hence, the earth doesn't move in a curved path around the sun because of gravitational attraction, but because space itself is curved in the vicinity of the sun.

Some idea of how curved space works may be gained from a two-dimensional analogy. Imagine that space is a horizontal sheet of rubber stretched tightly in all directions, like the mat of a trampoline. If we place a heavy ball on it to represent the sun, a hollow will be formed as the sheet stretches in response to the ball's weight. Another, lighter ball representing the earth will form a second, smaller hollow. If the two balls are close enough together, the 'earth' will roll into the hollow created by the 'sun'. And if we aim to roll the 'earth' past the 'sun', its path will curve around the heavier ball, and it may go into 'orbit' for a while. The smaller ball doesn't move towards the larger one because of an attractive force between them, but because of the curvature of the rubber sheet which is equivalent to the curvature of space.

Whether or not we can visualize curved space, the theory of relativity tells us that matter, gravity, space and time are intimately connected parts of an ultimate reality called spacetime. In other words, spacetime is not simply the arena in which the cosmic drama unfolds; it is all there is. It is the stuff of which the universe is made. There is nothing but spacetime and motion within it. Space and time affect, and are affected by, everything that happens in the universe. Mass, energy, gravity, space and time are not independent phenomena, but concepts we have created to help us make sense of the cosmos.

In the words of John Wheeler:
(We live in) a world whose properties are described by geometry, and a geometry whose curvature changes with time.

Quoted by Mathews (1991) p.61

Matter tells space how to bend; space tells matter how to move.

Quoted by Davies and Gribbin (1992) p.94

Spacetime is dynamic; stretching, rippling and curving as massive objects move, and forces act. Electromagnetism and gravity are waves in spacetime, and energy (mass) is curvature of spacetime. Since matter is concentrated energy, we experience it wherever spacetime is sharply curved. And when we feel drawn to the Earth's surface by gravity, and we imagine the Earth held in its orbit by the sun's gravitational attraction, we are actually experiencing a tight twist in spacetime following the curved lines of space around a small twist, which is itself moving around a larger curve. Or so we are told.

Just as there is no precise point at which an ocean wave ends, so the curvature in spacetime that is a material object does not have a clearly defined boundary. Rather, it merges imperceptibly with its surroundings, extending in some slight measure to all of spacetime. Thus, an object does not have a precise location in space, nor a clear identity independent of all other objects. These waves and curves and twists all overlap and interact with each other, forming an interdependent whole. Indeed, as J C Graves expressed it, there is "a sense in which each body is everywhere, and at the same time."[5] Similarly, electromagnetic and gravitational fields are not distinct. Electromagnetism has gravitational effects on particles, and gravitation has electromagnetic effects.

In Conclusion

We cannot shrug off relativity as a bizarre theory. Experiments have shown that it provides an astonishingly accurate picture of the real world. According to Roger Penrose: "general relativity (is) the most accurately tested theory known to science", and new experimental evidence continues to come in.[6] Nevertheless, there are rumblings, and one day its foundations may be shaken by new findings.[7] But until that day arrives we have little option but to accept Einstein's vision that objects do not exist at definite places, and events do not happen at definite times; that everything is somehow present everywhere and everywhen; that underlying the world of our senses is one in which all the objects and events we experience are mere ripples in the fabric of spacetime, and time itself is an illusion.

9 The Physics of the Very Small

If the concepts of relativity are mind-bending, then those of quantum physics are almost beyond belief. Yet, once again, they are experimentally well-proven and rapidly being incorporated into new technologies. So open wide the doors of your imagination ...

Fields

It is sometimes said that the countries of Central America are within the 'sphere of influence' of the USA. This means that events in those countries are somehow affected by the existence, actions and attitudes of the USA, even when there is no direct use of military, economic or other forms of power. Conversely, of course, a coup or revolution in Central America may influence what happens in the USA even though there is no obvious material connection.

Such political 'spheres of influence' are analogous to the idea of a 'field' in physics. A field is a region of influence in space. It is not made of matter, and cannot be seen or touched or weighed like matter. But fields interact with matter in various ways, creating forces between objects separated by empty space. Perhaps the best-known example is the magnetic field. You may have seen the existence of a magnetic field demonstrated by sprinkling iron filings on a sheet of paper near a magnet. The field is invisible, but it has the power to move the iron filings, and organize them into patterns that reflect the field's shape. And if you hold a piece of steel close to a magnet, you can feel a force pulling the two together, even though there is no visible connection – a force that we harness in myriad electric motors. Light and radio waves are other examples of the electromagnetic fields which permeate the whole universe.

Gravity is also a field, although we more often think of it simply as a force keeping us down to earth. We cannot see or touch the gravitational field, but it is nevertheless there, stopping us flying off into space as the Earth spins; and keeping the Earth in orbit around the sun. Indeed, it was the gravitational field pervading all space that drew matter together into stars and galaxies, thus forming the universe we know – a topic we'll discuss in more detail in Part IV.

Despite their elusive, mysterious nature, fields are very real. Indeed fields are a more fundamental aspect of reality than matter. But before discussing the nature of matter, I need to introduce another concept.

Solitons

Imagine throwing a stone into a calm pond, and watching as the ripples move outwards. The further out they get, the lower and more spread out they become until they disappear altogether. And if you stand at the stern of a boat moving through still water, you will see its wake gradually die away as it moves on. But not all waves fade away like this. In 1834, John Scott Russell was out riding near Edinburgh when he saw a boat moving along a narrow canal. Suddenly, the boat stopped, creating a surge of water at the bow which "rolled forward with great velocity, assuming the form of a large solitary elevation, a rounded, smooth and well-defined heap of water, which continued on its course along the channel apparently without change of form or diminution of speed." Russell followed it for two miles before losing it in the windings of the channel.[1] Solitary waves which retain their size and shape for long periods and over great distances are quite common, and have been named *solitons*. But how do they manage to maintain their identity when normal waves dissipate so quickly?

Despite what it looks like, a typical wave is actually made up of many superimposed waves, each with a different height between crest and trough, and a different length between crests. Waves of different sizes normally move at different speeds. So when you throw a stone into the water, the faster wave components leave the slower ones behind and the wave gradually spreads out, becoming longer and lower. In contrast, a soliton will form if all the component waves move at the same speed so that they stay together, and this happens when there is the right combination of wave size and water depth.

Solitons occur in many situations besides water. They can form in a liquid, solid, gas, or field. They have been found in planetary atmospheres, crystals, plasmas, optical fibers, nerves and electronic devices. They appear in electrical currents in superconductors, and may carry energy along DNA molecules.

Solitons not only persist for long periods and great distances, but also can be remarkably difficult to destroy. You might expect that when a large soliton overtakes a smaller one the two would merge, but in fact they separate again with their identities unchanged. Nevertheless, it is possible to destroy most solitons, but there are some that are truly permanent.

Imagine a strip of elastic stretched between two supports. An ordinary soliton can be formed by plucking the strip, like plucking a string on a guitar, as shown below. A permanent soliton can be made by rotating one end to put a twist in the elastic. The twist can move backwards and forwards along the strip, but cannot escape or be removed without unfastening one end. So a twist soliton will remain for ever. One example of a real-world permanent soliton occurs when there is an error in the regular stacking of molecules in a crystal. The resulting *dislocation* in the pattern can move around inside the crystal, but can never escape.

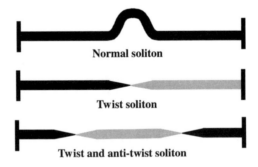

Figure 8   Solitons

Although a single twist soliton cannot be removed, it will be annihilated if it meets an anti-twist. Imagine taking hold of the elastic near the middle and rotating it so that there is a twist on either side of your hand, thus forming a twist – anti-twist pair. When you let go of the strip, the two twists will move until they meet and cancel each other out.

The Source of Matter

So what has all this talk of solitons to do with fields and matter? The answer is 'everything'. The water in the canal and the elastic strip are simple analogies to physical fields, and the solitons are analogous to the way matter arises from fields seething with creative energy like bubbles in a cauldron of boiling water.

The most fundamental field is the vacuum of what we normally think of as empty space. Far from being empty, however, most physicists believe this vacuum is an unimaginably intense energy field; so intense that one cubic centimeter of space contains far more energy than all the matter in the known universe.[2] This field cannot be observed directly, but it produces measurable physical effects on atoms and subatomic particles.[3]

The vacuum field is in constant movement, twisting and vibrating, and creating solitons as it does so. These solitons are particles of *virtual matter* which appear as particle-antiparticle pairs like the twist–anti-twist pairs in the elastic strip. And like the twists, these pairs of virtual particles exist only fleetingly, annihilating each other when they come into contact. But if, during its brief existence, a virtual particle is given enough energy from some other source, such as a black hole (See Part IV), it can become real.

These particles may be either matter or antimatter. Mind-blowing though the idea may be, antimatter is not confined to science fiction, but is created regularly in the particle accelerators physicists use to probe the nature of matter.[4] When a proton is smashed into an antiproton, the two particles annihilate each other to be replaced by a variety of other particles. At first, these were thought to be components of the proton and antiproton, but now it is recognized that they are actually new particles. When the proton and antiproton are destroyed, energy is released into the vacuum field, disturbing it in such a way that it creates these new particles. The effect is like a drop of water falling into a pool and splashing up several smaller droplets. The vacuum field has a component for each type of particle.

(I)t is easy to forget that ... we can never really know what is going on inside the atom. ... Our knowledge has been made by shadows.

John Briggs and David Peat (1984) p.79

Recall that, according to relativity theory, matter and energy are equivalent. Hence, each subatomic particle corresponds to a certain amount of energy equivalent to its mass – a packet of energy known as a *quantum*. For example, the electron is a quantum of the electron-positron field (the positron is the anti-matter equivalent of the electron); and the proton is a quantum of the proton-antiproton field. The spatial pattern of each field determines how likely it is that a particle of that type will be found at a particular point in space. Thus, it is the underlying fields that define both the spatial distribution and the form in which matter appears.[5]

As Albert Einstein himself wrote: "We may ... regard matter as being constituted by the regions of space in which the field is extremely intense ... There is no place in this new kind of physics both for the field and matter, for the field is the only reality."[6] Matter is like flecks of foam on the seething surface of the ocean of space.

Wave-Particle Duality

The particles that emerge from the vacuum field have split personalities. Sometimes they appear in experiments as solid lumps of matter with precise locations in space. And sometimes they appear as waves smeared out over a region of space. But these waves are unlike any others we've met so far. Imagine that the cross-section of a particle wave looks like the curved line in Figure 9. The height of the wave at any point represents the betting odds that the particle will be found there. Hence, it is most likely to be at A, and is about half as likely to be found at B. But it could be anywhere else in the region covered by the wave. This means that it is impossible to know the position of a particle precisely, and that the possible positions of different particles overlap. In the words of Fritjof Capra:[7]

> We can never say that an atomic particle exists at a certain place, nor can we say that it does not exist. ... the particle has tendencies to exist in various places and thus manifests a strange kind of physical reality between existence and non-existence. ... The particle is not present at a definite place, nor is it absent. It does not change its position, nor does it remain at rest. What changes ... (are) the tendencies of the particle to exist in certain places.

It is generally accepted that the particle and wave are complementary aspects of a single underlying reality whose nature is not yet understood. The particle aspect is revealed when the experimenter sets out to detect a particle, and a wave is seen when this is looked for. But it is impossible to observe both aspects at the same time. This is demonstrated clearly by the classic double slit experiment in which a point source of light shines on a screen with two slits in it. Light passing through the slits spreads out and illuminates a second screen with a pattern of dark and bright stripes called interference fringes. This pattern occurs where the beams from the two slits overlap, and is formed by the wave aspect of the light. Where the crests of the waves from the two slits arrive at the screen together, they reinforce each other to produce a bright patch; and where a crest and trough meet, they cancel out and there is darkness.

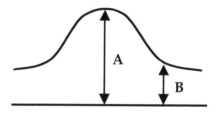

Figure 9   Probability wave

Now suppose the brightness of the source is reduced until light is being emitted one 'particle', or photon, at a time, and the impacts are recorded on a photographic plate. Each photon makes a spot on the plate as we would expect a particle to do, and yet as more and more photons arrive the pattern of interference fringes reappears. One possible explanation is that each photon passes through both slits at once and interferes with itself, the pattern gradually becoming visible as more photons arrive to brighten the image. An alternative explanation is that each photon somehow 'knows' about the other slit and the millions of other photons so that they can coordinate their behavior. So we can take our pick between an idea that is impossible to reconcile with everyday experience, and an example of interaction and connectedness that is hardly less mind-blowing in its implications, but at least is similar to the coherent behavior of billions of molecules in self-organizing systems as described in Chapter 5.

God does not play dice with the Universe!

Albert Einstein
Quoted by Davies and Gribbin (1992) p.208

Imagine now that a particle detector is placed so that we can tell which slit each photon actually passes through. This has the effect of smearing out the interference fringes until they are replaced by two superimposed patches of light – just as would be expected for two streams of particles passing through the slits. So tracking each photon changes the behavior of the light. We can observe either the wave aspect or the particle aspect, but not both at once.

This experiment was taken a step further by John Wheeler and Caroll Alley.[8] They delayed the decision whether to detect particles or allow the interference pattern to develop until the photon reached the screen. The results were the same. In other words, their decision determined whether the light was behaving as a particle or a wave when it passed through the slits *at an earlier time*. In this case, the choice was delayed by only a few billionths of a second, but in principle it would be possible to apply the idea to certain astronomical observations and control the nature of light that was emitted billions of years ago.

The double-slit experiment has been repeated many times with other subatomic particles and even whole atoms. In 2004, a team succeeded in demonstrating wave-particle duality with big fullerene molecules, and planned to do the same with even larger protein molecules.[9] According to quantum theory, every object, regardless of its size, has a wave aspect. So why don't we experience this in everyday reality? A recent suggestion is that in order to reveal the wave aspect, the system must be isolated totally from its surroundings in order to prevent any information about its state leaking out. The larger the object, the more difficult this is to achieve, and, at least with current knowledge and technology, it would be impossible with objects big enough to see with the naked eye.

Quantum Entanglement

We saw in the last section how the wave aspect of a particle spreads in all directions without a clearly defined boundary so that the waves of nearby particles overlap. Further, if two subatomic particles come close enough together, their waves merge to form a new and different particle. David Bohm concluded that "Ultimately, the entire universe ... has to be understood as a single undivided whole, in which analysis into separately and independently existent parts has no fundamental status."[10]

This unity has a still deeper dimension. Einstein's theory of relativity tells us that nothing can move faster than the speed of light. And so it may take eons for particles carrying news from a distant star to reach us. When we see a star explode, the readings on our instruments are caused by events that may have happened billions of years ago. In other words, distant parts of the Universe appear to be only weakly connected with each other.

By contrast, quantum physicists have discovered that when two subatomic particles interact they become linked. If something happens to one of them, the other will react instantly no matter how far apart they may have moved. Once again, this is not just speculation. Such action-at-a-distance has been observed in experiments, and applications are being developed in high-resolution imaging, superconductivity, computing and communications.[11] It was suggested recently that the vacuum field is filled with entangled particles, as they are called, and evidence is beginning to emerge that entanglement between atoms also may be possible. Even more bizarre are suggestions that entanglement across time may occur, thus bending the laws of cause and effect.[12]

> When you see light coming from a faraway star, the photon is almost certainly entangled with the atoms of the star and the atoms encountered along the way.
>
> **Thomas Durt**
> **Quoted by Brooks (2004a)**

Physicists are divided about the permanence of entanglement. When a subatomic particle interacts with a large system, such as a measuring instrument, it triggers a cascade of events which entangles all the particles in the apparatus. Roger Penrose argues that the particles become disentangled again when the cascade of events ends and the system settles into a particular state, such as a reading on a display. According to this view, there is a dynamic balance between entanglement and disentanglement which limits the extent of quantum connection.[13] However, many other physicists believe that entanglement is permanent. When we consider how many particles must have interacted since the big bang, this means that the universe is an interconnected whole, with particles in remote stars responding instantly to events on Earth; and particles on Earth resonating continually with distant cosmic events.

Many scientists are seeking to explain how action at a distance works. One possible explanation is the psi-field hypothesis put forward by Ervin Laszlo, which will be discussed in more depth later in this Chapter.[14] Laszlo suggested that particles are prevented from moving faster than light by resistance of the vacuum field. As they move, particles leave a wake in the field, like a ship moving through water. According to Laszlo, these waves are what mathematicians call *scalar* waves which do not experience resistance, and so can travel faster than light. The more matter there is nearby, the faster the scalar waves move; and where matter is dense, as at the surface of the earth, their speed approaches infinity. Just as the wake of a ship affects other vessels that cross it, so scalar waves generated by one particle affect the motion of other particles with which they interact. If this hypothesis is correct, then wherever matter is dense, every particle influences the movement of every other particle instantaneously.

All things by immortal power
Near or far,
Hiddenly
To each other linked are,
Thou canst not stir a flower
Without troubling of a star.

Francis Thompson

Consciousness and Matter

Classical science assumes that there is a physical reality 'out there' that exists in a well-defined state independently of any observer. But the double-slit experiment described above shows that reality is affected by the scientist's decision whether to look for a particle or a wave. Further evidence for the influence of mind over matter came in 1995 from an experiment showing that photons of light do not have a definite polarization until they are measured at a detector.[15] Many scientists still find such influence hard to accept, and several explanations for what actually happens have been suggested. But before discussing some of these, we need to probe a bit more deeply into the issue.

Imagine an electron in a box. If the electron's wave aspect fills the box, the particle could be anywhere inside it. Now imagine that we slide a partition across the box, dividing the space in two. Common sense tells us that the electron must be on one side or the other. But its quantum wave aspect still fills the box, and we cannot be sure which side it's on. We can remove this uncertainty by looking in the box to see where the particle is, but when we do so its wave aspect must suddenly contract to one side. The strange thing is that if we never look, the wave will never collapse. And so the behavior of the particle seems to depend on whether it is observed or not.

This may not be too hard to swallow when only subatomic particles are involved. But things get far stranger when we consider the interaction of quantum events with everyday reality. One of the best illustrations of this is a thought experiment called Schrödinger's cat.

Imagine a cat shut in a box with a bottle of cyanide. A mechanism is set up so that a chance event can trigger the fall of a hammer to smash the bottle. Suppose, for example, that the poison will be released if the ball on a roulette wheel falls into a red slot. If we cannot see the roulette wheel, we will not know whether the cat is alive or dead at any moment unless we open the box. But we can be sure that it is either alive or dead.

The situation changes, however, if the fall of the hammer is triggered when a lump of radioactive material emits an alpha particle. This is a random process, and so we cannot be sure at any particular time whether or not a particle has been emitted and the cyanide released. According to quantum theory, all possibilities coexist, and situations in which the cat is partly alive and partly dead are valid solutions. Thus the cat is neither alive nor dead, but in some strange sense is both alive and dead. In this case, when we open the box we do not simply discover what has happened, but actually cause the different possibilities to 'collapse' to a particular outcome. It is our act of observation which seals the cat's fate.

The situation becomes even stranger when we consider the world outside the box. The theory suggests that a quantum system takes on a definite state from amongst a range of possibilities when it is 'observed' by another system outside itself. But then the state of the larger system becomes indeterminate until collapsed into reality by a still larger system. And so on. Now suppose the scientist has looked inside and sealed the fate of the cat, but his colleague working in the next room does not know this. Does the whole room collapse to a particular state when he asks how the cat is? And is the state of the whole research institute only fixed when a reporter rings in to ask the result?

Physicists have struggled for many years to make sense of this seemingly absurd situation. Some believe that all the alternative possible quantum states are real, but that we are only aware of one of them. The others happen in parallel universes about which we can know nothing. Thus, every second the universe splits into myriad slightly different alternatives representing all possibilities. And each of these universes then evolves independently, splitting again and again and again. At each division, we, too, are split into slightly different alter egos each of whom experiences an alternative life history in an alternative universe.

As Marcus Chown expressed it: "The consequences are remarkable. A universe must exist for every physical possibility. There are Earths where the Nazis prevailed in the Second World War, where Marilyn Monroe married Einstein, and where the dinosaurs survived and evolved into intelligent beings."[16] For many years, this idea was little more than philosophical speculation. But in 2001 David Deutsch published a theory of multiple universes that even indicates how we could make use of other universes by means of quantum computers.[17]

If this is too much to swallow, then there are other possible explanations. The most popular is to assume that the collapse of possibilities is real, leaving us with only one universe, and the mystery of how and why such a collapse to a single state should occur. Roger Penrose suggested that gravity may cause all systems to be unstable and collapse to a defined state. The larger the system is, the faster this happens so that we never see the superposed quantum states at normal scales.[18] Many others believe that the collapse really is triggered by observation, but there is disagreement about what this means. For some, the observation could be by the cat itself, or even by an instrument such as a particle detector. But others believe that the chain of infinite regression to larger and larger systems can only be ended by observation by a conscious being. This then spins off into debate about the nature of consciousness which we will address in Part VI.

Let's assume for the moment that the collapse of the system to a defined state happens when it is observed by a conscious mind. As Paul Davies and John Gribbin put it: "It is as though through our observation we actually create, rather than explore, the external world."[19] This does not necessarily mean that consciousness can influence what nature does, but some physicists believe this is possible.[20] For instance, Frank Tipler argues that *active* intelligence will gradually spread throughout the cosmos until it controls the whole of nature.[21]

Intriguing evidence pointing towards this possibility comes from a variety of studies. In one, students of martial arts were able to influence the decay of radioactive particles *retrospectively*![22] In others, research by highly respected scientists has shown the effectiveness of prayer in healing.[23] And long-term studies have shown a relationship between the output of random number generators and global events such as the World Cup Soccer final or the crashing of an aircraft into the World Trade Center, apparently revealing the effect of coordinated global consciousness.[24] These studies are discussed in more detail in Chapter 20.

The idea that conscious observation 'creates' the real world has some interesting implications. In particular, it means that physical systems would evolve differently depending on whether or not they were observed by a conscious being. Roger Penrose argues that, since the weather is chaotic, it must be sensitive to quantum events. Hence, a planet without conscious beings would experience a mess of possible weathers, rather than any actual weather. More radically, we would have to believe that the universe did not exist in a concrete form before conscious beings evolved. Or assume that there is a Cosmic Consciousness which existed before the universe (just as the vacuum field is supposed to be pre-existent), and which is capable of collapsing the universe to a definite state. We will explore these ideas in more detail in Parts IV and VI.

The Nature of Ultimate Reality

When applied to subatomic scales, the smoothly curving space of relativity is at odds with the frothing, turbulent motion of the quantum world. And nonsense results when the two theories are combined to solve problems that involve both large and small scales, such as the evolution of the cosmos or the behavior of black holes (See Part IV). Hence, intense efforts are being made to develop a single, consistent Theory of Everything as it is often called. The front-runner in this stakes is string theory which has the potential to answer many outstanding questions.[25]

String theory suggests that the basic elements of matter are not sub-atomic particles, but even tinier 'strings' like infinitely thin rubber bands. These loops vibrate like guitar strings, and it is these resonant vibrations that give subatomic particles their characteristic properties. One of the most challenging aspects of string theory is that it requires a 10-D universe – as if 4 dimensions weren't difficult enough to understand! The extra 6 dimensions are thought to be curled up into minuscule tangled balls far too small to detect. And these balls of 6-D space have holes in them, each of which is associated with a family of potential particles. Whether or not string theory will prove to be the holy grail of physics remains to be seen, but its supporters must first overcome very challenging experimental and computational difficulties.

The rest of this section sketches some even more speculative theories about the nature of ultimate reality. David Bohm was a theoretical physicist who believed that the material universe is not simply the result of blind self-organizing processes.[26] He sought to reveal the deeper *implicate order* which is made manifest through a process of *explication* or *unfolding*. This process is analogous to a broadcast TV program that is encoded in a radio wave and then decoded by the TV set. Thus, the picture is *enfolded* in the *implicate order* of the radio wave, and the TV *explicates* or *unfolds* it. Bohm argued that the implicate order is holographic, so that the whole universe is encoded within each part of it, albeit with a loss of resolution.

We have seen how particles pop in and out of existence from the vacuum field. In much the same way, David Bohm argued that when a particle is observed, it is actually a rapid succession of similar particles that is unfolded from the implicate order, and then enfolded again. Hence, time is created by this flowing process of unfolding and enfolding of the material world around us. Bohm believed that the implicate order contains a memory of past events that guides the unfolding of the future. In the words of John Briggs and David Peat:[27]

> The guide wave or cosmic memory ... acts on the present moment. Then it acts on the next present moment, and the next. It acts to give a shape to the succession of moments. It guides whatever is unfolding in those moments of space-time. ... But as these moments fold again into the whole, they carry an imprint or afterimage of energy back into the implicate, where they affect the guide wave and subtly change the cosmic memory.

The concept of a guide wave from the implicate order meshes neatly with the ideas of biologist Rupert Sheldrake who is not convinced that the development of form in molecules, crystals and organisms can be explained adequately by self-organization. He claims that development is guided by *morphic fields* that have not been detected yet.[28] Since subatomic particles are formed by the vacuum field, and gravitational and electromagnetic fields organize matter at the cosmic scale, it seems reasonable that fields may be responsible for structures and forms at intermediate scales as well.

According to Sheldrake, morphic fields develop like habits. The first time something forms, it may take some time and be unpredictable. But, after thousands of times, it becomes a habit to do it a particular way. This can be illustrated by supposing that a newly discovered chemical crystallizes for the first time with a particular shape. In doing so, it creates a new morphic field that exists independently of the crystal. When a similar solution of the chemical is made later, it resonates with this field, and is guided to the same crystalline form. Each time crystals of this chemical are formed, the field gets stronger, and it becomes more likely that the same form will develop in future.

Rupert Sheldrake suggests that every material object has a holarchy of morphic fields associated with it, with field nested within field. Hence, if he is right, a living organism develops its characteristic form not only because of its genes and environment, as normally supposed, but also because it resonates with the morphic fields associated with its molecules, cells, tissues and organs, and the overall form of its species. (Development is discussed in detail in Chapter 14).

The theory of morphic fields implies that the supposedly absolute and eternal laws of nature may be no more than habits. Those which have been developing since the dawn of creation, such as the formation of atoms and common molecules, are very strongly established, and deviations are almost impossible now. Hence they appear to us as immutable laws. But the laws governing recent events should be less well established, and more readily changed.

Suggestive evidence in support of this idea again comes from the formation of crystals. The morphic field hypothesis suggests that a new compound may take some time to 'explore' alternatives before crystallizing for the first time. But after that, the process should become easier and easier because a strengthening morphic field exists for a specific shape. This is in fact what happens, as Rupert Sheldrake describes:[29]

> (N)ewly synthesized compounds are usually difficult to crystallize ... Moreover, generally speaking, compounds become easier to crystallize all over the world the more often they are made. This happens in part because chemists tell each other of the appropriate techniques. But the most common conventional explanation for this phenomenon is that fragments of previous crystals are carried around the world from laboratory to laboratory, where they serve as 'seeds' for subsequent crystallizations ... The carriers of the seeds are often said to be migrant scientists, especially chemists with beards, which can 'harbor nuclei for almost any crystallization process.' Or else seeds are thought to move around the world as microscopic dust particles in the atmosphere.

Bohm and Sheldrake agreed that morphic fields were not the same as the implicate order, but were an aspect of it.[30] Ervin Laszlo's psi-field hypothesis, introduced earlier in this chapter, provides a possible explanation for both.[31] Recall the suggestion that particles moving through the vacuum field leave a wake of scalar waves. Unlike water waves these do not die away but are permanent, each one being added to the waves created by previous particles to form an unimaginably complex and detailed memory of everything that has ever happened in the universe.

Subatomic particles are directly affected by these scalar waves, but large systems are not, just as supertankers do not react to small ripples on the water. However, living cells contain microscopic tubular structures which may respond to quantum events, potentially enabling organisms to interact with the psi field.[32] And when large systems become chaotic, they may be so sensitive that their behavior and development can be affected by resonance with these miniscule disturbances.

In Conclusion

The story of quantum physics is the latest in a long line of scientific attempts to explain the nature of ultimate reality. But this is not the end of the story. Research continues apace, and there is no reason to believe that this myth will be more enduring than earlier ones. Thought experiments have suggested that 'weak measurements' may be able to reveal the wave and particle aspects of a quantum particle simultaneously.[33] And experiments reported in 2004 appeared to detect both the wave and particle aspects of photons at the same time.[34] If verified, this could be the beginning of the end for wave-particle duality, along with the need for multiple universes or a reality frozen into existence by conscious observation. We can only wait and see!

Before leaving this chapter, let's reflect briefly on the deep similarities between the realities envisaged by quantum and relativity physics. Both see matter as springing from an underlying reality – the vacuum field or spacetime. Both reveal that matter has a wave aspect, that objects do not have a precise location, and that there are non-local interactions between matter. Such is the ultimate physical reality of the science of oneness.

Review of Part III

Approaching reality from the infinitesimal, we found a pulsating field of energy from which particles of matter appear, and into which they disappear like dust motes dancing in a sunbeam. And approaching from the cosmic scale, we found particles of matter in the guise of twists in rippling spacetime. In both cases, we discovered a weird world in which matter has no real existence, in which 'particles' overlap with no clear identity or location, and in which all matter is woven into a single whole across the vastness of the cosmos. We found a world in which time varies between observers; in which all time is present in every moment; and in which distances are elastic. Pressing on into more speculative realms, we explored the possibility that the development and evolution of everything in the universe, indeed of the very cosmos itself, may be guided by patterns encoded in underlying fields; patterns that learn and adapt in the light of experience.

Quite a journey!

Both quantum and relativity physics have sparked many a creative science fiction plot. But it is the complex, esoteric world of quantum theory that has most fired the imagination. Why is this? At one level it may simply be fascination with the bizarre and fantastic. But at a deeper level it may also be because quantum physics so clearly breaks free of the dead hand of classical science and opens up a vista of a new worldview – a worldview in which there is room for consciousness and spirit, meaning and purpose to re-enter our lives. This is why quantum physics is so often used, and misused, to justify spiritual ideas, New Age visions, and holistic thought and practice. And it is this space for subjective consciousness that most disturbs scientists wedded to the materialist view of reality.

In many ways, quantum physics is still in its infancy as a discipline, and it is by no means certain that all its bizarre aspects will survive into maturity. The validity of wave-particle duality is being challenged, and understanding of the relationship between the micro-cosmos of subatomic particles and the macro-cosmos of everyday life is gradually increasing. It is possible that one day in the not-too-distant future a deeper reality will be uncovered that sweeps away the paradoxes and shows us more clearly how the world of matter behaves.

However, many aspects of quantum theory provide reasonably solid intellectual ground. The interconnectedness of entangled particles and the unity of the cosmos that this implies are well established. A similar unity is revealed by relativity physics which approaches the issue from the opposite end of the size range. And in between, at the scale of everyday reality, we have seen that systems sciences also point to oneness. It is worth pausing for a moment to reflect on just how remarkable this consistency is. When coupled with the insights of mystics who have experienced oneness with the Cosmos or the Divine for thousands of years, we can have great confidence in the truth of oneness. We will see in more detail how this interconnectedness affects the reality of the world we experience as the rest of the book unfolds.

> The fate of any given particle is inseparably linked to the fate of the cosmos as a whole, not in the trivial sense that it may experience forces from its environment, but because its very reality is interwoven with that of the rest of the universe.
>
> **Paul Davies (1988) p.177**

As already noted, the role of observation in collapsing superposed states of quantum particles has been demonstrated in the laboratory. And even if wave-particle duality turns out to be due to incomplete knowledge of a deeper reality, there is no doubt that the decisions of the observer do affect what is observed in many experiments. But the apparent role of human consciousness in freezing normal reality from the sea of potential, as suggested by the story of Schrödinger's cat, may or may not prove to be correct. It may be replaced by the multiple-universe hypothesis, or by discovery of a cosmic Consciousness that continuously collapses possibilities into reality, or by some totally new explanation. We can only wait and see, but things will become a bit clearer as our journey continues.

Compared to quantum physics, little public attention has been paid to the relativity of time, and yet its implications are similarly deep. Perhaps this is because it is even harder to reconcile with our normal experience than quantum physics? It is difficult in our time-bound existence to comprehend that ageing is a relative process, and that every thing, every event is somehow present everywhere and everywhen. And yet, at another level, this should come as no surprise. Mystics down the ages have experienced not only the Oneness of the cosmos, but also a sense of timelessness; of all time being contained in this present moment. And experiences of other times and other places are quite common. Perhaps, through quantum entanglement or the tunnels of spacetime, our human awareness can travel instantly to the ends of the universe? We will explore these ideas in more depth in Part VI.

What of the implicate order, morphic fields, and the psi field? These mesh to form a conceptual whole in which the present and future evolution of the cosmos are guided by memories of past events. And they gently suggest that there is no pre-existent design, not even any fixed laws of nature, but rather a cosmic memory and cosmic habits which have evolved from the dawn of time. We will revisit these ideas later at various points.

A light-being riding a photon created in the big bang would see all of the past simultaneously with all of the future. There would be no before or after, no here and there. Only the eternal Here and Now.

Malcolm Hollick

Once again, this Part reveals the need to come to terms with the tension of opposites, and to face the fact that there are deep paradoxes at the heart of reality rather than a single truth. The wave-particle duality is the most obvious. But several other tensions also have been brought into the light: the relationships between matter and energy, mind and matter, space and matter, space and time, and between time and timelessness. At a deeper level, the paradox of determinism versus creative freedom has reappeared in the shape of the evolving morphic and psi fields. And at perhaps the deepest level of all, there is the tension between the fundamental scientific assumption that the universe obeys absolute and eternal laws of nature, and the radical alternative that the universe pulls itself up by its own bootstraps, and evolves its own laws.

If an electron exists at all it is because science tells me so – if they decide it's a particle today, a wave tomorrow and a clump of energy the day after, it makes no difference to frog being frog and cheese being cheese.

Geoff Mortlock (On reading the first draft of this book)

Science has not yet disclosed a single, incontrovertible truth about the fundamental nature of reality, and in some areas may never be able to do so. We once again face questions about ultimate reality that cannot be resolved by the methods of science. And once again we are free to make our own judgments; to choose what makes most sense to us based on our individual and collective inner knowing as well as on rational knowledge. We need not only to weigh the evidence before us, but also to seek inner wisdom. Greater clarity may come as we continue on our journey.

Reflections on Part III

The worlds of quantum and relativity physics defy normal, commonsense understanding. So in the first of these Reflections we explore what it means to understand something, and how we may approach these difficult concepts. We then move on to meditatively reflect on the experience of time, and the nature of reality.

Once again, choose a time to reflect when you will not be disturbed for a half hour, and have a notebook or other appropriate resources available to record or express your reactions.

Make yourself comfortable, breathe deeply and slowly a few times, and relax before you begin.

What Does It Mean to Understand Something?

Physicist J. J. Thompson said he couldn't really know what an atom was unless he could be one; and biologist Charles Birch claimed that "To really know is to be at one with that which is known."[1] These insights from two distinguished scientists reinforce the message of Chapter 3 about inner knowing, and knowing through relationship. And they mirror the words of the Buddha more than 2,000 years ago. As Zen Master Thich Nhat Hanh explains it:[2]

> The Buddha ... said that in order to understand, you have to be one with what you want to understand. ...
>
> The French language has the word comprendre, which means to understand, to know, to comprehend. Com means to be one, to be together, and prendre means to take or to grasp. To understand

something is to take that thing up and to be one with it. The Indians have a wonderful example. If a grain of salt would like to measure the degree of saltiness of the ocean, to have a perception of the saltiness of the ocean, it drops itself into the ocean and becomes one with it, and the perception is perfect. ...

Understanding means to throw away your knowledge. You have to be able to transcend your knowledge ... The technique is to release. The Buddhist way of understanding is always letting go of our views and knowledge in order to transcend. ... That is why I use the image of water to talk about understanding. Knowledge is solid; it blocks the way of understanding. Water can flow, can penetrate.

In the following reflections, you might like to try letting go of what you 'know' about reality. Try 'tuning-in' to your intuitive self. Allow yourself to identify with what you are contemplating; to feel what it is to be a particle, an energy field, or the whole of spacetime. Listen to the deeper messages of physics.

To help you do so, close your eyes, breath deeply and relax for several minutes before continuing.

Experiences of Time

The present is really the fringe of memory tinged with anticipation.

A N Whitehead[3]

We cannot sense time directly. We hear present sounds, not those of yesterday or tomorrow; and we see, taste, smell, touch and feel only present sensations, not past or future ones. Our sense of the past comes from memories, but memories are present experiences of brain activity whose images often differ markedly from the actual past. And when they happened, past events occurred in what was then the present. Thus we infer the past from present mental activity. In a similar way, we can never experience the future, only anticipations of it which are part of our present experience. In the words of Alan Watts:[4]

Remember the incident of seeing a friend walking down the street. What are you aware of? You are not actually watching the veritable event of your friend walking down the street. You can't go up and shake hands with him, or get an answer to a question you forgot to ask him at the past time you are remembering. In other words, you are not looking at the actual past at all. You are looking at a present trace of the past.

We create the sense of time when we draw a boundary between past and future. We spend most of our lives either remembering past events, or anticipating future ones. We regret past actions and dread future consequences. We brood anxiously on the future – searching for some way to make it better than the present. But all too often we forget to live in the only time we have – now. We actually live in the ever-changing present moment, but past and future are so absorbing that, as Ken Wilber put it "our present moment, the very meat of the sandwich, is reduced to a mere thin slice, so that our reality soon becomes all bread-ends with no filling."[5]

*   *   *

We think of time as passing at a constant, measured rate. But our subjective experience of time varies. Sometimes it flies with whole hours passing in a moment, and sometimes it drags, a few minutes stretching to an age. In the timeless present of the mystics, the 'eternal now' with no before or after, time becomes suspended in the beginningless, endless present.

The sense of time also varies between individuals. We all know someone who is 'laid back' – who is always late for appointments, never finishes tasks on time, and whose sense of time and the urgency of life is weak. And we all know others who are 'up tight', always on the go, unable to stop for a moment because of the pressures of life.

The subjectivity of time affects our perceptions in other ways as well. Evolution has equipped us to perceive events that take seconds, minutes, years and perhaps decades. But things which happen faster or slower than this are invisible to us without the aid of instruments. Thus we are unaware of events at the molecular level and below, not only because they are too small to sense directly, but also because they happen too fast. And we are oblivious of processes in geological or cosmological time because change is imperceptible during a human lifetime.

*   *   *

We live in a time-obsessed culture. We pore over past events in the news, biographies and histories; we run our days rigidly by the clock; and we are forever planning for the future. We spend time, buy time, use time, waste time, save time, make time, take time and are pushed for time. As a culture, we believe time is both short and valuable. But such attitudes to time are not universal, as these two stories told by John Frodsham reveal:[6]

> I recall the story of a visiting American who attempted on eleven different occasions to meet a Persian to conclude a business deal in Tehran. Finally, on the twelfth occasion, he fixed the time and place very definitely and explained that he was leaving Tehran and this would be his last chance to see this man, who very much wanted to conclude a business deal with him. Well, twelve o'clock came and his Persian friend hadn't arrived, so finally, half an hour later, he picked up the phone and said "Abdul – why haven't you come, you know we arranged to meet here?" "But it's raining," said Abdul querulously.
>
> In Afghanistan one gets even further afield ... A few years ago in Kabul a man appeared looking for his brother. He asked all the merchants of the market place if they'd seen his brother and told them where he was staying in case his brother arrived and wanted to find him. He did this for several years running. Finally one of the members of the American Embassy met him and asked if he had found his brother. The man answered that he and his brother had agreed to meet in Kabul, but neither of them had said where precisely, or what year.

*   *   *

Should we regard our subjective experiences of time as nothing but illusions, or are they part of its very nature? How do they differ from the concept of time in modern physics, as revealed by the following quotations?

To us who are committed physicists, the past, present and future are only illusions, however persistent.

Albert Einstein[7]

In the four-dimensional spacetime landscape, that moment (in memory) is still 'there', that past 'you' that so enjoyed that morning swim is still 'real'. The past does not cease to exist simply because your mind moves beyond it. No one moment is any more or less real than any other.

Darryl Reanney[8]

So does the future, in some sense, already exist "out there"? Might we be able to foresee events in our own future by changing our state of motion? ... it turns out that you cannot travel fast enough to see into your own future.

Paul Davies and John Gribbin[9]

*   *   *

The experience of eternity, of the present moment of the mystics, is vastly different to our normal perception of time, as the following quotations make clear.

These roses under my window make no reference to former roses or to better ones; they are for what they are; they exist with God today. There is no time for them. There is simply the rose; it is perfect in every moment of its existence ...

Ralph Waldo Emerson[10]

In this moment there is nothing which comes to be. In this moment there is nothing which ceases to be. Thus there is no birth-and-death to be brought to an end. Thus the absolute peace in this present moment. Though it is at this moment, there is no boundary or limit to this moment, and herein is eternal delight.

Platform Sutra[11]

(E)ternity is not everlasting time nor a split fraction of a second –
rather, it is timeless, a moment without date or duration, existing in
its entirety right now. This present moment, since it knows neither
past nor future, is itself timeless, and that which is timeless is Eternal.
... time is a vast illusion, and this timeless moment is Eternity itself.

Ken Wilber[12]

If you don't experience (eternity) here and now, you're not going to
get it in heaven. Heaven is not eternal, it's just everlasting. ...

Joseph Campbell[13]

*   *   *

One of the simplest ways to become aware of this present moment is
walking meditation. In walking meditation, attention is concentrated
on what is happening in every moment. How our bodies feel as our
hearts beat and as we breathe; the contraction and relaxation of
muscles as we slowly move our arms and legs; the sensation of the soft
wind in our faces and the warm sun on our skin; the sound of birds
and people, of wind and water; the light and shade and colors, the
forms and textures and patterns of the world around us. And we let
go of our thoughts about what happened yesterday, or last week, and
what may happen tomorrow or next week. We just let ourselves be
present. You might like to try it some time.

Walking meditation can be very enjoyable. We walk slowly, alone or
with friends, if possible in some beautiful place. Walking meditation
is really to enjoy the walking. Walking not in order to arrive, just
for walking. The purpose is to be in the present moment and enjoy
each step you make. Therefore you have to shake off all worries and
anxieties, not thinking of the future, not thinking of the past, just
enjoying the present moment. You can take the hand of a child as
you do it. You walk, you make steps as if you are the happiest person
on Earth. ... We have to walk in a way that we only print peace and
serenity on Earth.

Thich Nhat Hanh[14]

*   *   *

Some dark, clear night, preferably when you are in the depths of the country away from city lights and air pollution, try lying on your back and looking up at the stars.

Gaze at the stars and let your vision encompass the vastness of the heavens.

Become aware of your breathing, in ... and out ...

And as you breathe, gently relax your body and mind.

As you breathe in, let your anxious thoughts of past and future gently subside.

And as you breathe out, let the tiredness and tension flow from your muscles.

In ... Out ... In ... Out

And as you gaze at the sky, reflect that the light you are seeing now, in this present moment, began its journey at very different times.

Light from the moon set out only seconds ago, but light from the nearest star bears tidings of four years past.

And light from the most distant galaxies tells of the cosmic birth, thousands of millions of years ago.

As you gaze, you are experiencing the whole history of the universe in this one present moment of time.

Rest on that thought for a while.

And when you're ready, return to your body and this present time and place.

If you're with a friend, you might like to share what you experienced.

If you're alone, you might like to dance to the heavenly music of the spheres.

Or when you get home, you might like to express your experience some other way.

> Time present and time past
> Are both perhaps present in time future,
> And time future contained in time past.
> If all time is eternally present
> All time is unredeemable.
>
> T S Eliot *Burnt Norton*

The Nature of Reality

The material world that we sense is not ultimate reality. There is a deeper, mysterious Reality as spiritual leaders have taught through the ages. The nature of this Reality has always been hard to express in words, and modern physicists find it no easier than the mystics of old. The following quotations attempt to convey something of the essence of reality as seen by physicists.

> Which 'comes first', atoms or universe? The answer is 'neither'. The large and the small, the global and the local, the cosmic and the atomic, are mutually supportive and inseparable aspects of reality. You can't have one without the other. ... There is a unity to the universe ... It is a unity which says that without everything you can have nothing.
>
> Paul Davies[15]

> Wholeness in this case amounts to asserting that there is no here and there or that here is identical to there.
>
> John Briggs and David Peat[16]

> Far from dominating the activities of the cosmos, matter seems to assume an almost peripheral role. The main activity comes instead from the most insubstantial entities conceivable, a foam ... a froth of empty space ... And it is only by leave of the special properties of this foam that ordinary matter exerts the influence it does in the Universe today.
>
> Paul Davies and John Gribbin[17]

> (T)he proper starting point for the description of particles is all this (foam), all the time and everywhere, throughout space.
>
> John Briggs and David Peat[18]

> There are no basic particles; everything is built of everything else. Particles are made of other particles by binding forces that are themselves created by the exchange of particles among particles – the observed world lifts itself into existence by its own bootstraps.
>
> Ervin Laszlo[19]

Is spacetime really curved? ... Might not even the most perfect of clocks slow down or speed up, and the most perfect of rulers shrink or expand, as we move them from point to point ....? Wouldn't such distortions of our clocks and rulers make a truly flat spacetime appear to be curved?

Yes. ...

People ... who are unaware of the ruler's rubbery nature, and thus believe its inaccurate measurement, conclude that space is curved. However, people ... who understand the rubberiness, know that ... space is really flat. ...

What is the real, genuine truth? Is spacetime really flat ... or is it really curved? ... Both viewpoints ... give precisely the same predictions for any measurements ... Which viewpoint tells the 'real truth' is irrelevant for experiments; it is a matter for philosophers to debate, not physicists. Moreover, physicists can and do use the two viewpoints interchangeably ... They may regard spacetime as curved on Sunday ... and as flat on Monday.

Kip Thorne[20]

How do these descriptions make you feel? Perhaps you are confused? Frustrated? Disbelieving? Excited? Awed? Reverent? Or ... ?

Why do you think you feel this way?

*   *   *

The implicate order, morphic fields and the psi field are serious scientific hypotheses, but read more like the wilder fantasies of science fiction, or the reflections of religious mystics. You might like to pause here to consider their implications.

Relax once more in a comfortable position, and focus on your breathing for a few minutes before slowly reading the following questions. Pause between questions to ponder your answers, jot down your thoughts, and consider any other questions that may arise.

How does the idea that everything 'unfolds' from a pre-existent implicate order differ from Christian, Jewish and Islamic beliefs that the universe was created by God?

How does it differ from Taoist beliefs that everything flows from the Nothingness of the Tao? Or from other religious traditions?

If the implicate order is holographic, then each one of us somehow enfolds the whole universe. Perhaps, as the Quakers claim, there is that of God in everyone? Or perhaps, as Buddhists believe, the Buddha-nature is here inside each of us? Or ... ?

If each one of us is shaped physically and psychologically by resonance with the morphic fields of our kind, or by the 'memories' of the psi field, then perhaps we are in some sense 'reincarnations' of earlier beings. Is it possible that experiences that many people have of past lives arise from such resonance? How does this idea differ from belief in reincarnation?

If we are all intimately connected in space and time through the psi field, does this explain psychic phenomena such as telepathy and remote viewing?

Could the meaning and purpose of life be to further the creative evolution of the cosmos through our contribution to the psi field? Or are meaning and purpose simply illusions?

What other questions suggest themselves to you?

When you feel ready, return to this time and place. And if you wish, record or express your thoughts in some way.

Part IV:
The Evolution of the Cosmos

With the foothills of systems science and basic physics behind us, we are now ready to start up the mountain itself.

So far, we have discovered that the myriad objects and organisms we perceive are not truly independent, but are united at a deeper level into a single whole. It is as if the diversity of the material universe blinds us to this underlying reality, just as sunlight sparkling off the rippled surface blinds us to the water beneath. And we have seen how order, form and creative newness emerge from the processes of self-organization and evolution.

It is now time to explore in more detail how the many emerged from the primal One. At first sight the two visions may seem contradictory. How, you may ask, can the many connect to form One at the same time that the One differentiates to form the many? But, as we probe deeper, we will find that these are simply two facets of the same reality. They are like the two faces of a coin, so different in appearance and yet parts of the same whole.

Chapter 10 explains current theories of how the physical universe evolved after the big bang, revealing a remarkable ignorance. In order to shed a different light on these cosmic mysteries, Chapter 11 explores the possibility that consciousness rather than matter is the fundamental reality from which the cosmos has evolved, and discusses the questions of cosmic purpose and human destiny. Chapter 12 then goes on to examine the case for cosmic design, and the possibility that our universe evolved from other universes. Finally, this Part closes as usual with Review and Reflections sections.

Enjoy the climb!

10   A Brief History of the Universe

The Challenges and Limits of Cosmology

Everything we know about the universe comes from observations made at a single point in space (Earth), and a single moment of cosmic time (the last few centuries).[1] And it depends on information which travels to us across space and time rather than being gathered by active exploration. As a result, much of what we 'know' relies on theory without the support of direct observation. For example, as we will see, cosmologists have deduced that the matter we can detect is only a small part of the whole, and they have little idea what the majority of matter actually is.

Early humans gazed in wonder at the heavens with their naked eyes, but could learn little more than to recognize the constellations and their movements. Simple observatories made it possible to track these movements more precisely, but a closer look had to await the telescope. Later, we realized that visible light carries only a small part of the information reaching Earth, leading to radio astronomy and particle detectors. Today, we can observe more closely and more precisely than ever before, and have found things that were beyond the wildest stretches of earlier imaginations. And as we try to make sense of them, so our theories suggest new things to look for. Who knows what surprises still lie in store?

However, there are limits to our knowledge, no matter how sophisticated the technologies we develop. Radiation takes time to reach us across the vastness of space even at the speed of light. So we cannot see anything that happened further away than the distance light has traveled since the birth of the cosmos. We are thus at the center of an observable sphere which is slowly expanding. But it seems very unlikely that this horizon is actually the edge of the universe, for that would put the Earth back at the exact center of creation, as it was believed to be before the days of Galileo.

Indeed, many cosmologists believe the universe is infinite, so that what we can observe, despite its vastness, is no more than the tiniest local fragment of the whole.

A slightly different limit is imposed by the expansion of space. When we look at the universe, all other galaxies appear to be moving away from us, and the further away they are the faster they are moving. Hence, there may be very distant objects that are receding faster than light can carry information towards us. At first sight this may appear to conflict with Einstein's theory that nothing can travel faster than light. However, this is not the case as the objects are not moving through space; it is spacetime itself which is expanding as will be explained later. Nevertheless, radiation from such distant objects can never reach us, and when cosmologists speak of events at the edge of the universe, they really mean events close to this horizon. We have no way of knowing what lies beyond.[2]

Our theories create conceptual horizons too. The law of gravity applies throughout the known universe, but does that mean the universe might be attracting objects outside itself? We cannot tell. Similarly, the second law of thermodynamics predicts that every part of the universe must eventually decay, but is this necessarily true of the universe as a whole?[3] For all we can tell, the universe may be self-maintaining in some way.[4] Further, the 'big bang' which we believe gave birth to the universe is a mathematical singularity – a point at which current theories break down. Thus, what we call the origin of the universe is really no more than the limit of validity of our present model.

These boundaries mean that cosmology is limited in a way that other sciences are not. We cannot directly observe most of the universe, and hence we can only partially test any model of it. And there is only one universe that we can observe, so we cannot derive laws by studying the behavior of universes in general. As a result, we can never have a complete and perfect theory of the universe. Despite these challenges, cosmologists have created a coherent picture of the evolution of the universe by combining the insights of quantum and relativity physics. However, many aspects of this history are controversial, and there are still huge gaps in knowledge as we shall see. This chapter summarizes the most widely accepted story of the creation and evolution of the cosmos; and the next two chapters delve into some of its deeper mysteries.

The Moment of Creation

It is flat-out strange that something – that anything – is happening at all. There was nothing, then a Big Bang, then here we all are. This is extremely weird.

Ken Wilber (1995) p.vii

The moment of creation, what preceded it, and what caused it are beyond the scope of all but the most speculative theories. But many cosmologists suggest that our universe appeared out of nothing with no cause. In the words of Heinz Pagels:[5]

> The nothingness 'before' the creation of the universe is the most complete void that we can imagine – no space or time or matter existed. It is a world without place, without duration or eternity, without number ... Yet this unthinkable void converts itself into ... existence ... What 'tells' the void that it is pregnant with a possible universe?

Stephen Hawking argues that such speculation on what came before the big bang is futile:[6]

> (E)ven if there were events before the big bang, one could not use them to determine what would happen afterward, because predictability would break down at the big bang. Correspondingly, if, as is the case, we know only what has happened since the big bang, we could not determine what happened beforehand. As far as we are concerned events before the big bang have no consequences, so they should not form part of a scientific model of the universe. We should therefore cut them out of the model and say that time had a beginning at the big bang.

Similarly, John Hick claims from a spiritual perspective that it is meaningless to ask what preceded the big bang; we should simply accept that it is sheer mystery.[7] Nevertheless, it is human nature to be curious, to want to know, and the creation myths of many cultures and spiritual traditions provide intriguing insights. We will explore some of them in the next two chapters.

And so back to consideration of the moment of creation. Recall that, according to quantum physics, the vacuum of space is not truly empty, but is an incredibly intense energy field. One of the most popular theories today is that this vacuum field existed before the big bang. But it did not exist in empty space, nor for all time before creation, since time and space and matter had not yet come into being. This field simply was. But it was alive with potential, pulsing gently with quantum fluctuations. And from these fluctuations emerged a stream of tiny ephemeral 'bubbles' of spacetime, appearing and disappearing like bubbles in frothy water. Then (if then has meaning in timelessness) one fluctuation triggered a self-organizing, evolving process. Just the right kind of bubble appeared by chance, and it grew and developed and evolved into the universe we know today.[8] As Darryl Reanney described it:[9]

> Science says the universe 'began' in what it calls, in its own mythic language, a quantum fluctuation. But when we ask 'What is this?' we run headlong into a situation where language falters on the brink of breakdown. I could speak of a point that is not a place and a moment that is not a time, of a micro-bubble of nothing, but words would distort the very phenomenon they were seeking to define. At the barrier between existence and non-existence we approach a place where truth can only be glimpsed in paradox: *an emptiness that yet is full.*

How could the universe possibly create itself out of nothing? We can perhaps get some feel for this by imagining the vacuum field as like the surface of a big lake. When it is calm, the water fills the lake's basin to a certain level. When it is stormy, the surface becomes rough and undulating. But the water which mounds up into waves is drawn from the troughs between, and the average level remains the same because the total amount of water doesn't change. Now consider the gravitational energy of the water. The potential energy of any object increases with height above the Earth. So the water gains energy as it is lifted up to form a wave crest, and releases it again when it falls into the trough. But, despite the surging energy at the surface, the total energy of the lake is the same when it is stormy as when it is calm.

Before the big bang, the vacuum field was like an ocean covered with waves. And the big bang itself was like one of those rare occasions when several waves just happen to coincide to produce a very large one, appearing as if from nowhere and out of nothing.[10] Although the big bang was an explosion of unimaginable power, David Bohm pointed out that, according to quantum theory, it was actually little more than a ripple on the surface of the underlying vacuum field.[11]

Recall that matter is concentrated energy. It turns out that the energy locked up in the matter of the universe exactly balances the energy used to blow it apart against the force of gravity. Thus, just as the energy of storm waves is drawn from the troughs leaving the total energy of the ocean unchanged, so the total energy of the vacuum field is the same as it was before the big bang. Energy and matter did not have to be created from nothing, since the sum of matter and energy is zero and all is still in balance.[12]

The Evolution of the Cosmos

Once that first tiny bubble of spacetime has come into existence, quantum and relativity theories can trace its growth and development into the universe we know today, although there are still many unresolved questions. Currently, the favored picture is that during the first 10^{-32} seconds (about a millionth of a billionth of a billionth of a billionth of a second!) the bubble expanded with unimaginable speed to form a symmetrical, homogeneous, empty and timeless region of space about the size of a grapefruit.[13] At the end of this *inflation*, enormous amounts of energy were released irreversibly as radiation and particles of matter. Time emerged from 4-D space, and the universe began its journey with no possibility of return.

15 billion years after the big bang, the galaxies are still flying apart. However, it is important to understand that the big bang did not occur at a point in space, and matter is not expanding to fill a pre-existing empty space. Rather, space itself was created in the big bang, and it is space itself which is expanding. This idea can be illustrated with a simple example. Imagine that the universe is a balloon, and that the galaxies are spots painted on its surface. As the balloon is inflated and it stretches, so the 'galaxies' move apart. Now imagine sitting on one of these galaxies and looking out. All the other galaxies move away from you as the balloon expands. Another image is that of raisin bread dough left to rise. As the dough expands, the raisins move away from each other. This would be true no matter which galaxy you sat on. Thus, the fact that the rest of the universe appears to be moving away from Earth, does not mean that we are at the center where the big bang occurred.

In its early moments, the universe was so hot that matter could not exist. All was pure energy. After a few seconds, subatomic particles began to form, and at around one minute protons and neutrons began to fuse into helium nuclei. Originally, too, there was only one force, but as the temperature dropped this divided into gravity and electromagnetism, followed by the so-called weak and strong nuclear forces. It was only much later, after about a million years, that nuclei and electrons came together to form atoms.

Initially, equal quantities of matter and anti-matter particles were created, and should have annihilated each other totally when they came into contact. So why do we have a matter universe at all? Both matter and anti-matter particles have limited lives, dissolving back into the vacuum field which spawned them. But, in a mysterious asymmetry, it appears that anti-matter decays slightly faster than matter. Hence, in the first second of the universe, a tiny imbalance occurred in which perhaps one matter particle remained for each billion pairs that annihilated each other.[14]

Once matter had been created, gravity formed it into the structures we see today – the planets, stars, galaxies, and clusters of galaxies. If matter had been spread completely evenly through space, the pull of gravity on each particle would have been the same in all directions, and there would have been no resultant movement. But small variations in density meant that particles experienced a net pull towards regions of higher density, and thus, over astronomically long periods of time, matter was drawn together. And the more it was drawn together, the stronger became the force of gravity pulling it closer still.

If gravity had been the only force at work, the concentration of matter would have continued until it disappeared into black holes – remnants of stars where matter shrinks to nothing leaving empty space behind, and which we'll discuss further in Chapter 12. In practice, however, the repulsive forces of electromagnetism provided a counter-balancing push. Thus, according to physicist Paul Davies: "gravity is the fountainhead of all cosmic organization. Way back in the primeval phase of the universe, gravity triggered a cascade of self-organizing processes ... that led, step by step, to the conscious individuals who now contemplate the history of the cosmos and wonder what it all means."[15]

As stars and galaxies formed, so they began to forge the diversity of atomic nuclei we know today, and to spew them into space. Planets and other inter-stellar objects harvested these atoms and froze them into crystals and molecules of ever-increasing complexity. Earth emerged from the nebula of dust and gas about the sun soon after its formation 5 billion years ago. And the process of ever-growing complexity continued, and still continues, on Earth with the emergence of life and consciousness.

Dark Matter and Dark Energy

The afterglow of the big bang, known as the *cosmic microwave background radiation*, has now cooled to within $3°$ of absolute zero. In the last decade, very precise measurements of this radiation have shown that the early universe was more uniform than predicted. As a result, cosmologists have concluded that there must be far more matter in the universe than previously estimated in order for gravity to create the structures we see today. Similar conclusions have arisen from observing the movement of galaxies. They must contain far more matter than we can detect or they would have been torn apart by the centrifugal forces due to their rotation. These observations have led to the conclusion that the universe consists mainly of invisible *dark matter*.

No-one knows for sure what this dark matter is. Some of it may be 'ordinary' matter in forms that we cannot see such as black holes (See Ch. 12) or brown dwarfs (dim objects larger than planets but smaller than stars). But much of it may consist of new types of particles which only interact with ordinary matter through gravity. In a touch of cosmic humor, cosmologists have dubbed these two sources of dark matter MACHO's and WIMPs (or Massive Compact Halo Objects and Weakly Interacting Massive Particles). Black holes are bizarre objects whose strangeness has already entered our mythology through popular accounts and science fiction. But WIMPs are a new concept, and yet it is possible that billions of these particles are whizzing through our bodies every second. Nobody knows what they are, but theoretical candidates abound. And there are still those who claim WIMPs do not exist, and propose alternative solutions for the 'missing' mass of the universe ranging from massive black holes to a revised theory of gravity.[16]

Some people find the idea of WIMPs metaphysically challenging, posing a new threat to humanity's self-image. Not only has astronomy evicted us from the center of the universe, but also cosmologists now tell us we are made of different stuff to most of it. We appear to be an insignificant epiphenomenon, and the 'real' universe is something else entirely!

As if the idea of dark matter weren't challenging enough, cosmologists have found the need to postulate a mysterious *dark energy*. Without it, estimates of the age of the universe suggest it is younger than many stars in our galaxy. And recent measurements of the movement of distant galaxies have shown that the expansion of the universe is accelerating. This contradicts predictions made from measurements of the cosmic background radiation, and from the theory that the universe inflated rapidly in the first fraction of a second after the big bang. In order to reconcile the various theories and observations, a previously unknown source of dark energy has been postulated that draws matter apart in opposition to gravity. Possible candidates for this role are the vacuum field of space, an unknown form of matter with negative gravity, or something else. However, predictions of the energy of the vacuum field made in this way differ from those of quantum physics by a factor of 10^{60} or even 10^{120}. As Leonard Susskind wryly observed "this is by far the all time worst discrepancy between theory and experiment."[17] The only way out of this dilemma is to assume that different contributions to the vacuum energy cancel out with extraordinary exactness, as described above in the section on the moment of creation.

As already noted, matter and energy are interchangeable according to the theory of relativity. Hence it is possible to combine dark matter and dark energy into a single property, conveniently expressed as the average density of energy or matter in the universe. Intense efforts are being made to measure this density more accurately. Recent estimates published by NASA are that the universe is 73% dark energy, 23% cold dark matter, and only 4% atomic matter. We are in the extraordinary situation that 96% of the universe is in a form that has never been detected directly![18]

Where Will It End?

What is the ultimate destiny of this self-organizing universe? In one sense, it doesn't matter apart from satisfying our intellectual curiosity and validating one belief or another. Whatever the end, it is so far in the future that it is irrelevant to the prospects for humanity; and whenever it occurs, it may well mean extinction for our evolutionary descendants despite optimistic speculations about ways of escape that we'll look at in the next chapter. But, in another sense, it is of great importance, potentially affecting our whole worldview as Bertrand Russell's despairing reaction in the text insert below indicates.

A few years ago, there seemed to be three possible fates. The favorite was unending expansion to a cold and featureless nothingness. But expansion followed by contraction to a 'big crunch', or a delicate balance in which expansion eventually stops were both on the cards. Today, we can choose from a smorgasbord of other possibilities as well, including cycles of cosmic death and rebirth, sudden transformation of the quantum vacuum into something else, or a 'big rip' which tears the universe apart. It all depends on the nature of that mysterious dark energy.[19]

> … that all the labours of the ages, all the devotion, all the inspiration, all the noonday brightness of human genius, are destined … to be buried beneath the debris of a universe in ruins – all these things … are yet so nearly certain that no philosophy which rejects them can hope to stand. Only … on the firm foundation of unyielding despair, can the soul's habitation henceforth be safely built.
>
> **Bertrand Russell (1957) p.107**

The best guess at present is that dark energy will remain constant, in which case the universe will continue to expand and slowly sink into heat death, all galaxies disappearing after about 200 billion years. However, if dark energy weakens and then becomes negative, the universe could collapse to a big crunch, the reverse of the big bang, as soon as 25 billion years. On the other hand, if dark energy becomes stronger, the universe could be torn apart in a 'big rip' in 40 billion years or more.[20] But don't hold your breath while you wait to find out!

And long before we need to start worrying about the fate of the universe, our little corner of it will become distinctly uncomfortable. In about 6 billion years, the sun will swell into a red giant, making Earth uninhabitable. And even before that, our Milky Way galaxy will be torn apart by a collision with the Andromeda galaxy.

In Conclusion

From media publicity and popular science books, we could be forgiven for thinking that the big bang theory of the universe is proven, with only the details to be filled in. Nothing could be further from the truth. In May 2004, a group of 33 scientists from 10 countries issued a hard-hitting public statement arguing that the theory is seriously flawed and that insufficient resources are being given to researching viable alternatives. In mid-2005, they convened a conference on the crisis in cosmology.[21] In their own words:[22]

Big bang theory relies on a growing number of ... things we have never observed. Inflation, dark matter and dark energy are the most prominent. Without them, there would be fatal contradictions between the observations ... and the predictions ... In no other field of physics would this continual recourse to new hypothetical objects be accepted ...

What's more, the big bang theory can boast of no quantitative predictions that have subsequently been validated by observation. The successes claimed ... consist of its ability to retrospectively fit observations with a steadily increasing array of adjustable parameters, just as the old Earth-centered cosmology of Ptolemy needed layer upon layer of epicycles.

Strong stuff! But not without support. For example, in early 2004, measurements of the cosmic background radiation by a few research groups were reported as sowing "Seeds of doubt for theory of inflation" – a central plank of big bang theory.[23] Later in the year, observations of distant quasars indicated that either the universe is expanding slower than estimated, thus implying that it is older than was thought, or that there is far less dark matter than calculated.[24] And new models emerging from string theory suggest that the big bang may be a cyclic cataclysm every few trillion years, rather than the beginning of it all.[25]

All these have never yet been seen –
But scientists who ought to know,
Assure us that they must be so …
Oh! Let us never, never doubt
What nobody is sure about!

Hilaire Belloc

11 Cosmos, Consciousness and Destiny

In the last chapter, we looked at what science has to say about the origins and evolution of the physical universe. It is now time to pause and ask ourselves: Is this all there is? Or does the cosmos hold even deeper mysteries?

We have seen at a number of points on our journey how non-physical qualities are an important part of reality. Even simple systems display creativity, autonomy and purpose. And consciousness plays a key role in the quantum world. We will discuss the evolution and nature of consciousness in Part VI, but it is pertinent to explore its deeper origins here. Is consciousness simply an epiphenomenon that arises when the organization of matter and energy reaches a certain level of complexity, as most scientists believe? Or could it be qualitatively different from matter, and have evolved alongside matter from the big bang? Or could it even be the primal 'stuff' of the universe, from which matter emerged? And what would such ideas mean for human existence?

These are the subjects of the first section of this chapter, which draws on insights from Gnostic traditions going back to ancient Babylon and Egypt. The second section goes on to explore scientific speculations about the ultimate fate of humanity as the universe ages, and the relationship between these speculations and the Gnostic insights.

Conscious Evolution

We saw in the last chapter how mathematical models break down at the singularity of the big bang, making it impossible to know what came before. We are asked to accept that the universe simply created itself out of nothing; to accept that what came before is sheer, unknowable mystery. This conclusion is based on the limitations of mathematics, and yet it leaves us exactly where religions have always left us: faced with the necessity of believing in and accepting the insoluble Mystery of God, the Ground of Being, or the Vacuum Field as it is now called. Unlike past religions, however, science asks us to believe that creation is a purely mechanical, purposeless process; that consciousness just happens to have emerged from matter; and that life is without meaning or purpose.

The most beautiful experience we can have is the mysterious. It is the fundamental emotion which stands at the cradle of true art and true science. Whoever does not know it and can no longer wonder, no longer marvel, is as good as dead, and his eyes are dimmed.

Albert Einstein (1954)

John Hick divides the creation myths of the great religions into two broad groups.[1] The philosophical roots of modern cosmology lie in the mainly western religions which tell of creation by God from nothing. But whereas scientific creation is purposeless, God is said to have wanted to share existence with beings upon whom He could lavish his love, and by whom He could be loved. The ultimate end is the eventual perfection of humanity in the presence of God. The second group of mainly eastern religions believe that everything is God, or within God, without beginning or end, and that the purpose of life is to transcend the individual self and merge with the Divine. In Hinduism, for example, the beginningless and endless life of the universe is Shiva's dance of continuous destruction and creation. The dance is an end in itself, danced purely for its own sake.

At this point I want to introduce a spiritual model of cosmic evolution based on Gnostic sources as described by Timothy Freke and Peter Gandy.[2] I have chosen this particular model not because it is true in any absolute sense, nor because it is necessarily better than others. Rather I have chosen it because it makes sense to me, and complements the current scientific model of physical evolution by providing a place for consciousness and purpose. Also, being relatively unknown, it carries less 'baggage' than other religious models, hopefully making it easier to approach with an open mind. I find it helpful and stimulating, and I never cease to be amazed at the sophistication of these ancient philosophers. But please feel free to interpolate your own preferred religious model if that is more helpful for you.

Gnostic philosophers faced the same challenge as modern cosmologists in trying to work out the source of the universe. There is a singularity beyond which human thought cannot take us. Hence, the ultimate reality is an utterly unknowable Mystery. In the words of Basilides:[3]

> Nothing existed. Not even nothing. The truth, naked of opinion and conceptualization, is that there was not even the One. And when I say "there was", I don't mean that anything was. When I use the expression "There was absolutely nothing" I am just giving some sort of indication of what I am trying to talk about. Nothing was. Neither something nor absence of something. Not the One. Not the impossibility of complexity. Not the imperceptible or the inconceivable. ... there was not anything for which human beings have ever found a name.

We will return to the mystical experience of being united with this cosmic Mystery in Part VI, and here I want to focus on the rest of the Gnostic model. The Mystery is somehow pregnant with potential; but unlike the vacuum field which is pregnant with space and time, matter and energy, the Mystery is pregnant with Consciousness (with a capital C to distinguish it from human consciousness).

In contrast to the scientific worldview, the fundamental essence of reality for Gnostics is consciousness not matter. Matter springs from Consciousness, not the other way round – an idea that makes sense when we consider our own experience. It is not the atoms of which our bodies are made that makes life real for us, but our consciousness; our awareness of the world around us and of our own inner selves. Without the conscious experience of living, we would be unaware of existence; nothing would exist for us, and we would not exist for ourselves. Our lives would be of no consequence to us, and we certainly wouldn't be seeking answers to such difficult questions! And so it makes absolute sense from this perspective to consider Consciousness to be the fundamental reality.

According to the Gnostics, the Mystery from which all existence springs is one undivided whole. It follows that the Mystery cannot actually be Consciousness because consciousness requires an object of which to be conscious. I, the subject, am only conscious when there is something, an object, of which I am aware. Hence, Consciousness only came into being when the primal Mystery divided into two – Consciousness and its object. Before that, the Mystery can perhaps best be likened to the potential consciousness of our minds in deep sleep.

How this potential became actual is unknown. But the Gnostics suggest *why* it happened. This Mystery, this ocean of potential, 'desired' to be conscious of itself – whatever that may mean in this context – and hence split its unity, making itself both subject and object. The unknowable became aware of itself as both knower and known, observer and observed, witness and experience. In mythical language, this is the split between God and Goddess, Consciousness and Psyche, Spirit and Soul, which can be expressed only as polarities similar to yin and yang, and 'the ten-thousand' things of Taoism as discussed in Chapter 3 (See Table 3).

God	Goddess
Essence	Appearance
Mind	Thought
Potential	Manifested reality
Pure consciousness	Physical and mental cosmos
One	Many
Existing	Becoming
Spiritual Knowledge	Journey towards
(Gnosis)	Gnosis
Eternal perfection	Evolution over time

Adapted from Freke and Gandy (2001)

Table 3   Attributes of God and Goddess

Following that primal split, Consciousness thought, imagined, dreamed the universe in all its complexity and diversity, thus shaping the potential of the Mystery into the matter and life of the cosmos – the original meaning of which is 'beautiful order'. The process by which this happened is similar to the evolution of the holarchy of nature as described in Chapter 6. At each stage there first comes differentiation from one to two, followed by synthesis and integration into a greater whole before the cycle is repeated. In this way, the One manifested itself as the many.

Figure 10 below illustrates this process in terms of the levels of human consciousness. At the first level, the Mystery divides into Consciousness and Truth – the image that Consciousness has of itself. As humans, we experience this duality as our inner 'I', the witness or observer in our minds, versus our experience of existence. Experience in turn splits into two aspects, our private inner world of imagination, thoughts and feelings, and the outer world of material reality, or sensation. And objective experience splits again into the part which we identify as ourselves, and everything else.

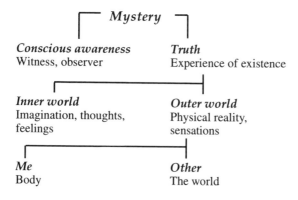

Figure 10   Evolution of Consciousness

We don't need to go further into the evolution of consciousness and matter here, but it is worth looking a little deeper into the Gnostic view of the purpose of the universe. As already noted, Gnostics believe that the Mystery wishes to know itself, or achieve *gnosis*. But in order to do so it had to divide itself between subject and object, knower and known, God and Goddess. In the guise of God, Consciousness, or the One, it achieves gnosis instantly, being aware of and embracing all that is. But as Goddess, or the multiplicity of the manifest cosmos, it is plunged into ignorance.

To clarify why this is, imagine the Mystery as an empty space within which Consciousness arises as a central point. From this center the myriad individual consciousnesses of creation radiate out, each radius ending at a different point on the circumference of a circle. These points represent the infinite diversity of the physical cosmos. Consciousness at the center is aware of all this multitude of forms and consciousnesses, and identifies with each one. Hence, instead of seeing the unbroken line of the circumference, it sees an infinity of separate points. Instead of there being the undivided whole of the circle, there is an infinity of separate individuals. In order for the Mystery to achieve complete self-knowledge, every conscious being must awaken to its true nature as part of the whole that is both the circumference and the center. Or, in more religious terminology, we must each return to God, be Redeemed, or achieve Enlightenment.

When we reach the still point in the center, there is only Being; no becoming, no evolution, no time, no matter. In the words of Timothy Freke and Peter Gandy there is only "The Mystery knowing itself. Love making love with itself. Beauty delighting in itself. Truth true to itself. Being being itself." And "We are co-creating this shared dream which reflects our wisdom and foolishness back to us to help us wake up."[4]

> At the still point of the turning world. Neither flesh nor fleshless;
> Neither from nor towards; at the still point, there the dance is,
> But neither arrest nor movement. And do not call it fixity,
> Where past and future are gathered. Neither movement from nor
>   towards,
> Neither ascent nor decline. Except for the point, the still point,
> There would be no dance, and there is only the dance.
>
> **T S Eliot** *Burnt Norton*

We will return to these ideas on cosmic purpose later in this chapter and particularly in Part VII.

Cosmic and Human Destiny

Scientists are reluctant to speculate about what came before the big bang, but show no such reluctance about the far distant future of the aging universe and ultimate human destiny. Perhaps it is simply that their crystal balls are not blacked out by mathematical singularities as they gaze into the future. But for some at least it is a search for meaning and purpose, and an escape from the existential despair expressed by Bertrand Russell; a despair rooted in the sheer pointlessness of existence if all humanity's creativity and passion, all our love and suffering will ultimately come to nothing in the death of the cosmos. We are now in a position to set this response to the second law of thermodynamics in the context of modern science, and to ask: Was Russell's angst justified? In addressing this question, we need to look at three issues. What is it that is so important to preserve? When and how is it threatened? And is there no hope of escape?

Few people would seriously lament the loss of the human body as such. What we fear is the passing of the uniquely human combination of mind, body and spirit – but what exactly is this? I believe it is the ability to love, laugh, and empathize; to feel awe and wonder, fear and courage, joy and sadness, compassion and anger; to imagine and create both beauty and utility. It is the yearning for fulfillment and self-transcendence; the spirit to rise above adversity, to seek adventure and novelty, and to slake our curiosity. It is self-awareness, self-reflection, spirituality, and whatever unsuspected realms of consciousness are still waiting to unfold. Perhaps, above all, what we would regret is the loss of potential; the sense of an unfinished journey, and being cut off in our prime. It is also, perhaps, a projection of our own fear of death onto humanity as a whole.

Before going further, let's set the issue in an appropriate time frame. Early hominids emerged 5 million years ago, and Homo sapiens less than 0.5 million. Civilization began 5,000 years ago, and our scientific and industrial culture goes back less than 500 years. By contrast, on even the most pessimistic assumptions, the Earth will remain habitable for another 2-3 billion years – 5,000 times the life of humanity so far, 500,000 times longer than human civilization, and 5 million times longer than our technological civilization. The universe will be habitable for at least 10 times longer still; several times the current age of the Earth. For all practical purposes, we have infinite time to play with, provided we survive more immediate threats.

The most immediate threats come from our own actions here and now, and in the next century or two. We truly have no future unless we can learn to live at peace with each other and our planet; unless we can learn to use the fruits of science and technology wisely; and unless we choose to foster the higher human values instead of pandering to materialist greed and selfish individualism.

Assuming we avoid overt self-destruction, we face a more subtle threat. Given the explosive and accelerating advance of science and technology, it seems almost inevitable that we will transform ourselves through genetic engineering and non-biological enhancements within the next couple of centuries. The process has already begun, and resistance to tinkering with our genes or implanting silicon-based technologies is crumbling. Step by little step, it seems certain that we will choose to fix our frailties and enhance our abilities.[5] And even if we decide against such a course, the natural process of evolution will transform us within a few million years unless we take heroic steps to prevent it. So it seems likely that within a blink of the cosmic eye neither our bodies nor our brains will be recognizably 'human'. Perhaps no-one will seriously mourn their passing any more than we mourn Homo erectus. But unless we're very careful, side effects of these 'enhancements' will destroy the very qualities we wish to preserve.

An on-going threat to survival is posed by asteroids. There have been four occasions in the last 600 million years when many species have become extinct in a short space of time, the most famous being the death of the dinosaurs 65 million years ago. These mass extinctions are most commonly attributed to massive environmental changes following the crash of an asteroid into the Earth. Further collisions every few million years or so seem inevitable unless we develop the technology – already possible – to detect and divert large space rocks on a collision course with Earth.

And the wild card in the pack. What if a new race of intelligent aliens arrives on Earth? Whether beneficent or not, any aliens who visit us in the foreseeable future will be technologically more advanced than we are, and human civilization, perhaps humanity itself, would never be the same again. Given the rate at which astronomers are finding planets circling neighboring stars and the possibility that there was once life on Mars, I believe this scenario is becoming more and more likely. Indeed, it is possible that aliens are already here and are trying to communicate through crop circles which are far more than ingenious hoaxes. In particular, one appears to be a response to the 1974 radio message transmitted from Arecibo to tell the galaxy of our presence, and another looks like a picture of an alien face.[6]

In the light of these more immediate threats, it seems misguided neuroticism to worry about the fate of the universe. Our challenge is to survive the next century, and then the next millennium. If we can do that, then a whole new future opens up. A future in which we may indeed escape from the universal catastrophe.

Let's imagine that we make it through the next 1,000 years. Already by then our technical and scientific progress will have made unimaginable things possible. We will be able to travel through space at close to the speed of light, and to preserve life on centuries-long voyages into the galaxy. We may well live far longer than we do now, and will have enhanced brains as well as bodies. It seems likely that we will be able to engineer ourselves to flourish on alien planets with less oxygen or more gravity, as well as to 'terraform' other worlds. Even if we could travel at only 1% of the speed of light, it would take a 'mere' 10 million years to cross the galaxy. If we paused on our way to colonize 100,000 planets, taking 200 years over each, we could still inhabit the galaxy in just 30 million years – leaving almost untouched the available 2-3 billion years![7]

Science fiction writers mostly envisage that we will carry our greed and power-hunger across the galaxy, spreading war, political intrigue, aggressive competition, corporate dominion and environmental destruction in our wake. This would be nothing short of a disaster, and I pray we learn wisdom and transform our inner selves before embarking on this journey. My point here is simply that we have ample time. Anything is possible. And our present challenge is to ensure that we don't destroy ourselves and our planet, or transform ourselves into psychopathic cyborgs.

And now let's take a leap into even more distant times, looking first at the possibility that the universe will go on expanding forever, and then at the scenario where it reaches a maximum size followed by contraction to a big crunch.

If the universe is open, there will eventually come a time when sentient beings will face a cosmic challenge. As far as we know, mental processes need energy. And, as the universe runs down, there will be a new energy crisis hundreds of billions of years in the future. How might our far-distant descendants cope? One possibility is to create more favorable local conditions by shifting stars around using chaos and small nudges. Another is to develop information processors that don't need energy – perhaps by becoming disembodied consciousnesses. A third possibility is that we are wrong about the death of the universe. We may yet discover that the evolutionary process towards ever-increasing complexity does not increase entropy, and that the second law of thermodynamics is balanced by a complementary law of complexification. If this turned out to be true, the future would be unlimited.

Some more intriguing ideas relate to our subjective experience of time. The faster our minds work, the more experience we can cram into a given period. Thus, beings on a neutron star using nuclear instead of chemical processes might experience years in one human second. In the reverse direction, as the universe cools and slows down, perhaps we could slow our consciousness to match the available energy. After all, if infinite time is available, the pace of experience doesn't matter.

This idea becomes even more significant when we look at the possibility of a big crunch. We still have lots of time – at least 25 billion years. But, before it disappears completely, the universe will become too hot for anything to survive. So how could sentient beings possibly escape? One idea is to encode all our memories and information in radiation and transmit it through the new big bang – provided, that is, that the universe 'bounces' into a new existence.[8] Another idea explored by physicists John Barrow and Frank Tipler[9] is that a kind of immortality might be experienced as the mind sped up with rising temperature until infinity could be experienced in finite time. In the words of Paul Davies:[10]

> What might a brain of unlimited capability do? According to Tipler, it would not only be able to deliberate on all aspects of its own existence and that of the universe it had engulfed but ... it could go on to simulate imaginary worlds in an orgy of virtual reality. There would be no limit

to the number of possible universes it could internalize in this way. Not only would the last three minutes stretch to eternity but they would also permit the simulated reality of an infinite variety of cosmic activity.

How does this scientific vision of an infinite mind differ from the religious vision of God? How does the capacity of infinite mind to simulate possible universes differ from Consciousness dreaming our cosmos into being? Is the emergence of infinite mind a return to the Consciousness of Gnosticism, or the Enlightenment of Buddhism?

There are still a couple of steps to go in this speculative exploration. Yet another possible escape route from our doomed universe is to move to another one through a 'wormhole' in space; or to learn how to create a new universe for ourselves. These are not the dreams of fantasists, but serious theoretical possibilities proposed by physicists, as we'll see in the next chapter. And if it might be possible for us to learn how to create a new universe, how do we know that this one was not deliberately created by sentient beings as an experiment, a demonstration, an escape route from another dying universe, or perhaps as a way for Consciousness to know itself?

Physicist Paul Davies has this to say about cosmic purpose:[11]

> If there is a purpose to the universe, and it achieves that purpose, then the universe must end, for its continued existence would be gratuitous and pointless. Conversely, if the universe endures forever, it is hard to imagine that there is any ultimate purpose to the universe at all. So cosmic death may be the price that has to be paid for cosmic success.

In Conclusion

What lies beyond the creation of the cosmos? Did consciousness emerge from matter, or matter from Consciousness? Does the cosmos have a purpose? What is the destiny of humanity and the cosmos? These are big questions. And probably unanswerable. In such arenas, the insights of mystics are at least as valid as the theorizing of physicists and philosophers.

We must choose the answers that make most sense to us, based on our own intuition and inner knowing. And, if we care about the future of humanity, the answers that maximize our chances of survival.

12 Cosmic Design and Evolution

The last chapter touched on the theoretical possibility that advanced civilizations could create new universes. But how could we tell if our particular universe was a product of design rather than chance evolution? The best evidence so far is the extraordinary precision of the laws and fundamental constants of physics. If they had been different by even the minutest amount, life in general, and intelligent life in particular, would not exist. Indeed, they are so finely tuned that it seems impossible at first sight that they could have arisen by chance. In this chapter we explore these mysteries and some of the explanations which are emerging.

A Designer Universe?

The universe is full of unlikely coincidences, a few of which will serve to illustrate its mystery.[1]

- The complexity of the cosmos is possible because subatomic particles can combine into stable atoms of many types. The first nuclei to form were hydrogen and helium, but both of these are inert, and there was not enough energy in the early universe to fuse them into heavier elements. Nature's answer was to provide a catalyst in the form of carbon. But how could large amounts of carbon have formed from hydrogen and helium in the first place? The creation of diverse atoms turns out to depend on helium, beryllium, carbon and oxygen having exactly the right energy levels.
- Matter accounts for only about one billionth of the energy of the universe, but this tiny amount is exactly what is needed to permit the evolution of life. A little bit more and there would be more stars,

and more collisions between them, potentially destroying life-bearing planets.

- If the strong force binding atomic nuclei together was a little bit weaker, the stars could not shine. If it were slightly stronger, the stars would inflate and might explode.

- If the neutron was not heavier than the proton, stars would burn out in a few hundred years.

- If the electric charges of the electron and proton did not balance exactly, matter would be unstable, and the universe would consist of nothing but radiation and a mixture of gases.

- If the electric charge of the electron had been only slightly different, stars would either have been unable to burn hydrogen and helium, or else they would have exploded.

- The rapid inflation of the early universe lasted just long enough, and was at just the right rate to produce a universe in which stars and galaxies could form. A shorter inflation would have left the universe too 'lumpy', and it would have collapsed again. Longer or stronger inflation would have spread the stuff of the universe too thinly for gravity to draw it together.

These mysteries led cosmologists to propound *the Anthropic Principle*. In essence, this states that "we see the universe the way it is because we exist."[2] Another way of expressing this idea is that we would not be here to observe it if the universe was different. Three forms of this principle have been proposed: weak, strong and final. In its weak form, the Anthropic Principle argues that, in a large and long-lived universe, conditions will vary from place to place and over time. Hence, conditions suitable for intelligent life are likely to occur in certain parts and at certain times, and intelligent beings should not be surprised to find themselves in a region that supports their existence. We observe such conditions because we exist; if conditions were different, we would not exist to observe them. This version of the Principle can be used to 'explain', for example, why the big bang occurred 10-20 billion years ago. It simply takes that long for intelligent life to evolve.

The strong Anthropic Principle claims that the universe was designed to produce intelligent life, and its evolution has been guided by that purpose. This idea is controversial because it implies the existence of a purposeful creator, or creative intelligence, outside the cosmos and beyond the big bang. Traditionally this creator has been called God, Allah, Brahma,

Consciousness and a host of other names. But bear in mind the possibility raised in the last chapter that the creative intelligence could be sentient beings in another universe.

The biologist, Professor Charles Birch, claimed that the strong Anthropic Principle reflects a misunderstanding of the process of evolution. Rather than creating an environment fit for particular types of organism to inhabit, evolution forces organisms to adapt to fit the environment that happens to be available.[3] From this perspective, it is quite possible that very different types of life may exist in quite different conditions elsewhere in the cosmos, or in other universes. On the other hand, there is nothing in principle that forbids a creative intelligence from establishing a basic framework of cosmic laws and constants that facilitate the evolution of intelligent life.

Some cosmologists have proposed a third, or final version of the Anthropic Principle. They argue that there must be a reason for the universe to be designed to suit intelligent life, and this purpose would be reflected in some way in the lives of these sentient beings. Any such purpose would be negated if intelligence were to die out soon after emerging, and before it had a significant influence. Hence, they claim, intelligent life will never die out once it has evolved.[4] This belief could help to inspire confidence and hope in the future. On the other hand, it would be very damaging if it was used to justify actions that would jeopardize human survival.

Cosmic Evolution?

Many attempts have been made to explain the cosmic mysteries without recourse to purpose and design by developing models that make our type of universe more likely to develop than others. One such is the theory that the early universe inflated very quickly to the size of a grapefruit. This rapid inflation would have smoothed out any irregularities which existed at the moment of creation, like wrinkles disappearing from the surface of an inflating balloon. Thus wide variations in conditions at the Big Bang could still result in the uniform state we observe today.[5] While inflationary theory makes our universe more probable, it is still not the only type that is possible. Amongst the more intriguing theories is the suggestion that the universe may have evolved through many generations, and that ours is the type most fitted to survive.[6] But before we can discuss this idea it is necessary to describe the nature of black holes.

Black hole is the evocative name given to an object that is so dense that not even light can escape its gravitational field.[7] Black holes are thought to be created when stars larger than about three times the mass of the sun die and collapse inwards under the pull of their own gravity. As their matter becomes packed into a smaller and smaller space, so their gravitational field increases in strength until all matter, and even light itself, is trapped within them. In order to turn the earth into a black hole in this way, it would be necessary to shrink it to the size of a pea. Since nothing can escape them, we cannot see black holes, or detect them directly in any other way. We can only deduce their existence from the effect that their intense gravity has on other objects. For instance, many astronomers believe the Milky Way galaxy is held together by the gravitational pull of a massive black hole at its center.

Once a star has passed the light-trapping stage, it continues to collapse, shrinking away virtually to nothing, and leaving 'empty' space apart from the gravitational effects due to intense warping of spacetime. Anything which falls into a black hole simply disappears. But how? Theoretical studies suggest that rotating black holes may act as 'tunnels' in spacetime, connecting our universe with others. But if we somehow traveled through this tunnel, we could never return because re-entering the tunnel at the other end would not bring us back home, but take us to a third universe. Thus, according to Paul Davies and John Gribbin, a rotating black hole "is connected to an infinite sequence of universes, each representing a complete spacetime of possibly infinite extent, and all connected through the interior of the black hole."[8]

Now imagine that you live in another universe at the other end of the tunnel. What would you see? According to some theories, there would be an explosive outpouring of matter from its mouth. And this has led to speculation that exploding objects in our universe may be the ends of spacetime tunnels, pouring matter from other universes into this one. However, other theories indicate that the tunnels would be blocked. Yet other theoretical speculations suggest that it may be possible to use black hole tunnels to travel quickly between remote parts of our own universe, and possibly even to travel in time – but these are beyond the scope of this book.[9]

We are now in a position to explore the idea that our universe may be the product of an evolutionary process. Imagine blowing up a balloon. As it inflates, a tiny pimple appears on the surface, and gradually expands to form a small bubble attached by a slender neck. This neck then gradually becomes pinched off until the bubble can continue expanding without affecting the original balloon. In a similar way, baby universes may be born from an existing one, linked by a black hole at the neck. Eventually, the black hole would evaporate away, and the young universe would become independent of its parent. According to this theory, every black hole may be the seed of a new universe, and our own universe may be the offspring of a black hole in another one. We thus get an image of an infinite, frothing foam of universes.

The greatest problem with this model is that the natural size for a universe is unimaginably small – about 10^{-35} meters (a millionth of a billionth of a billionth of the size of a hydrogen atom) – so how could a universe as large as ours come into being? Evolution could select bigger and bigger universes given two conditions. First, the laws of physics must be able to vary slightly between the parent universe and its offspring. For example, gravity might be a bit stronger or a little weaker. Second, the larger the universe, the greater the number of black holes it must be able to create, and hence the more baby universes it could spawn.

Now imagine that a new universe just happens to grow a bit bigger than usual, and collapses to two black holes instead of one. Each of these creates another new universe in which the laws of physics may again be slightly different. One offspring might lose its ability to grow larger than usual, but the other might grow larger still, giving birth to more black holes and universes. In other words, large universes reproduce better than small ones, and hence would become the most numerous type. In the end, most universes would be as big as they could be without disrupting the black hole method of reproduction. Our universe matches this description, and is thus the logical outcome of 'universal' evolution. According to John Gribbin, "We live in a Universe which is exactly the most likely kind of universe to exist if there are many living universes that have evolved in the same way that living things on Earth have evolved."[10]

A slightly different picture emerges from Ervin Laszlo's psi-field hypothesis described in Chapter 9. If the laws of nature are eternal, preceding space and time, then the values of the physical constants must be defined precisely at the big bang, giving rise to the cosmic mysteries already described. However, if the laws of nature are habits, the constants can emerge within the developing cosmos by evolution. Laszlo's model envisages an endless sequence of expansion and contraction cycles from big bang to big crunch. The 'experience' of each cycle becomes imprinted in the psi field of the vacuum, thus helping to guide subsequent cycles. In this way, the universe learns and evolves from cycle to cycle, gradually refining its laws and constants to produce more and more complex structures, and, ultimately, sentient life itself.[11] This suggests that the implicate order which David Bohm believed underlies the world we experience is not itself eternal and unchanging, but has evolved over many generations of the universe.[12]

Not surprisingly, these ideas are controversial. Paul Davies claims that there are big theoretical difficulties with the idea that the cosmos might 'bounce' at the big crunch into a new big bang.[13] He argues that this would be physically possible only if no information was transferred from one incarnation to the next. But without such information transfer, evolution of the cosmos from generation to generation is impossible. However, as noted earlier, more recent research based on string theory suggests that information could be passed through a big bang.[14]

The multiple universe model is consistent with traditional evolutionary theory. It suggests that the laws of physics act like genes, passing on particular characteristics from parent to offspring. And like genes, the laws mutate. Gradually, those laws which reproduce most successfully become dominant, determining the form of most universes. Thus cosmic evolution appears to have pursued the goal of producing our type of universe, although as far as we can tell there may be many alternative 'reproducing' forms.

By contrast, Ervin Laszlo's model implies evolution of the laws of physics by learning, a process which is more similar to the inheritance of acquired characteristics in organisms (see Part V). Laszlo argues that the evolutionary process is accelerated and given apparent direction by the feedback of information from the memory held in the psi field.

In either case, the evolutionary process is stabilized by the fact that the laws of nature cannot change randomly. If they do not form a consistent set in which all the 'laws' are compatible with each other, the universe will self-destruct.[15] And so, just as a mutation in nature may render an organism unable to survive, a mutation to a natural law may not be viable.

In Conclusion

Was the universe designed? Perhaps. But not necessarily by the traditional creator God. Did the universe evolve from other universes? Quite possibly. And evolution could account for the signs of design. Does the universe have a purpose? Maybe. But the appearance of purpose could be the result of the feedback of information between 'generations'. Once again we are faced with questions that are beyond the ability of science to answer. And scientific theories in this field are as much acts of faith as the inner knowing of the mystical traditions.

We need to tune in to our own inner selves, and choose the alternatives that make most sense to us.

Review of Part IV

Despite almost universal acceptance of the big bang theory, cosmology is in flux. Hardly a month goes by without some new finding that challenges accepted ideas or opens up new questions. Data gathered in the last decade, often with the help of space observatories, have yielded a far more precise picture of the universe than ever before, in some cases confirming theoretical predictions and in others creating new theoretical challenges. The result is that cosmologists seem to be less sure about the nature of the universe now than they were 10 years ago! We are in the extraordinary situation of not knowing what 96% of the universe is made of. And one of the key ideas about the development of the early universe – the theory of inflation – is being called into question by some researchers. It will come as no surprise if radical changes occur in the next decade, possibly including replacement of the big bang theory.

Whatever the story that ultimately emerges, it seems likely to be stranger than any science fiction writer would dare to dream, and no easier to swallow than many traditional creation myths. Protagonists of the big bang ask us to believe that the universe created itself from nothing with no cause. And the leading alternative theories claim that the universe has always existed and always will. But one thing all the scientific theories have in common is the belief that consciousness is an accidental by-product of complex physical systems. However, as we've seen in the case of Gnosticism, an equally cogent case can be made that Consciousness is the fundamental reality, and that the material universe is an outcome of conscious processes. This approach could resolve the challenges faced by quantum physicists in understanding the interaction of mind and matter, and help to explain many mysterious aspects of consciousness which we will examine in Part VI.

Whether or not the unfolding of the universe is a conscious process directed by a cosmic purpose towards some final goal is unknown, and probably unknowable in an absolute sense. The dominant scientific view that the cosmos is mechanistic and purposeless is as much a belief as the minority view that it is purposeful. At this stage, the controversy is partly about what we mean by purpose. Does purpose have to be associated with the pursuit of precisely defined goals? Or can it be a more general sense of direction? As an analogy, consider our vague notion of the 'pursuit of happiness'. Most of us regard this as purposeful even though we are uncertain about what would make us happy, and change direction occasionally as we learn. In a similar way, the evolution of the cosmos appears to pursue increasing complexity and consciousness but may not have a predetermined goal.

It is clear that once again we are faced with alternative theories that are equally valid, and that the choice depends on our own worldview. We can accept the dominant scientific view that the cosmos is without meaning or purpose. Or, without invoking a creator God or Cosmic Consciousness, we can take the emergence of complexity and consciousness as evidence for an evolutionary direction built into the laws of nature – a weak kind of purpose. It is also possible to believe that Cosmic Consciousness exists and created the universe without negating the findings of science. The evolutionary process might have been left open-ended because there are many ways in which the goal of Cosmic self-knowledge could be achieved. Indeed, it can be argued that freedom to act as the participants choose is essential to true knowing in any relationship. Hence, a precise prescription of the outcome which left no creative freedom for conscious beings would obviate the aim of Cosmic Consciousness.

Cosmologists have puzzled for many years over the mystery of why the universe has precisely the natural laws which enabled intelligent life to evolve. Some believe that the answer lies in design, and hence a designer. Others have suggested that the universe itself may have evolved through many generations, and that the cosmos we see around us is the type most fitted to survive. A variation on this idea is that the universe has gone through many expansion-contraction cycles, being reborn each time in a new big bang. With each cycle, it learns from its forebears, gradually producing more and more complex systems, and ultimately intelligent life. At present, there is no scientific way to choose between these alternatives, and they are all potentially compatible with the Gnostic view of the primacy of Consciousness.

And so we are left with a vision of a universe most of which consists of mysterious matter and energy, which may or may not have a design and purpose, which may or may not have a Designer and Creator, which may or may not be the result of cosmic evolution, which may or may not have originated in a big bang, and which may or may not obey immutable laws of nature. But it is a universe which always forms a self-consistent, integrated whole.

The cosmos is still full of mysteries, and may forever remain so.

Reflections on Part IV

In these Reflections I have gathered together quotations from a variety of spiritual and scientific sources that I find both inspiring and thought provoking. I suggest you choose a time when you can relax undisturbed for a while, and read them slowly and reflectively. Pause to meditate longer on any which particularly draw you. Once again, you may wish to record or express any insights that come.

Before the Beginning

There is a reality even prior to heaven and earth,
Indeed, it has no form, much less a name;
Eyes fail to see it; it has no voice for ears to detect;
To call it Mind or Buddha violates its nature,
For it then becomes like a visionary flower in the air;
It is not Mind, nor Buddha;
Absolutely quiet, and yet illuminating in a mysterious way,
It allows itself to be perceived only by the clear-eyed.

Dai O Kokushi[1]

Before the world was
And the sky was filled with stars ...
There was
    a strange, unfathomable Body.
This Being, this Body is silent
    and beyond all substance and sensing.
It stretches beyond everything
        spanning the empyrean.
It has always been here, and it always will be.

Everything comes from it, and then
   it is the Mother of Everything.
I do not know its name. So I call it TAO.
The Body of The Tao
      is a mist beyond your eyes.
Tao of No Body,
       and yet within it is All Creation.
Like a seed in the dark, and a dim light
And from it, comes everything.

Lao Tzu[2]

(God) is neither a soul, nor a mind, nor an object of knowledge; neither has He opinion, nor reason, nor intellect; neither is He reason, nor thought ...; neither is He number, order, greatness, littleness, equality, inequality, likeness, nor unlikeness; neither does He stand nor move, nor is He quiescent; neither has He power, nor is power, nor light; neither does He live, nor is life; neither is He being, nor everlastingness, nor time, nor is His touch knowable; neither is He knowledge, nor truth, ... nor wisdom, nor one, nor one-ness, ... nor goodness; ... neither is He darkness, nor light; nor falsehood, nor truth; neither is there any entire affirmation or negation that may be made concerning him.

St Dionysius[3]

And I'm sure that today St Dionysius would transcend his gender-ridden language, and say that God is neither male for female.

Philosopher of science, M. K. Munitz, uses the term 'Boundless Existence' for what others call God. He claims that we cannot know Boundless Existence intellectually, or through the senses, observation, imagination, or concepts, and that it cannot be described or communicated. Rather, it must be apprehended directly and personally. For those who experience it, that apprehension is powerful and self-validating, not requiring any other justification. Following is his attempt to describe what he claims cannot be described.

Boundless Existence can only be "characterized" in negative terms. It is not a complex whole composed of parts. It is not "in" space or time, nor does it have any spatial or temporal structure. It has no history: neither a past, present, nor future. Nor is it spread out in space or have any location or shape. It does not have any structure at all. ... (Boundless Existence) does not refer to an individual that lies beyond the range of observation, since it is not an individual at all. It is not an agency, cause, source, generator, producer, or creator. It has no mind, purpose, value. Nor is it a mind, purpose, or value. Nor is it anything material, decomposable, or transformable into something else. It has neither extension nor thought, nor any other attributes, properties, or qualities. ... There are no laws, generalizations, or patterns that can be confirmed as holding of it. Boundless Existence is so totally unique ... that all similarities with anything in our ordinary experience must fall short and be inadequate. ...

Boundless Existence "shines through" the known universe ... but is not identical with it. ... In a way that is wholly unintelligible and unknowable, Boundless Existence "underlies" the existence of the universe and all that it contains.

M.K. Munitz[4]

The Creative Act

(A theory) is just a set of rules and equations. What is it that breathes fire into the equations and makes a universe for them to describe? ... Why does the universe go to all the bother of existing? Is the unified theory so compelling that it brings about its own existence? Or does it need a creator, and, if so, does he have any other effect on the universe? And who created him?

Up to now, most scientists have been too occupied with the development of new theories that describe what the universe is to ask the question why. ... If we find the answer to that, it would be the ultimate triumph of human reason – for then we would know the mind of God.

Stephen Hawking[5]

(T)he vacuum whispers to us softly, very softly, that the universe began as *a thought that dreamed itself awake.*

Darryl Reanney[6]

And God said, 'Let there be light'.
The Bible, Genesis

According to cosmologists, "let there be light" is a scientifically accurate description of the opening moments in the cosmic drama.
Willis Harman and Howard Rheingold[7]

The cosmos is like a growing organism, forming new structures within itself as it develops. Part of the intuitive appeal of (the Big Bang) story is that it tells us that everything is related. Everything has come from a common source: all galaxies, stars and planets; all atoms, molecules and crystals; all microbes, plants and animals; all people on this planet. We ourselves are related more or less closely to everyone else, to all living organisms, and ultimately to everything that is or that ever has been.
Rupert Sheldrake[8]

(F)rom hydrogen, the simplest of all the atoms, have come symphony orchestras, diamonds, the glimmer of dew on spiderwebs at dawn, the underwater flight of dolphins, the colors inside abalone shells, Voyager spacecraft, fractal images on computer screens and the laser light of cognitive awareness. This is not a mechanical progression from simplicity to complexity; it is a creative act of stupendous proportions.
Darryl Reanney[9]

The creation theories of science have grown up within the Judaeo-Christian (culture), with its ... conception of a beginning, a Fall, a historical progress towards the end of history, and an end that in some sense re-establishes the beginning. The theory of the Big Bang and the modern doctrine of universal evolution bear a striking resemblance to this fundamental myth of our culture.
Rupert Sheldrake[10]

(T)he expansion-contraction cycles hypothesized by contemporary cosmologists were prefigured with astonishingly accurate descriptions in the ... Hindu (scriptures) ... which described cosmic cycles known as "the breath of Brahma." These cycles not only corresponded to "fiery," explosive phases and eventual contractive phases, but were predicted to be on a time scale ... within an order of magnitude of the time scale hypothesized by modern astrophysicists!

Willis Harman and Howard Rheingold[11]

Now the physicist himself, who describes (the world) is ... himself constructed of it. He is, in short, made of a conglomeration of the very particulars he describes, no more, no less, bound together by and obeying such general laws as he himself has managed to find and to record.

Thus we cannot escape the fact that the world we know is constructed in order (and thus in such a way as to be able) to see itself.

This is indeed amazing.

G. Spencer Brown[12]

Cosmic Design

If we accept the evidence of cosmic design, we must then ask: Who was the designer? It could be God, or Cosmic Consciousness, or an advanced alien civilization in another universe that has learned how to create universes and travel between them. But why go to all the trouble of creating an actual universe? Surely such an advanced civilization would have learned to simulate consciousness as well as physical and biological processes. So why not simply create a virtual universe? If the computer were powerful enough, the virtual world would appear real to the 'beings' within it.

> The key question is this: do the simulated people exist? As far as the simulated people can tell, they do. ... There is simply no way for the simulated people to tell that they are "really" inside the computer, that they are merely simulated, and not real. They can't get at the ... physical computer, from where they are inside the program. ... There is no way for the people inside this simulated universe to tell that they are merely simulated, that they are only a sequence of numbers being tossed around inside a computer, and are in fact not real.
>
> Frank Tipler[13]

But this whole discussion begs the obvious question: how do we know that we ourselves are "real", and not merely a simulation inside a gigantic computer?

Paul Davies[14]

Nick Bostrom argues that there's no way to be sure, but that it's highly likely we are virtual people.[15] And John Barrow has thought about how we could discover if we are a simulation, with some disturbing conclusions.[16]

So what might we learn from these ideas about the purpose of the universe?

And about the meaning and purpose of our own lives?

How is the idea of being in a virtual cosmos created by an advanced civilization different to being in a real world created by God?

In the end, does it matter if we are virtual beings?

What difference would it make if we knew?

Part V:
Life

Our current creation myth tells us that the big bang gave birth to subatomic particles and simple atoms. Gravity then drew them together, forming the stars in which the elements we know today were forged. The next step up the great holarchy of nature was the emergence of a rich diversity of molecules, thus paving the way for the evolutionary leap into life.

Despite dramatic advances in the biological sciences, life still holds many unexplained mysteries. What is it? How did it begin? How do complex organisms develop from a single cell? How do living things evolve? And is the Earth a living organism? We explore these questions in this Part, discovering that the commonly accepted scientific answers are often half-truths at best.

Chapter 13 attempts to define the nature of life, and then explores ideas about how it may have begun. This is followed in Chapter 14 by discussion of how the 'seed' cell of a higher organism such as a flowering plant or mammal develops into its mature form. The process by which life has evolved from its simple origins to its present diversity is the subject of Chapter 15, followed by an introduction to ecology and the idea of a living planet in Chapter 16. As usual, the Part closes with Review and Reflections sections.

Once again, the climb is often steep and the going rough. But the struggle is rewarded with fresh vistas.

13 The Nature and Origins of Life

In September 2004, New Scientist magazine listed the top 10 unresolved issues in the life sciences.[1] Amongst them were the questions: "What is Life?" and "How did life begin?" These are the subjects of this chapter.

Life is so extraordinary in its properties that it qualifies for the description of an alternative state of matter.

Paul Davies (1998), p.xviii

What Is Life?

Most of us feel we can tell the difference between living and non-living things quite easily. A tree is obviously alive, whereas a rock is not. And a dead dog is very different to a live one. But it is actually very difficult to draw a clear distinction between living and non-living. Indeed, in the words of New Scientist: "The best minds of biology and philosophy have tried for decades, and failed, to agree on a universal set of criteria for life."[2] So what is the difference between an inanimate object and a living organism?

When my son was at University in the mid-1990's, his biology text gave four criteria for life.[3] First, an organism must be composed of one or more *cells*. A cell is a complex collection of molecules enclosed within a membrane that protects what happens inside from disruption by events outside. More generally, an organism must have a boundary that separates it from its environment.

The second criterion is that an organism must use *energy* from outside to grow and maintain itself. Plants get energy from sunlight, animals derive it from food, and many microbes make use of chemical reactions. Whatever its source, this energy is accompanied by flows of nutrients and wastes. When an organism dies, the only immediate change is that the flow of energy stops.

According to most biologists, the third sign of life is *reproduction*. Since all organisms eventually die, living things must either reproduce or be continually re-created from inert matter. But reproduction means more than simple duplication. In order to qualify as living, a system must be able to pass on through *heredity* not only its original characteristics but also changes to them. This enables the species to evolve to meet the challenge of change in its environment.

If you took a vote today, the most popular definition (of life) would probably be … a self-sustaining chemical system capable of evolving through Darwinian natural selection. … but critics worry that, broad as it is, it may not be broad enough to encompass absolutely everything we would want to call life.

Bob Holmes (2004)

However, these criteria are not universally accepted. Humberto Maturana and Francisco Varela point out that cross-bred animals such as the mule cannot reproduce but are very much alive. They argue that the sole hallmarks of life are growth and self-maintenance.[4] James Lovelock also rejects reproduction as a criterion in light of his hypothesis (to be discussed in Chapter 16) that planet Earth is a single living organism – one that clearly cannot reproduce in any normal sense of the term.

The dividing line between life and non-life is fuzzy as well as disputed. A virus is a fragment of genetic material encased in protein which can enter a cell and use it to supply its needs.[5] Viruses often are classified as non-living because most of them cannot survive or reproduce independently. But all organisms depend on their environment for the energy and nutrients they need, and are parts of an interconnected ecosystem. Hence, viruses differ from other organisms only in that their environment is a living cell rather than an ecosystem.

Phospholipid molecules are another step away from life as normally defined. When dissolved in water they form hollow spheres which may trap energy-processing molecules. They can grow by absorbing 'food' molecules from their environment and subdivide when they get big. Most biologists regard them as non-living because they have no mechanism of heredity, but they are self-organizing systems which meet the other criteria for life.[6]

Still further removed from normal living things are the protein molecules called prions, one of which causes BSE, or 'mad cow' disease. Copied using the animal's genes, these prions occur in two forms which have the same chemical composition but different shapes. One form is found in all cells, but the other only occurs in diseased brains. A molecule of the normal prion changes shape if it comes in contact with an abnormal one, and the abnormal form cannot be broken down by the cell. Hence, just one abnormal molecule can lead to infection of the whole brain.[7] This is a successful form of reproduction, but prions do not meet the other criteria for life.

We may conclude that there is no clear dividing line between living and non-living. Rather, there is a continuum of systems, and where we choose to make the division is largely arbitrary.

(L)ife and non-life are words like hot and cold. They are positions on a scale graduated from simple (no-life) to complex (life).

Charles Birch (1990), p.79

As well as being controversial, the four criteria for life fail to capture its richness. Non-living things have little flexibility and autonomy. But an organism defines its own form, size and functions, and maintains and reproduces itself. This reflects a will to exist as an integrated whole, and to continue to exist beyond death. But whilst science is rapidly revealing the mechanics of life, the source of this 'life force' remains as mysterious as ever.

Not only are organisms autonomous, but also they can adapt creatively to change. Even simple organisms such as amoebae learn from experience, whilst higher animals teach what they learn to their offspring. And it is becoming apparent that organisms can direct their own evolution, as we shall see in Chapter 15. The variations that enable organisms to evolve also mean that each one is different in subtle ways from all others of its kind, and has its own unique identity and characteristics – unlike machines or molecules of a given type which are all identical.

Without the membrane to hold its contents together, cell metabolism would be impossible. And without metabolism to maintain it, the membrane would disintegrate. Similarly, the genes specify the proteins needed to create and maintain the cell, and are themselves repaired by cellular processes. These processes replace most of the cells in my pancreas every 24 hours, and those of my stomach lining every 3 days; my white blood cells last about 10 days, and 98% of the protein in my brain is replaced in less than a month. Thus, an organism's identity resides not in the matter of which it is made, but in the dynamic unity and integrity of the processes by which that matter is organized.

The importance of process compared to structure becomes obvious when an organism dies. Immediately after death, the form and composition of a dead cat are unchanged from when it was alive. But something clearly is different. The flow of energy that powered the cat's growth and movements, and maintained its body tissues has stopped. But this scientific explanation of the mechanics of life fails to tell us why this energy flows in the first place, or why it ceases. And it doesn't tell us what gives the body its particular form, since the same energy and matter take many forms as they pass from sunlight to grass, to cow, to human. According to all non-scientific cultures, there is an essence which leaves the body at death, variously called the life force, breath, spirit, soul, subtle body, vital factor, or organizing principle.[8] Perhaps scientists are missing something?

Arising from this discussion, it is tempting to add yet another concise definition of life to the pile. But I can't think of one that captures all the subtle nuances discussed above, and excludes everything that is clearly non-living, such as computer software that creates itself.[9] We must accept that life is life, and refuses to submit to our desire to put it in a neat box.

The Miracle of Life

So extraordinary is the phenomenon of life that Paul Davies called it *The Fifth Miracle* – the first four being God's creation of the universe, light, the heavens and dry land.[10] One reason life seems so miraculous is scientists' belief that it must have evolved by chance. Rough calculations show that the probability of a virus forming by chance in a billion years is about $10^{2,000,000}$ to one against – an impossibility.[11] Yet fossils of blue-green algae far more complex than any virus were laid down 3.5b years ago, just 1b years after the Earth formed. So it seems that something other than chance was at work. We will explore what that might have been in the next section.

A second reason life appears to be a miracle is the complexity of even the simplest forms. How could the leap possibly have been made from matter to life? This complexity is demonstrated well by the chemical processes that occur in every cell. You may find it difficult to absorb all the following ideas and technical terms, but most of them are relevant to later sections.

Feel free to refer back when you need to!

At the heart of each cell are the spiral-shaped DNA molecules that encode the recipe for life. If we imagine untwisting the spiral, the molecule would look like a very long ladder with rungs made of the four *bases*, A, T, C and G. Each base connects with one other to make 4 types of rung: A-T, T-A, C-G and G-C.

Now imagine that we pull the ladder apart, breaking each rung in the middle. Each half is a template which grows into a complete new DNA molecule as appropriate bases attach themselves to the broken rungs. This process is facilitated by a variety of specialist molecules called *enzymes* that unzip and rejoin the DNA, and guide the right bases into position. For the organism to survive, this copying must be incredibly precise, and higher organisms have evolved specialized proof-reading enzymes to fix almost all errors.

Taken three rungs at a time, combinations of these four bases form 64 'words' representing the different *amino acids* from which all *protein* molecules are made. As all organisms use only 20 different amino acids, there is more than one 'word' for some of them but the reason for this is not known. It also remains a mystery how this code could have evolved by trial and error since changing the amino acid associated with just one 'word' would change all the many proteins containing that particular amino acid, normally with fatal results. This inflexibility may be one reason why the code has endured unchanged for an astonishing 3.5b years.

Protein molecules are long chains of 100 or more amino acids which form the basic building blocks of life. Every cell in higher organisms contains around 30,000 different proteins. The recipe for each protein is 'read' from its *gene* in the DNA molecule by a *messenger RNA* (mRNA) molecule which takes it to one of thousands of *ribosome* 'factories'. Here, the proteins are assembled from amino acids gathered as needed by *transfer RNA* (tRNA). Once the chain of amino acids is complete, the protein folds into a precise 3-D shape. How it chooses its shape is also a mystery, but that is another story.[12]

We've only touched upon some aspects of the awe-inspiring complexity of cellular chemistry here. But I hope it is sufficient to demonstrate just how miraculous the processes of life really are.

A third mystery of life is how the information encoded in the genes came into existence. The proteins in each cell of a higher organism represent about 30 million bits of information, and even a typical bacterium stores the equivalent of a large book. However, the second law of thermodynamics states that information cannot spontaneously spring into being any more than an object can spontaneously heat up. So where did the vast treasure trove of genetic information come from? Paul Davies compares the origin of this information to tipping out a jar of coffee beans and forming a specific random pattern.[13]

DNA is often portrayed as a super molecule that controls life for its own selfish ends. But DNA is impotent without the cooperation of millions of specialized molecules within each cell – the bases, amino acids, RNA, ribosomes, proteins, enzymes and so on. Many of these molecules are themselves enormously complex, are found nowhere else, and perform an amazingly intricate dance without a choreographer. In the words of Paul Davies: "Our cells contain sophisticated chemical repair and construction mechanisms, handy sources of chemical energy to drive processes uphill, and enzymes with special properties that can smoothly assemble complex molecules from fragments ... As fast as the second law tries to drag us downhill, this cooperating army of specialized molecules tugs the other way."[14] And so we are left to wonder how such an interwoven whole could come into being. Which came first: DNA, RNA, or proteins? And how were the first constituents made in the absence of the supporting cast?

Possible answers to these mysteries are examined in the next section.

The origin of life appears … to be almost a miracle, so many are the conditions which would have had to be satisfied to get it going.

Francis Crick (1981), p.88, discoverer of the structure of DNA

The Origins of Life

We don't yet know how life began, but at least three steps were involved. First, complex organic molecules must form. From these, simple self-replicating molecules, such as proteins or RNA, must emerge that have the potential to evolve into life as we know it. And finally, a complex soup of these chemical reagents must be isolated from its surroundings to facilitate self-organizing reactions.

Much has been made of the fact that amino acids, the building blocks of proteins, can be created in the laboratory. And the presence of amino acids in deep space has fuelled the idea that Earth could have been seeded from elsewhere[15] – an answer which is no answer to the mystery of how life began. But amino acids form relatively easily because the reaction gives off energy. By contrast, the next step of combining them into short *peptide* chains is more challenging because appropriate energy must be supplied. Nevertheless, chains of up to 4 amino acids form naturally in volcanoes.[16]

Imagine for the moment that peptides are available. It is now necessary to create proteins from them. One of the challenges here is an embarrassment of riches since the 20 amino acids can be assembled into 10^{130} different chains of 100 amino acids, of which life uses only a tiny fraction. The odds against this particular set of proteins being selected by chance have been estimated at 1 in $10^{40,000}$ – as near to impossibility as can be imagined. So how were the right ingredients chosen?

Part of the answer lies in the facts that what a molecule does is determined by its shape as well as its composition, and that many molecules of different compositions have similar shapes. For instance, the 10^{130} amino acid sequences may fold into as few as 100 million shapes.[17] Thus, life did not have to wait for exactly the right molecules, only ones of the right shape. This reduces the problem from looking for a needle in a haystack to sifting the one useful molecule from a few thousand.

Once complex molecules were available, the next step may have been the emergence of simple self-replicating systems with the potential to evolve into life as we know it today. One candidate for this role is a virus that uses RNA molecules instead of both DNA and enzymes. In one experiment, an RNA virus called Q_β replicated its RNA in the laboratory, but not its protective protein coat. In another experiment, self-replicating RNA molecules formed spontaneously in a broth of raw materials that included an enzyme that the virus uses to facilitate replication. Exciting though this is, it does not demonstrate the emergence of life since the enzyme had been extracted from the virus, and almost certainly would not have been available in the primeval soup. Further, some scientists argue that RNA evolved relatively late, and hence is unlikely to be the original self-replicating molecule. An alternative possibility is a simple protein molecule since some of these can not only replicate but also correct errors.

Whichever self-replicating molecules were used by early life, self-organization seems certain to have played a major part in producing them. Computer simulations by Stuart Kauffman show that a diverse mixture of organic molecules can not only organize itself, but also reproduce.[18] And, if the diversity is high enough, an avalanche of new molecules may be unleashed. This suggests that sufficient diversity may be all that is needed for the dynamics of self-organization to produce life.

Attractive though this idea may be, self-organization is not the whole answer because life is different to other complex chemical systems. Self-organizing complexity, such as chemical clocks, can be described by simple laws and formulae, and hence contains little information. By contrast, the genetic code is extremely rich in information that cannot be reduced to simple models in this way.[19] In practice, the protein molecules are specified by the information-rich genes, and self-organization then plays a major role in arranging them into cells, tissues, organs and organisms, as we will see in the next Chapter.

The source of this genetic information remains obscure, although two possibilities have been suggested. Some scientists believe there is an evolutionary trend towards ever-greater complexity. They argue that an as-yet-undiscovered law of complexity garners information from the environment and stores it in self-organizing systems. Others believe that cells may act as quantum computers that have the ability to generate information.[20] Similarly, the psi field, if it exists, could inform living things about what has happened in other parts of spacetime, enabling genetic information to be shared across the universe.

The final factor needed for the emergence of life is some means of protecting the self-organizing chemical reactions from disruption by their environment. Simple theoretical models show how this might have happened. Imagine that the chemical stew includes molecules that can form a membrane around a 'bubble' of liquid, and a catalyst that facilitates this reaction. Possible candidates for the membrane include phospholipids or iron sulfide which is abundant around deep-sea volcanic vents. Also imagine that some catalyst molecules are trapped inside when a membrane forms. Hence, if the membrane is punctured, it will be repaired whenever a catalyst molecule tries to escape. Such bubbles can grow by absorbing 'food' molecules, and reproduce by dividing when they reach a critical size. If the chemical composition inside each bubble were slightly different after division, a process of evolution would begin towards bubbles that grow and reproduce more efficiently.

Attractive though this theory is, it is not the complete answer. There also must have been an environment conducive to the formation, survival and reproduction of the bubbles, and that not only brought the ingredients together, but also protected the reactions from disruption. Yet, when life began, the Earth was almost completely covered in hot water which surged back and forth in huge tides. The atmosphere was dense and unbreathable, thick clouds hid the sky, and UV radiation was intense. Fissures in the

Earth's crust spewed forth molten rock, and asteroids frequently smashed into the fledgling planet. Hardly conditions conducive to life as we know it!

But in the last few decades, scientists have been discovering that life is far tougher than had been thought. The most inhospitable regions of the Earth are replete with microbes dubbed *extremophiles*. Some use hydrogen sulfide for energy, eat concrete or thrive in high acidity. Others enjoy high alkalinity, the salt of the Dead Sea, the heat around volcanic vents, the intense cold beneath the Antarctic ice sheet, or pressures of a thousand atmospheres in deep ocean trenches. Still others thrive on nuclear waste, and bacteria have even survived two years on the moon.

It seems almost certain that the earliest living things were heat-loving extremophiles that emerged near volcanoes or fissures in the Earth's crust. But even these tough organisms needed a habitat protected from asteroid impacts, UV radiation, extremes of climate change and other destructive events. The most likely place is in the pores of spongy under-water basalt rocks near fissures in the crust. Today, deep-sea volcanic vents teem with life where molten rock meets cold water and creates a rich chemical brew. Most species of microbes here are ancient archaea, and conditions are probably little changed since life began – including pitch darkness, high temperatures and high pressures. Complex ecosystems containing thousands of species of microbes also have been discovered up to 7km deep in rocks. Whilst it is possible that they migrated there from the surface, it seems more likely that they are remnants of early life. Hence, we can have confidence that life on Earth will survive the most violent catastrophes. Even if surface life is extinguished, the basis will remain for evolution to start again.

In Conclusion

The origin of life is the key to the meaning of life.

Paul Davies (1998) p.3

Despite intensive research, the boundary between living and non-living things remains ill-defined. And, although currently unpopular amongst scientists, the idea that there is some unidentified life force or vital essence refuses to die.

Most scientists continue to believe that life emerged by chance. Yet the odds against this are astronomical, and there is a growing body of opinion that the evolution of life, followed by consciousness, is an inevitable consequence of laws of nature that we don't yet understand. If this is true, we should expect to find life throughout the cosmos. This difference of opinion is more than a normal scientific debate. It represents a major philosophical shift away from belief in a random, meaningless universe towards a belief that evolution has a definite direction towards complexity, life and consciousness.

14 Development

Each of us developed from a fertilized ovum to a mature human; a journey similar to that of every multi-cellular organism. And whilst this process is not quite so extraordinary as the origin of life, it still seems miraculous that the 'seed' or 'egg' cell develops of its own volition, with no more than energy and nutrients from the outside world.

As this primal cell divides and re-divides, differences appear. And these differences increase as the young organism grows, until what was a mass of similar cells becomes organized into roots, stems, leaves, and flowers; or the complex tissues and organs of an animal. Eventually, the organism reaches maturity with all its parts taking on their appropriate sizes and forms in the correct relationships with each other.

This development process is very robust. Half a sea-urchin embryo develops into a normal adult; and two fused embryos become a single organism. Similarly, a chicken grows normal wings even after the wing buds are surgically removed from the embryo. Mature organisms also show remarkable abilities to repair themselves. Plants can be grown from cuttings or even single cells. A flatworm can be chopped up, and a slice will redevelop into a complete organism. Salamanders re-grow whole limbs that are as complex as those of humans. And even humans regenerate lost skin and liver cells, replace shed blood, knit broken bones, heal wounds and re-grow nerves.

With all the hype about genetics, non-biologists could be forgiven for believing that genes are all-powerful determinants of this development process. And yet the genes are the same in every cell of an organism, and do no more than specify the amino acid sequences of a list of proteins. So how does each cell know which proteins to produce? How does it organize these proteins into a functioning whole? How does the developing organism ensure that the various cell types are produced in the right places and arranged correctly in relation to one another? And how, for example, does a human embryo know it is not supposed to turn into a chimpanzee, given that there is only 1% difference in their DNA?

This Chapter investigates the twin processes of *differentiation* into a variety of cell types, and then *morphogenesis,* or the organization of cells and tissues into specific forms. It reveals that the idea of a genetic 'program' or 'blueprint' is not sufficient to explain the process of development, and self-organization, *morphogenetic fields* and *archetypes* also may be involved.

The Process of Differentiation

As already noted, the genes are identical in every cell of an organism, and every cell is derived from the original 'seed' cell. Yet adult humans, for example, have 256 different types of cell. This diversity results from inactivation of many of the genes in most cells, leaving switched on only those needed for a particular cell type, such as skin or muscle. But how does this happen?

In the earliest stages of development, every gene is active in each cell, and hence they have the potential to become any type of cell in the organism. In humans, these are the *stem cells* that medical researchers hope will cure a range of diseases, and which became a controversial focus of the 2004 US Presidential campaign. At present they are extracted from human embryos, thus raising ethical and religious concerns, but in time ample supplies will doubtless become available by cloning or 'de-differentiation' of the patient's cells.[1]

Differentiation begins when self-organizing reactions concentrate certain chemicals at each end of the 'seed' cell. Hence, when this cell divides, its two daughter cells have slightly different chemical compositions. Genes are switched on and off by special molecules, and so a slightly different set of genes may be active in each of these cells. The first obvious sign of differentiation in animals is the emergence of three types of tissue: *endoderm*, which develops into glands and the digestive system; *mesoderm*, which becomes muscles, bone and the circulatory system; and *ectoderm*, which forms skin and the nervous system.

One theory is that differentiation is controlled by a small group of 'master' genes, but this idea has some serious shortcomings. First, like all others, the master genes are the same in every cell. Hence, they are in no better position than any of the other genes to know which proteins are needed at this particular location and stage of development. Second, experience with computer software suggests that a hierarchical control system such as this would not be stable enough to produce reliable development, particularly as genes are being continuously deleted, moved, duplicated and changed.[2] Third, the response of developing organisms to injuries indicates that their parts communicate with each other, changing the type of cell they become as necessary to repair the damage whilst continuing development. Such a process cannot be centrally controlled.

In the absence of a genetic blueprint, it appears that differentiation is coordinated in stages, so that a cell only has to choose between 2 or 3 alternative paths at any time. For example, the first 'decision' for animal cells is between endoderm, mesoderm and ectoderm. Once made, this choice is irrevocable. Hence, at each parting of the ways, the history of earlier decisions defines the options that are available for the next stage. In this way, the cells are channeled through a cascade of self-organized bifurcations towards increasingly specific types as more and more genes are switched off. Eventually, each cell in the mature organism produces the specific proteins for just one type of cell.

Further insights into this process come from the network simulations of Stuart Kauffman. Imagine a network of 10,000 interconnected genes, each of which may be active or inactive. As we saw in Chapter 5, such a network has an almost infinite number of possible combinations of gene activity. In practice, however, it settles quickly into a stable cycle of only 100 patterns, each of which represents a different cell type. We may conclude from this that only a few of the myriad possible combinations of genes in any organism will produce viable cell types. However, those combinations that are viable are very stable – a conclusion that is supported by the robustness of the development process in practice. Finally, the network of genes does not have to go through the tedious process of searching for viable combinations amongst the vast number of possibilities, but is attracted rapidly towards these stable patterns.

The Process of Morphogenesis

The process of cell differentiation is only half the story of development. We also have to consider how the various cell types are assembled and organized into the specific form of the mature organism. Once again, the idea of a genetic blueprint is inadequate.

Each gene is simply the recipe for a specific protein which may affect one or more traits of the organism, either on its own or in combination with proteins coded by other genes. In general, therefore, a single gene may affect several traits, and a particular trait may be produced by the interaction of many genes. Hence, an organism with a particular gene has a greater chance of developing an associated characteristic, but is not certain to do so. And to make the situation even more complex, the effect of a gene may be modified, or even reversed, by environmental conditions. For example, the sex of Mississippi alligators is determined by the temperature of the eggs in their unattended nest.

Richard Dawkins is an outspoken advocate for the view that genes and Darwinian evolution can explain all the characteristics of life. Nevertheless, he stresses that development is a holistic process. For instance, a gene can affect the number of fingers only if an arm is being formed, and an arm will be formed only if there is a whole developing embryo. The development of the embryo in turn depends on the interaction of myriad genes with each other and environmental factors, and on the complex interplay of processes in cells, tissues, and the organism as a whole.[3]

Dawkins also stresses that the genes are more like a recipe than a blueprint. This recipe specifies what ingredients to use, and how to combine and cook them, but does not say what shape the finished cake should be.[4] In similar vein, Brian Goodwin claims that "a genetic program can do no more than specify when and where in the developing embryo particular proteins and other molecules are produced, and we know that molecular composition isn't sufficient to explain physical form."[5] A more complete explanation requires the power of self-organization and morphogenetic fields.

The processes of development and regeneration both suggest that cells somehow 'know' where they are in relation to the whole organism, and respond accordingly. Long ago, this led to the idea that there must be an internal coordinate system, but it is only in the last couple of decades that a possible mechanism has become apparent.

Imagine that there is a high concentration of some chemical at one end of an organism, and a low concentration at the other end. If cells could distinguish between high and low concentrations, they would be able to tell if they were at the front or the back. Now suppose that a second chemical has a low concentration at the two ends and a high concentration in the middle of the organism. By combining these two pieces of information, cells could determine which quarter they were in. Additional gradients, or greater sensitivity to chemical concentrations, would enable the cells to locate themselves more precisely. And similar chemical gradients could provide coordinates in the other two directions as well. If these chemicals also acted as gene switches, it is possible to envisage the coordination of complex patterns of cell differentiation.[6]

In Chapter 5, we saw how chemical reactions with both positive (catalytic) and negative (inhibitory) feedback loops can generate complex spatial and temporal concentration patterns. Such patterns could guide both the formation of an organism, and the sequence of stages through which it develops, as shown by two examples.

An animal's coloration comes from the combination of two skin pigments. Hence, complex patterns can be produced by varying the activity of just two genes. Simulations of self-organizing reactions that produce gene-switching chemicals on the surface of an embryo have successfully reproduced many patterns that occur in nature. If the reaction happens when the embryo is very small, the animal develops a plain color, but patterns emerge if the reaction is delayed until the embryo is larger. Broad bands of color appear first, then stripes and spots, followed by large blotches separated by narrow strips, which finally merge into a single color. Thus, this one process can explain the plain coat of a mouse, leopard spots, zebra stripes and giraffe patches. More complex patterns, such as butterfly wings, can be produced by similar processes, some of which are sensitive to environmental factors such as temperature.[7]

Further support for the role of self-organization comes from studies of fruit flies. As a fruit fly develops, the original cluster of cells divides into compartments, each of which contains cells of one particular type. There are several such compartments in the wing of an adult fly, and computer models of self-organizing chemical reactions in a wing-shaped dish produce similar patterns and sequences of development. In some cases, these chemical patterns are highly stable, remaining unchanged over a wide range of conditions.

Self-organization also helps to explain the robustness of development. If an embryo is damaged, the chemical patterns around the wound change, and hence the remaining cells develop differently. For instance, if the wing buds of a chick embryo are amputated, the cells nearby no longer have wing cells next to them. Hence, conditions are more like those before the wings started to grow, and they respond accordingly.

As noted in the last chapter, it takes little information to define a self-organized pattern. Thus, self-organizing processes require few genes, helping to explain how the complexity of the human body can be created with only 30,000 instructions. Fractal patterns are similarly efficient.[8] They are generated by the repeated application of a simple formula, and are widely used in nature. An example is the Koch curve which can be generated from a straight line by repeating three simple instructions:

1   Form an equilateral triangle on the central third of the line;
2   Erase the original base line of the triangle;
3   Repeat for all resulting line segments.

The first two cycles are shown in Figure 11. When repeated many times, this produces a line of potentially infinite length within a finite area. In three dimensions, it creates a huge surface area within a finite volume similar to that required for gas exchange in the lungs. Other examples of biologically-useful fractals are the branching patterns that enable blood vessels to reach every part of the body without taking up the whole volume, and the self-similar shapes of fern leaves.

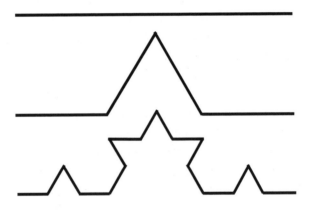

Figure 11   Koch fractal

We have seen that the differentiation and arrangement of cells in space probably is not controlled by a genetic blueprint, but by self-organizing chemical patterns that 'tell' each cell where it is in relation to the whole, and what nearby cells and tissues are like. This information can determine which cell type is produced where, but still does not explain fully how this mass of cells is transformed into a fully functioning organism. For this we also need to invoke the old idea of *morphogenetic fields* as reinvented by Brian Goodwin and others.

Much of their work involved a beautiful little green alga called *Acetabularia acetabulum* that grows in shallow waters around the

Mediterranean. Its slender stem, up to 50mm long, is anchored to the rocks at its base, and tipped with a delicate fungus-like frilled cap. Before the cap forms, the stem grows a whorl of fine leaf-like elements which later wither and fall off. Perhaps the most remarkable thing for an organism of this size and complexity is that it is a single cell. How could such a structure possibly form?

Detailed studies have revealed a fascinating story.[9] *Acetabularia* is formed by a morphogenetic field resulting from interactions amongst the concentration of calcium in the cell fluid, the pressure inside the cell, and the cell's structure. Wherever there is more calcium than normal, the cell wall softens and stretches in response to the internal pressure. And where it stretches, calcium is released from storage thus increasing the concentration further. This positive feedback results in further softening and stretching, but negative feedback kicks in to limit the process before the cell ruptures.

Brian Goodwin simulated the first stages of regeneration after the stem was cut in two. He correctly predicted closure of the cut, growth of a new stem with a smaller diameter than the original, and the appearance of 'buds' where the whorl of 'leaves' forms. To the surprise of the investigators, these results were easy to produce, suggesting that this form is robust, and would arise readily in nature. However, the model predicted a large bulbous end to the stalk rather than the delicate cap that actually grows, indicating that this shape is less robust.

The arrangement of leaves on plant stems provides another example of the way morphogenetic fields determine form. The 250,000 species of higher plant all arrange their leaves in one of three ways: singly, on opposite sides of the stem; as whorls of two or more leaves at nodes in the stem; or, most commonly, as a spiral with leaves spaced at fixed angles around the stem. These arrangements arise naturally as the layer of cells on the surface of the growing tip stretches in response to pressure from new tissue underneath. A simple model of this process revealed that two key factors are involved: the outward pressure of growth, and repulsion between leaf buds. If the leaves grow slowly, single leaves form on opposite sides of the stem. But as growth speeds up, spiral patterns appear, with the most common angle of 137.5° emerging when growth is fast. Whorls form if more than one leaf grows at a time.[10]

Some important insights emerge from these examples. First, neither the form of *Acetabularia* nor the arrangement of plant leaves are specified by the genes. Second, these forms have not been selected to improve the organisms' prospects for survival as the theory of evolution holds – a topic that we will discuss at length in the next chapter. Rather, the forms emerge from the interaction of cell chemistry and structure with the changing shape of the organism and the characteristics of its environment. This is a holistic process in which genes, biochemistry, physical structure, environment, shape and the organism as a whole are involved at every stage. The genes define the starting point, and strongly influence what is possible for the organism. But it is the complex field dynamics, in which the genes are only one player, that determine the outcome. This suggests that related organisms, such as different species of *Acetabularia*, follow development paths towards accessible, stable, generic forms, and that the genetic variations between them simply add the details of shape, size and color that distinguish each species.

Brian Goodwin concluded that "the main ... characteristics of organisms – hearts, brains, guts, limbs, eyes, leaves, flowers, roots, trunks, branches, to mention only the obvious ones, are the robust results of morphogenetic principles. There is also a lot of variation of these structures in different species, and it is in these small-scale differences that (genetic mutations) and natural selection find a role."[11]

Archetypes and Wholeness

Before leaving the topic of development, it is worth taking a look at another approach derived from Plato's belief that every organism is an expression of a pre-existent ideal form, or archetype. With the dominance of mechanistic molecular biology and Darwinian evolution on the one hand, and emerging understanding of self-organization and morphogenetic fields on the other, it is tempting to see Plato's approach as a redundant legacy from the past. But it is a mistake to dismiss it too easily as it serves to open up different perspectives on what it means to be an organism, or to be a 'whole'.

Brian Goodwin's approach is holistic in the sense that the morphogenetic field depends on the complex, integrated interaction of all aspects of the organism. Indeed, the totality of the integrated process *is* the organism. And the organism as a whole has no existence or meaning beyond this process. But this approach to holism arises out of the analytical, mechanistic method. Its perception of wholeness is founded on the study of separate components and processes, and emerges from their necessary interconnection and integration. There is no ideal form, no genetic blueprint or program, no formative field that exists before the organism comes into existence, and that is independent of and external to it.

As discussed in Chapter 3, there is another approach to holism that starts with the whole, and explores how diversity emerges from this unity. The application of this approach to biology is exemplified by Goethe's perceptions that all the parts of a plant are transformations of the leaf; and that each plant is an expression of an archetypal Plant. More recently, a similar approach to the development of form was taken by D'Arcy Thompson in the early twentieth century, and Wolfgang Schad half a century later.

From a wide-ranging study of the forms of organisms, D'Arcy Thompson showed that, within broad groups, they could be understood as simple geometrical transformations of one another.[12] Similarly, Wolfgang Schad claimed that the diversity of mammalian forms could be explained by variations in the balance between their constituent systems.[13] Of the three major groups of mammals, he argued that rodents are dominated by the nerves and senses; carnivores by respiration and circulation; and ungulates by metabolism and limbs. He explained variations within each of these groups by the relative strengths of the other two systems and environmental factors. As an example of his analysis, Schad noted that whilst all mammals have the same types of teeth, those of rodents are dominated by incisors, whilst carnivores put more emphasis on the canines, and ungulates on the molars. Much earlier, Goethe had observed that horns are a transformation of teeth, and hence any animal with a full set of upper teeth cannot have horns.

These studies provide a complementary perspective from which it appears that organisms use the mechanisms of genetics, self-organization and morphogenetic fields to transform themselves, and hence explore the range of forms available to them. Whether or not these transformations are *active*, with organisms purposefully exploring a pre-existent archetype of potential forms cannot be proved. But Brian Goodwin, reflecting the ideas of David Bohm discussed in Chapter 9, suggests that an organism's morphogenetic field enfolds an implicate order that is unfolded during development. This is tantamount to stating that there is an archetypal form that guides development, and comes close to Rupert Sheldrake's concept of morphic fields (See Chapter 9). According to this hypothesis, the development of every organism is guided by resonance with morphic fields associated with similar organisms – in effect a self-organized and evolving archetype.

In Conclusion

The impression is often given that the development of organisms is determined by a genetic blueprint. But the discussion in this chapter has shown that this is simplistic, and that the truth is much more complex, creative and holistic. It seems that development results from the interplay of many factors and processes including genes, self-organized bifurcations and patterns of chemical concentrations, structural dynamics and environmental factors. These mechanisms seem to leave little space for the old Platonic idea that the form of an organism is an expression of an underlying archetype, or Ideal Form. Nevertheless, this idea persists, and provides enriching perspectives on what it means to be a whole organism.

15 The Evolution
of Life

Once life had emerged, it faced the challenge of surviving as asteroid impacts blanketed the sky in dust, the climate swung from 'snowball Earth' to sweltering tropics, the atmosphere was poisoned by oxygen, and the intensity of sunlight grew. Some species managed to remain unchanged for billions of years, holding out in stable habitats such as the deep oceans. But most life evolved new forms.

Darwinian theory claims that the evolution of life is due to the chance occurrence of genetic variations, or *mutations*, which are then sifted by natural selection to eliminate those less fit to survive. But, as we will discover in this chapter, this simple statement hides a far more complex scene.

Variation

For evolution to happen, natural selection needs inheritable differences to work upon. Traditionally, three causes of chance genetic mutations have been recognized: external environmental factors, internal biochemical errors, and sexual reproduction. More recently, another two mechanisms have been discovered: the expression of latent characteristics, and experimentation by the organism. These are more controversial because they open the door to the active involvement of the organism itself – an idea that was heretical until recently. This section discusses each of these sources of variation, and then takes a look at another heresy – non-genetic inheritance.

The risk of genetic damage by toxic chemicals and radiation in the environment is a common concern in our industrialized world. Less well known is that some viruses can insert genes into their host cells – an ability often harnessed by genetic engineers. However, many mutations are due to internal errors rather than external factors. DNA molecules need constant repair, and genes are copied every time a new cell is made or an old one is replaced. These processes are amazingly accurate, but nevertheless errors do occur.

Scientists are still debating the evolutionary value of sexual reproduction, but its main benefits are probably to increase variation by mixing the genes of individuals, and to replace faulty genes with healthy ones. However, the story is more complex than this suggests. Some thriving organisms are asexual, and others switch backwards and forwards between sexual and asexual reproduction. Also, bacteria don't have sex, but nevertheless exchange DNA fragments amongst themselves, including between different species. So it seems that neither strategy is a clear winner in all situations.

Sex has two key disadvantages. First, an organism with really good genes can't pass them on without mixing them with those of a mate who is likely to have less good ones. Second, males don't have offspring! The importance of this can be illustrated by imagining that every reproducing organism has two young. A sexual pair will simply replace itself, but two asexual individuals will have four offspring. Hence, all other things being equal, the asexual should inherit the Earth. So why haven't they? Clues come from two sources: studies of a small fish that switches between sexual and asexual reproduction[1]; and research on the constant race to outwit disease organisms. These suggest that sex has the edge when the future is uncertain and hence genetic diversity is of high value. But asexual reproduction works better in extreme, but stable, environments where it is important to keep genes that work.

A fourth source of variation is gene expression. In the last chapter, we saw that many genes are normally switched off. Which genes are active, or *expressed*, may be affected by the cell's metabolism, the surrounding tissue, or the organism's environment.[2] An oft-quoted example of the effect of environment is the English peppered moth which occurs in a pale and a dark variety. Before the industrial revolution, the pale moth was far more common than the dark-colored one. But when pollution blackened the trees on which it rested, the dark-colored moths were less visible to predatory birds, and their numbers increased greatly. When pollution controls were introduced, the balance swung back again.[3]

The importance of gene expression becomes apparent when we consider that only 1% of human genes are different to those of the mouse.

This suggests that the differences between species may be due largely to differences in gene expression. Indeed, it is now thought that control of expression may be as important, or even more important, than actual mutations. Gene expression is strongly influenced by 'jumping genes' or *transposons* that move spontaneously from one site to another in the DNA molecule, inserting copies of themselves wherever they fancy. Until recently they were regarded as useless junk, but research is revealing that transposons are important agents of change that organisms actually use to modify their own genes.[4]

When a transposon creates a new gene, the organism often keeps the original as well. These duplicates sit in different positions in the DNA molecule, and thus experience different chemical environments which may turn them on and off at different times or in different tissues.[5] This may be sufficient in itself to change the organism significantly, but even more important is the fact that duplicates can be altered without endangering survival. For instance, some plants can revert to a gene from their grandparents if a mutation is not successful.[6] This ability enables risk-free genetic experimentation.[7]

The importance of active genetic tinkering is shown by the evolution of color vision. After the green receptor had emerged, a duplicate gene was modified repeatedly until the red receptor was discovered. This strategy is far more efficient than waiting passively for the right chance mutation, and is similar to that used by scientists to discover new drugs. However, there is often a price to pay for this creativity. For instance, the genes that impart resistance to malaria have duplicates that are associated with genetic diseases such as sickle-cell anemia.[8]

Research is revealing that the activities of transposons are not random. Mutations to sections of DNA where they are harmful are weeded out by natural selection, whilst those in areas that can benefit from change survive. In this way, the genome 'learns' to steer mutations towards sections of DNA where increases in fitness are likely. Thus, the immune system has learned to focus mutations where they help to create innovative antibodies, whilst protecting proven means of eliminating identified pathogens. This led Lynn Caporale to ponder: "if natural selection is a teacher and genomes can learn, should we think of them as intelligent?"[9]

One of the most fiercely defended tenets of modern biology is that characteristics acquired during the lifetime of an organism cannot be passed on to its offspring. This position is being challenged by the research on transposons, duplicates and focused mutations. And the following examples provide clear evidence of the inheritance of acquired characteristics, as well as revealing that genes are not the only mechanism of heredity.

Barry Hall starved a species of bacterium, and then gave it food it could not digest. Two mutations were required simultaneously to overcome this challenge, and these occurred in the laboratory 100 million times more often than expected by chance.[10] This does not necessarily mean that the bacteria 'knew' what they needed and set about producing it, but at the very least they appear to have actively shuffled their genes to find useful new combinations that were then inherited by subsequent generations.

Another fascinating series of experiments is described by Rupert Sheldrake.[11] Fruit fly eggs were treated with ether, thus producing mutants with two pairs of wings. The eggs of these mutant flies were again treated with ether, resulting in an increase in the percentage of mutants with each generation. After several generations, untreated eggs continued to produce a higher percentage of flies with extra wings, indicating that this acquired characteristic had been inherited. Further, eggs from flies whose ancestors had never been treated with ether also showed a higher proportion of mutants after the experiments. Sheldrake's explanation was that the experiments had strengthened the morphic field for the mutant form, thus increasing its incidence in all fruit flies. This implies that morphic fields may provide a second, non-genetic mechanism of inheritance. (At this point you may need to refer back to the description of morphic fields in Chapter 9.)

In yet another challenge to traditional theory, there are hints of a third mechanism of inheritance. *Paramecium* is a single-celled organism about 0.1mm long, with little hair-like protuberances called *cilia* that propel it through the water. In an amazingly delicate operation, Tracy Sonneborn cut out a patch of cilia and replaced them the wrong way round. When the organism divided, this modified feature was reproduced even though the genes were unchanged. In other words, the structure of the parent cell was used as a template, and the form was not determined by genes alone.[12] This reinforces the conclusions of the last chapter on the role of non-genetic factors in morphogenesis.

This section has revealed several sources for the inheritable variations on which natural selection works. Mutations caused by environmental factors, errors in copying or repairing DNA, and sexual mixing are chance variations as traditional evolutionary theory holds. But, contrary to established theory, organisms appear to influence gene expression and experiment with mutations, leading to the inheritance of some acquired characteristics. It also appears some traits may be inherited via morphic fields and structural templates rather than through genes. Slowly, a more holistic and less mechanistic model of evolutionary variation is emerging.

Natural Selection

Variation alone cannot drive evolution. There also must be some means of choosing which variants survive and reproduce. This is the role of natural selection, the basic theory of which is very simple. Any hereditary trait that helps an organism to reproduce better than others of its kind will gradually spread through the population. And over many generations, the accumulation of tiny variations can lead to significant changes. Thus, natural selection works like a ratchet, locking in beneficial variations and eliminating harmful ones. But once again things are not quite so simple in reality.

The process of natural selection is often likened to movement in a *fitness landscape.* Imagine a species of animal in which eyesight and skin color determine its fitness to survive. In principle, its scores on these characteristics can be marked as a point on a 3-D graph in which eyesight and color are plotted on the horizontal axes, and fitness is plotted on the vertical axis. Plotting many such points produces a surface like a hilly landscape. The higher an animal is in this landscape, the fitter it is to survive. This idea can be extended mathematically to multi-dimensional 'landscapes' that include many different traits. Animals that are high in the landscape raise more offspring. Hence, as the generations pass, the average fitness of the population increases, and it moves upwards.

If each gene produced a single trait, the landscape would have only one peak. But most traits are determined by the interaction of several genes, and the resultant landscape has many peaks. Hence, as a species evolves, it may find itself climbing a relatively low peak. It cannot switch to a higher hill without temporarily becoming less fit in order to cross the valley in between – something that is not possible according to evolutionary theory. So how can it continue to evolve?

One strategy would be to take a big jump, and hope to land on the other side of the valley. But large mutations usually reduce fitness. A better strategy revealed by computer models is a combination of small steps uphill, with occasional moderate jumps that cross small valleys without the risks associated with large mutations.[13] Sexual reproduction is one way of doing this. If each parent is on a different fitness hill, then combining their genes has a good chance of landing on a third one.

So far we have assumed that the fitness landscape is fixed. But, when species interact in an ecosystem, any change to the fitness of one changes the fitness landscape of the others as well. Thus species are trying to move uphill in constantly shifting terrain, like a group walking about on a slack trampoline. Hence, there is no fixed optimum fitness that a species can achieve, and the best it can do is move in a positive direction whenever possible.

Using computer models, Stuart Kauffman found that a stable ecosystem emerges when most species have reached local fitness peaks, and hence cannot evolve further unless the landscape changes.[14] At this point, the landscape 'freezes' because so many species are stuck. Occasionally, however, a mutation can trigger an avalanche of change, leading to the extinction of some species, and the emergence of new ones. These results support the idea that evolution occurs in bursts as observed in the fossil record and discussed in Chapter 7. They also show that major extinction episodes are normal features of ecosystem behavior, and do not need to be triggered by a catastrophe such as an asteroid colliding with the Earth.[15]

Kauffman also found that simple ecosystems tend to become rigid and unable to respond to change, but that too much complexity leads to chaos and the loss of any coherent structure. He concluded that natural selection may be able to optimize complexity even though it works on individual organisms. In other words, when the fitness of a species increases, it also improves the fitness of other species with which it is linked, and hence improves the fitness of the whole system.

Powerful though natural selection is, evidence is accumulating that it is not the only force driving evolutionary change. In the last chapter, we saw how *Acetabularia* develops a temporary whorl of leaf-like projections on its stem. Darwinian theory would explain this as a redundant legacy of the evolutionary past that has not yet been removed by natural selection. But computer simulation revealed that the whorl is formed by the self-organizing dynamics of the organism. Similarly, the different ways that leaves are arranged on the stems of plants are equally effective at gathering light, and arise naturally from the dynamics of bud growth. Natural selection does not appear to be involved.

A third example is the evolution of the eye. Explaining how it could have arisen from the accumulation of small mutations has taxed the ingenuity of biologists for many years. And yet similar structures have evolved independently in 40 different lineages, suggesting that there must be an easy pathway. Brian Goodwin suggested that the basic form of the eye may result from the interaction of calcium metabolism with cell structure, growth, deformation, and movement. Hence, he argued that natural selection did not create the primitive organ, only refined it.

Emergence of New Species

The metaphor of a landscape illustrates how the fitness of an existing species can improve, but does not explain how new species emerge. One way this happens is when environmental change divides a population into two separate groups which experience different conditions, and hence evolve along different paths. For instance, climatic change may reduce an extensive forest to smaller, isolated pockets; or rising sea level may turn a mountain range into an archipelago. Until the last decade, biologists thought this was the commonest mechanism, but now they recognize that a species in a uniform environment frequently splits into two. The puzzle is why the whole species does not evolve in the same direction when some individuals become fitter.

Imagine a change in food supply alters the fitness landscape of a species so that it finds itself in a saddle between two hills. Due to slightly different traits, one group may climb one side, while another sets off towards the other peak. Thus, the two types evolve in opposite directions to occupy different ecological niches, and avoid direct competition. For instance, the beak of one bird may grow shorter, and the other longer so that they become adapted to different foods. Such changes can be rapid, with a significant increase in the length of a bird's beak happening in just a few generations.[16] In laboratory experiments, bacteria differentiated into several strains when given an environment with alternative niches. And, when faced with competition for food, bacteria created a new niche occupied by a species that fed on by-products.[17]

The Role of Organisms

According to Darwinian theory, organisms are passive, caught between the rock of chance mutations and the hard place of natural selection. But closer examination shows that species are actually active agents in their own destiny. It is the species-in-its-environment which survives, and organisms often create their own environments as the examples in this section show.

Plants alter the composition of the air and change the local micro-climate. Their roots make channels through which water flows, they engage in chemical warfare with creatures that eat them, and they regulate forest fires. Fungi break down dead organic matter and weather rocks. Worms create fertile soils. Bacteria decompose wastes, and make plant nutrients available. Birds spread the seeds of plants on which they feed. And so on. Hence, the habitats in which most species live were created by the collective efforts of their ancestors, and all the other organisms in the ecosystem. And the species themselves have co-evolved in association with each other and their physical environment.

Mathematical models show that this process of niche construction is a major evolutionary force. Through it, organisms acquire a non-genetic ecological inheritance; a legacy of enduring environmental changes and modified natural selection pressures. This inheritance may enable apparently damaging mutations to survive and spread; eliminate seemingly useful variants; open spaces for new patterns of gene expression; and even enable populations to resist external selection pressures.[18]

In a similar way, any learned behavior that increases reproductive success and can be taught to offspring effectively becomes a cultural inheritance. This process is strongest in higher animals, particularly in humans for whom science and technology have become the key determinants of evolutionary fitness. But cultural transmission also occurs in some lower animals, including several species of coral reef fish which have preferred migration paths between feeding and resting sites.[19] Further, resonance with behavioral morphic fields may influence evolution of species that cannot teach learned behaviors directly. For instance, tits throughout the UK rapidly learned to pierce foil milk-bottle tops once birds in one area had discovered the trick; and isolated groups of monkeys began to throw grain in water to separate it from sand once this skill had emerged in one place.[20]

Competition and Cooperation

Natural selection is often portrayed as a bitter struggle in which only the fittest survive. It is the contest between herbivore and plant, predator and prey, parasite and host, with the one looking for a square meal and the other trying to avoid being eaten. And it is the competition for resources between members of the same species, or with other species that occupy the same niche. But this is only one side of the picture, and nature is also deeply cooperative as we have seen with regards to niche construction. So important is this alternative view, that most of this section is devoted to exploring it in depth.

Competition between species leads to *co-evolution*. If a moth becomes better camouflaged, then the birds that eat it must evolve better ways of detecting it. But better means of detection apply pressure for the evolution of better camouflage, and so it continues. Similar 'arms races' occur between the chemical repellants of plants and insects' strategies to avoid them; and between carnivores' weapons and herbivores' defenses. But there is a price to pay for this relentless improvement in armaments – it diverts energy and nutrients from other purposes such as reproduction and feeding young. Hence, arms races eventually end either when one species loses and becomes extinct, or neither side can afford to continue.

Competitive races also occur between members of the same species. The trees in a forest which grow fastest and tallest get the most light; and animals with the most elaborate threat displays get the biggest territories and best mates. In most cases, the individuals would be better off with a peace treaty under which everyone stayed short, respected boundaries, or gave up elaborate dress to attract the opposite sex. But no-one knows how to get out of the competitive cycle.

The competitive struggle described above gets most publicity, but is only half the story. We have noted already that it is the species-in-its-environment which survives, and hence "the fittest for survival are those who fit in best."[21] From this perspective, it is not surprising that even apparently competitive relationships often conceal mutual benefits.

From the point of view of individual animals, the predator-prey relationship looks like a life and death struggle. But from the perspective of the ecosystem, it helps keep the whole system in balance. Without the prey species, the predator would die out; and without the predator the prey population might explode unsustainably. And such changes, in turn, would affect the parasites of the two species, and the growth and distribution of plants, sending ripples throughout the ecosystem. Very similar relationships exist between herbivore and plant, and parasite and host. Grazing animals eat grass plants, but also control the growth of shrubs, thus helping grasses in general to flourish. And a parasite may debilitate its host, but in doing so help to maintain the balance between species. At the ecosystem level, therefore, the competition between species looks more like cooperation to maintain their environment and genetic fitness.

The focus on competition also draws attention away from myriad, mutually beneficial interactions as a few examples will demonstrate. Rhizobium bacteria turn nitrogen from the air into fertilizer for the plants such as peas and beans which nurture them. Mycorrhizal fungi make soil nutrients available to the trees from whose roots they draw their nutrition. Cows and sheep feed bacteria in their gut which help them digest the grass they eat. And we humans rely on friendly bacteria to digest our food and make B-group vitamins. Leaf-cutter ants cultivate the fungi that feed them; and other ants farm aphids, carrying them from plant to plant, and eating their sugar-rich excreta. Plants feed many animals in return for pollination of their flowers or distribution of their seeds. And so I could continue.

So important are these relationships that the loss of one partner may threaten survival of the other. For instance, in rainforests where bat populations are declining, the trees on which they feed are endangered as well.[22] And in Western Australia, a root fungus which kills many species of plants is endangering some native animals which feed on them.

Turning now to the relationships between members of the same species, we again find that the appearance of competition conceals a deeper cooperation. Establishing territories limits population and controls food distribution, thus keeping the species in balance with its environment. Individuals not only compete for mates, but also cooperate to produce and raise offspring. Even sex may be seen as cooperation to improve the species' prospects of survival.[23] Animals band together to defend themselves from predators, sometimes sacrificing their lives for the good of the group. And some predators cooperate to catch their prey. Dolphins, elephants and bats as well as humans help each other give birth, and care for their young.[24]

Social cooperation pervades simpler forms of life too. Social insects such as termites and bees cooperate for the good of the whole colony or hive. The queen, workers and soldiers assume specialized body forms to meet collective needs, and often are unable to survive without the hive, just as the hive cannot survive without them. So close is this cooperation that the colonies are best regarded as single organisms with a collective intelligence, and an ability to cope with change that is far greater than that of any individual.[25]

An even simpler organism, the coral polyp, forms colonies which build the largest structures made by living things – coral reefs. The polyps often share networks of nerves and reproductive functions to such an extent that it is difficult to consider them as individuals. And when two polyps of the same species have room for only one, the smaller disintegrates. As Lewis Thomas put it: "He is not thrown out, ... not outgunned; he simply chooses to bow out."[26]

Bacteria are amongst the most cooperative organisms known, forming slimes almost everywhere there is water. Until recently these were thought to be collections of individual bacteria, but it has been discovered that they are complex structures, like miniature cities, in which different types of bacteria collaborate to get food, dispose of wastes, resist antibiotics, and defend against predators. Such biofilms can be 1500 times more resistant to antibiotics than a colony of a single species of bacterium.[27] And, in the words of Mark Buchanan, bacteria "use a bewildering range of chemical messages not only to attract mates and distinguish friend from foe, but also to build armies, organize the division of labor and even commit mass suicide for the good of the community. Some experts even talk about "microbial language" with its own lexicon and syntax."[28]

Multicellular organisms probably evolved from cooperative associations such as these. But even when different types of cell are integrated into a single body, close cooperation with independent organisms may be critical. Amazingly, over 99% of the cells in the human body are micro-organisms living on the skin, in the gut or elsewhere.[29] Even more startlingly, the cells of all higher organisms probably began as a cooperative venture between more primitive organisms. In animal cells, one organism became the nucleus containing the genes, and another became the energy-generating mitochondria which still have their own independent genes. In plants, the chloroplasts responsible for photosynthesis form a third member of the cooperative.[30]

But cooperation runs still deeper. Individual cells in our bodies disintegrate when their DNA is damaged beyond repair rather than produce faulty copies of themselves. But first they ensure that their contents are in a suitable form for digestion by their neighbors. Cells which have lost this ability to commit suicide for the good of the whole may be the cause of many cancers.[31] And within cells, molecules cooperate to check, repair and duplicate DNA, and 'chaperone' molecules help proteins to fold into the right shapes.[32]

Richard Dawkins argues that it is not species which survive, but cooperatives of genes.[33] But genes cannot survive without an organism. And organisms cannot survive without ecosystems. Hence what survives is DNA-in-cell, cell-in-organism, organism-in-nest, -hive or –society, and species in environment.

A New Theory of Evolution?

The Darwinian theory of evolution portrays the combination of chance genetic variations with natural selection as an all-powerful creative force. But the theory has large gaps and inconsistencies, and there is a growing body of contrary research. Numerous studies have concluded that chance mutations cannot explain the evolution of life. And the idea that natural selection fits organisms to their environments biases perception towards clever adaptations and away from poor ones. Hence, we tend to see how well the beak of a humming bird matches particular flowers, and ignore the design flaw that requires light entering the eye to pass through the nerve cells before striking the light-sensitive retina. Similarly, evolutionary biology is replete with wonderfully imaginative 'Just-So' stories to explain how particular features benefit the organism and the steps by which they could have evolved.

In this chapter and the last, we have seen how:

- Organisms influence which genes are expressed, and engineer their own genetic modifications;
- Acquired characteristics and non-genetic variations may be inherited;
- Natural selection does not create new forms but chooses amongst those made possible by the self-organizing dynamics of the organism;
- The emergence of new species may happen within a few generations;
- Fitness to survive is a quality not of the individual but of the organism within its environment;
- Cooperation is as important as, or more important than, competition in determining fitness;
- Changes in behavior may be as important for survival as genetic mutations in species that learn and teach their young.

All these processes contradict traditional evolutionary theory, and it seems that we are on the threshold of a new theory of evolution in which the genes and natural selection remain important but no longer dominate the stage. We are beginning to see the outlines of a world in which organisms no longer wander aimlessly around a fitness landscape, but dance between order and chaos, move towards stable states and away from unstable ones, form alliances for mutual benefit, and actively explore their potential; a world in which natural selection does no more than choose amongst alternative possibilities presented by the species-in-its-ecosystem, and fine-tune the results.

Another valuable perspective on the emerging theory comes from Wolfgang Schad.[34] His studies of vertebrates convinced him that evolution has a direction and purpose, contrary to Darwinian theory. He claims that vertebrates have evolved increasing independence from their environment by progressive internalization of life functions. Fish breathe through gills, but amphibians have lungs and are thus not dependent on water, although they still need a damp environment. Reptiles have a closed fluid system, and hence are free of this constraint but depend on the environment for warmth. Birds introduced independent temperature control, and placental mammals internalized the development of offspring. Most mammals are still limited by specialized limbs, but humans have adaptable arms and hands that are not needed for support or movement. Thus, the human is the most advanced species because it is the least specialized and most flexible, and hence is the least dependent on its environment.

This idea of increasing independence of the environment contradicts Darwinian theory which holds that natural selection progressively improves the 'fit' between the species and its environment. Thus organisms become more specialized and dependent unless conditions are changing rapidly in which case less specialized and flexible organisms may have a survival advantage.

In Conclusion

The impression is often given that Darwinian theory is the last word on the evolutionary process, and that only the details remain to be filled in. This chapter has shown that nothing could be further from the truth, and that many new facts and theories are challenging the dominance of Darwin.

All organisms have similar needs for food, shelter, and reproduction, but there are myriad ways in which these can be met. We do not know what is possible, and, even if we did, we could not tell which paths evolution will choose. Its future course cannot be predicted. We cannot know what wonders still lie in store.

16 Ecosystems and Gaia

We saw in the last chapter how organisms cooperate, often co-creating niches for themselves. A cooperative of organisms in a particular place is an ecosystem. Ecosystems are often thought of as higher levels of organization than individual organisms or species. But the discussion of the holarchy of nature in Chapter 6 revealed that they are actually on the same level, with ecosystems forming the social environment of the organisms within them. In this chapter, we explore the characteristics of ecosystems, and particularly the global ecosystem, often called the *biosphere* or *Gaia*.

Ecology

Ecosystems vary dramatically from place to place, ranging from Arctic ice to tropical rainforests, from sandy deserts to coral reefs, and from high mountains to the ocean depths. Despite these obvious differences, the boundaries between them may be blurred, with one merging gradually into another. And at smaller scales, the boundaries between local ecosystems are often defined by subtle variations in soils, vegetation and fauna. As with all systems, exactly where the boundaries are drawn depends on human perception and judgment.

Ecosystems vary greatly in complexity, from the richness of tropical forests to the sparse life of the desert. But they all share certain characteristics and processes. Energy and materials flow through *food webs*, as shown in the diagram below. A tiny fraction of the solar energy reaching the Earth is used by plants to make organic chemicals by *photosynthesis*. About half of this fixed energy is used by the plants to meet their own needs, and the rest is turned into plant material, or *biomass*. *Herbivores* eat plants to supply their energy, and to build and maintain their bodies. *Carnivores* in turn eat herbivores, as when a lion eats a gazelle. And second order carnivores eat other carnivores, as when a flea bites a lion, or a hawk catches an insect-eating bird. At each stage, only 10-20% of the food energy is converted to body tissue, and the rest is lost as heat. Hence the total body mass, or *biomass*, and usually the number of animals, that can be supported decreases with each level of the food chain.

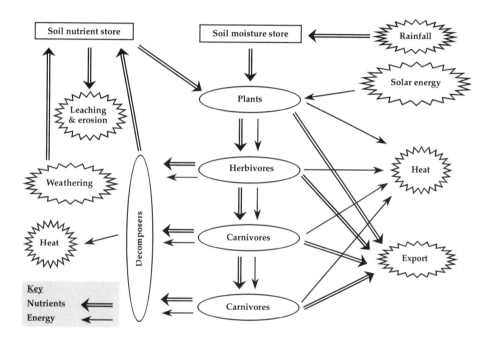

Figure 12   Food web diagram here

Wastes and dead organic matter are produced at each level, which feed decomposer organisms such as bacteria and fungi. These break down organic molecules into simpler chemicals which can be used again as plant nutrients. By the time this nutrient cycle has been completed, all the solar energy originally captured by the plants has been converted to heat.

In practice, food webs are more complex than this, with several species at each level, and some that span two or more levels. Humans, for example, are herbivores (eating grain, fruit and vegetables), primary carnivores (eating herbivores such as sheep and cattle), and secondary carnivores (eating chickens and ducks which eat insects, and shark which eat other fish). From the perspective of individual organisms or species, this food web is a competition for scarce resources. But to the ecosystem, it is a story of cooperation and harmony through which life creates and maintains conditions for life.

Ecosystems are dynamic – growing, developing and evolving over time. Hardy plants such as Marram grass colonize coastal dunes, stilling the blowing sand. Other plants then gain a foothold where they are protected from sandblasting. These plants provide food and habitat for herbivores, and then carnivores, whilst dead organisms and excreta provide a niche for decomposers. Thus, over many years, a sandy wasteland, derelict urban land, or an old mine may develop into an ecosystem of steadily increasing complexity.

These colonizing organisms actively shape their environment, as we saw in the last chapter. Plants halt erosion, bacteria break down toxic chemicals, and lichens release chemicals that weather rocks to soil. In a similar way, large ecosystems influence regional conditions. Forests hold water in the landscape. Conifer trees warm their cold lands by shedding snow and exposing their dark color to the spring and autumn sunlight. And coral reefs create sheltered lagoons, and form clouds to keep themselves cool.[1] Thus life and its physical setting form an interactive, co-evolutionary whole.

Old growth forest is born of catastrophe.

Leigh Dayton (1990)

It used to be thought that an undisturbed ecosystem would eventually reach a 'climax' state in which no further change would occur. Now, it is recognized that ecosystems remain dynamic, responding to disturbances of many kinds: good years and bad, pests and diseases, droughts and floods, fires and tempests, human exploitation, and invasion by new species. Ecosystems often follow long natural cycles, and evolve slowly compared with human time scales. For instance, many forests don't regenerate continuously, but mature for hundreds of years before the old trees are destroyed by fire or storm, and a new forest springs up. We tend to see such long-term changes as harmful, and try to prevent them whilst often failing to perceive slow degradation. Thus, most people in England barely noticed the loss of cheerful, chattering flocks of sparrows over a few decades. And the insidious increase in water weeds due to fertilizer pollution often goes unnoticed until a wetland chokes with toxic algae.

> The only constant factor in our forest ecosystems is change.
>
> **Shea and Underwood (1990)**

Like organisms, ecosystems are born, develop, grow old, die and are reproduced. The physical environment is equivalent to the skeleton; each species corresponds to an organ or tissue; and the flows of energy and nutrients are analogous to circulation, breathing and digestion. The biggest difference between an ecosystem and an organism is that there is no genetic inheritance. But each successive ecosystem inherits the physical environment created by its predecessors, as well as any species which survive the 'death' of the earlier system. Many people balk at the idea that ecosystems are super organisms, but they are undeniably super constructions, co-created by their species with such structural and functional integrity that they give the impression of being alive.[2]

> We used to think that a forest was basically a group of species that had evolved together or been together for a very long time. Now what we find is that a forest community is a group of species that may have recently migrated together and then, in the future, might migrate in separate directions.
>
> **Chad Oliver**
> **Quoted by Dayton (1990)**

The more species an ecosystem has, the more efficiently it can harness solar energy because each species has different ways of using light, nutrients and water. Hence, the total production of biomass is higher in diverse systems, even though the productivity of particular species, such as wheat, may be lower. Complex systems also suffer less from stress, and recover faster than simple ones. This was clearly demonstrated by fertilizing plots of tall-grass prairie in the USA. The fertilizer benefited species that could best make use of it, and they crowded others out, thus reducing the diversity of the plots. After 5 years, there was a severe drought. Production was reduced by half on the plots with most species, but fell to one eighth on those with the fewest. Furthermore, the species-rich plots had recovered three years later, while the others were still below their normal productivity. Another study showed that grazing by herds of zebra, gazelle and wildebeest in Tanzania removed three-quarters of the vegetation on less diverse plots, and only a quarter of it on the more diverse ones.[3]

There are two schools of thought about the relationship between diversity and stability which can be explained by imagining that each species is like a rivet in an aircraft. One group believes that losing even a single rivet weakens the whole structure, and losing more than a few may cause it to fall apart. The alternative view is that there are more than enough rivets so that many can be lost as long as key ones remain, such as those holding the wings on. Whilst there is probably some redundancy in diverse systems, ecologist Stuart Pimm warns that "There is no wide plateau of species richness where species do not matter. Species become important quickly."[4]

Stability is not just a matter of complexity, however. Complex technological systems, such as computer software, are often less stable than simpler ones. The stability of natural ecosystems owes much to co-evolution that has woven species, environment and processes into an integrated whole. In consequence, artificial ecosystems, created for example in mine rehabilitation, may be less stable than the natural systems they replace even if their diversity is as high. So effective is evolution at producing stable systems that computer programmers are starting to allow software to evolve rather than writing it in full.[5]

Modern farming systems are the ultimate in simplicity, with a single crop variety, or cattle grazing one species of grass, over large areas. Such systems collapse if disease strikes. In a natural ecosystem, by contrast, a disease rarely wipes out more than one or two plant or herbivore species, and others are often able to take over their function.

Diversity probably increases production and stability at the regional and planetary levels as well as in local ecosystems. But conservation of global diversity presents huge challenges. As an example, the Northern Spotted Owl of North America weighs some 600 grams and stands less than half a meter tall. But each pair needs around 900 hectares of ancient forests to live and raise their young.[6] The US Fish and Wildlife Service argued that 2.2 million hectares (an area 22 km by 10 km) should be set aside from logging to protect it.[7] Even larger areas are needed for big carnivores such as grizzly bears.

Conservation reaches truly global dimensions with migratory birds. Swallows fly from Europe to Africa for the winter; and some water birds escape the Siberian cold by flying south to Australia. If such species are to survive, not only must the feeding and breeding grounds at each end of their journeys be protected, but also their resting places en route. Large marine mammals such as whales and elephant seals have similarly long migrations.[8] It is clear from such examples that the Earth's ecosystems form one interactive whole, and elimination of a key species from any ecosystem may have a domino effect around the planet. But we don't know which are the key ones; the loss of any species could be critical.

Planetary systems are also connected physically, as illustrated by variations in Pacific Ocean currents known as El Nino. In 1982-83, falls in sea level in the western Pacific destroyed the upper layers of coral reefs on many islands. Corresponding rises in levels and water temperatures in the mid and eastern Pacific reduced fish populations. As a result, sea birds on Christmas Island abandoned their young to seek food, and huge numbers of fur seals and sea lions died along the Peruvian coast. The southern USA, Ecuador and northern Peru suffered massive floods, Hawaii and Tahiti experienced unaccustomed typhoons, and Indonesia and Australia were hit by drought and forest fires. The loss to the world economy was estimated to be over $8 billion, with a much larger toll in human suffering.[9] Since then El Nino has become a widely recognized and frequent event.

Gaia

In 1974, James Lovelock and Lynn Margulis sparked a long-running controversy by arguing that the planet is a self-organizing, evolving whole – a living organism that they named Gaia after the Greek Goddess of the Earth. Gaia was "gentle, feminine and nurturing, but also ruthlessly cruel to any who failed to live in harmony with the planet."[10] The Gaia hypothesis emerged from evidence that life itself created the conditions in which it now flourishes, and has stabilized them for hundreds of millions of years. In human terms, there have been major fluctuations between ice ages and warm periods. But compared to what might have happened, these waves of change are mere ripples.

Gaia originated with the emergence of life. The earliest forms, similar to modern bacteria, dominated life for 2 billion years. They filled the oceans and lakes, formed a scum wherever the land was damp, and created a truly global system that dramatically changed the fate of the Earth. Without life, Earth would probably have become like Mars or Venus; and without bacteria, the atmosphere would never have become suitable for oxygen-breathing organisms.

James Lovelock stresses that the Gaia hypothesis does not mean that the planet is conscious, or that Gaia's actions are purposeful. Rather, the planet is simply a giant example of a self-organizing system. Others, however, prefer to marry the science of Gaia with ancient religious traditions which honor the Earth as sentient and purposeful.

Many people still find the idea that Earth is a giant organism hard to swallow. So what is the evidence? The starting point for James Lovelock was a project for NASA to devise ways to detect life on other planets. He hit on the idea of studying the composition of the atmosphere, arguing that it would be changed by life. On a dead planet, all possible chemical reactions would have taken place over eons, and the atmosphere would consist almost entirely of molecules that do not react with each other – a state known as *chemical equilibrium*. By contrast, the atmosphere on a planet with life would contain many reactive chemicals released in metabolic processes.

When Lovelock compared the atmospheres of Mars, Venus and Earth, he found that the first two are close to equilibrium, but that Earth's atmosphere is a cocktail of reactive gases such as oxygen and methane. Apart from 1% of inert gases, Gaia's air is the product of living organisms.[11] Not only can we breathe it, but also it protects us from harmful ultraviolet radiation, and keeps us warm. Without the blanketing *greenhouse effect*, the temperature of the Earth's surface would be about -19° C – too cold for most forms of life.

The idea that life could control the environment of the planet is hotly disputed by many scientists, including Richard Dawkins, whose work on genetics and evolution has been cited in earlier chapters. He raises three arguments against the Gaia hypothesis. First, natural selection works on the genes, and there is no way that evolution could favor cooperation between organisms at a global scale. Second, regulation of the planet would require foresight and planning – "a meeting of committees of species to negotiate next year's temperature."[12] And third, Gaia cannot be an organism because there is no possibility of the Earth reproducing, and hence evolution by natural selection amongst planets cannot occur. A powerful counter-argument is provided by the story of Daisyworld.

Over the last few billion years, the energy output of the sun has gone up by 25%. Other things being equal, we should have been cooked. Instead, the temperature of the Earth has remained constant except for the relatively minor ups and downs of ice ages and warm interludes. In order to demonstrate how life could control the global temperature, James Lovelock created a computer model of Daisyworld. On this planet grow two daisies, one pale-colored and the other dark, with the bare soil an intermediate shade. Both daisies can grow at temperatures between 5°C and 40°C, but grow best at 22°C. And Daisyworld's sun is getting hotter every year.

At first, the planet is too cold for daisies to grow. But there comes a year when Daisyworld is warm enough for a few plants to germinate. The dark daisies grow better than the pale ones because they absorb more sunlight, and hence are warmed. They also absorb more light than the paler-colored soil, and hence warm their surroundings too. So year by year the dark daisies spread, and warm the planet. But when the temperature reaches the ideal 22°C, it stops rising.

Stabilization occurs because the pale daisies reflect sunlight, and hence have a cooling effect. As the sun continues to get hotter, so the pale daisies spread, reflecting more and more light to keep the planet cool. In this way, the temperature is kept at the ideal level for life until the planet is covered with pale daisies. Eventually, however, the sun gets so hot that even the pale daisies cannot control the temperature, and the planet dies.

This model is an extreme simplification, and doesn't mean that the color of the vegetation controls the Earth's temperature, although it is an important factor. But Daisyworld demonstrates that temperature regulation by the cooperative interaction of species is possible without purpose or planning, and that a living planet can evolve without having to compete for survival with other planets. More complex versions of Daisyworld have been tested with more colors of daisy, rabbits, foxes, and plagues that destroy 30% of the daisies. None of these seriously changed the daisies' ability to regulate the temperature.

We've seen that all organisms change their environment to a greater or lesser extent. The way the colors of Daisyworld's flowers varied to optimize the temperature demonstrates that natural selection can favor organisms that make their environment more livable. Using this simple idea, it is possible to start constructing the story of how the conditions we enjoy on Earth today were created and are maintained by life. There isn't the space here to recount how life regulates the composition of the atmosphere and oceans; how it prevented the loss of water from the young planet; or how it influences the great geological cycles of erosion, sedimentation, uplift and continental drift. James Lovelock's books provide fascinating and detailed accounts if you would like to explore further.[13] Here, we must be content with a short excursion into some of what is known about the control of climate.

I have already mentioned that the greenhouse effect keeps the earth warm enough for life. The present concern is not about the greenhouse effect as such, but about the increase that is occurring due to human activities. Most attention is centered on carbon dioxide, the concentration of which is rising due to burning fossil fuels and other causes. Carbon dioxide has a strong 'blanketing' effect, allowing solar energy in, but preventing heat escaping back to space.

Carbon dioxide is a source of carbon for plants which release the oxygen as a waste product. If Earth only had plants, more and more carbon would be locked away in organic matter, both living and dead. Eventually growth would be limited to the rate at which carbon dioxide is released from volcanoes. Also, the oxygen level in the atmosphere would rise until the vegetation was destroyed by fierce fires. Thus decomposer organisms are needed to break down dead organic matter and recycle the carbon. But if carbon was recycled perfectly, the concentration of carbon dioxide in the atmosphere would rise steadily due to volcanic emissions. Hence, there needs to be some means of removing carbon permanently from the cycle. The main mechanism is the burial of carbonates and carbon in sedimentary rocks. In concentrated form, the latter appears as coal, oil, and natural gas. Hence burning these as fuel returns stored carbon dioxide to the atmosphere.

Living organisms play an important part in geological storage of carbon. Plants raise carbon dioxide concentrations in soil pores to 10 – 40 times those in the atmosphere. The resulting acid greatly accelerates the weathering of rocks, the chemical products of which percolate to the ocean. There they either precipitate as limestone or are used by algae, together with carbon dioxide dissolved from the atmosphere, to make their skeletons. When the algae die, the skeletons sink to the sea bed to form sedimentary rocks, thus removing carbon dioxide from the atmosphere.

Over geological time, the accumulated weight of limestone may have played a key role in the break-up of the original landmass into the continents we know today, and their movement across the globe. The arrangement of the continents, in turn, affects the circulation of water in the oceans which is important in the regulation of climate.

Marine algae play another critical role by emitting the gas dimethyl sulfide (DMS), which is converted to sulfuric acid droplets in the atmosphere. These droplets provide nuclei for the condensation of water vapor over the oceans, without which clouds wouldn't form. As temperatures rise, algae grow faster, and evaporation of water from the sea surface increases. Hence cloud cover is likely to increase, producing cooler, less sunny conditions which would decrease the growth of the algae again.

Rainfall on land mostly comes from clouds formed over the ocean. And sulfur is an essential nutrient which is washed from the soil faster than it is replenished from weathering rocks. So the release of DMS both waters and fertilizes the land, affecting the growth of land plants, which in turn affect global climate.

It is not yet clear what benefit the algae get from releasing DMS. It may be a by-product of chemical reactions which help them survive in the salty water of the sea. Or the cloud cover may shield them from ultra-violet radiation, or increase the strength of winds, thus mixing the surface waters and making more nutrients available. Recent research has shown that coral reefs also release DMS, particularly when stressed by high temperatures or UV radiation. So this may be a way they keep cool by forming clouds.[14]

These examples illustrate how life can control climate, but the full story is far more complex and still not well understood. Amongst major uncertainties are the effects of clouds, which in some cases can cause warming rather than cooling; the release of methane, a potent greenhouse gas, from soils and ocean sediments; and the effects of industrial smog. In the last few years there has been an alarming increase in the number of positive feedback mechanisms identified, which could lead to runaway warming and a Venus-like Earth. However, the stability of the climate over hundreds of millions of years suggests that there are as-yet undiscovered negative feedbacks that will eventually regain control. But temperatures could first climb too high for human civilization to survive.

In Conclusion

We are only just beginning to understand the wonderful, intricate ways in which the web of life controls conditions on Earth. Ways that are powerful and resilient enough to have survived for billions of years, and through the many catastrophes that have extinguished countless species. But however far our understanding increases, we may never be able to predict and control global processes due to the sensitivity of the chaotic and self-organizing processes involved.[15]

Many people worry that we may kill the living planet. But the diversity of life and Gaian history suggest that this is unlikely. As we saw in Chapter 13, there are many extremophiles that live in totally different conditions, isolated from the world we know in deep ocean trenches, and even in the heart of terrestrial rocks. And there are bacteria and other simple organisms that have evolved to feed on the toxic wastes we produce with such abandon. So I believe that Gaia herself is safe from humanity's threats. But that doesn't mean we humans will necessarily survive. If we continue to act as a cancerous plague, we may be eliminated as just another unsuccessful experiment. In that case Gaia will continue her evolution without us.

Review of Part V

It's time once again to pause for breath and look back over the country we've traversed. Have we progressed towards our goal, or are we wandering lost in impenetrable thickets? Do the lands we've crossed form a coherent landscape, or are they separated by geological fault zones? And what does this landscape look like through the eyes of the science of oneness?

At the end of Part II, we concluded that non-living systems may grow, develop, adapt, maintain themselves, and evolve; and that they display elementary forms of purpose, autonomy, creativity and cooperation. These perceptions led us to ask: Why is matter self-organizing? Why does it assume certain forms and not others? Is creative emergence of diversity and complexity simply what matter happens to do? Is the apparent direction of evolution an illusion? Does the self-created world have meaning, value and purpose? Is there a law of nature that drives matter towards ever greater complexity? Might the cosmos itself be conscious and purposeful, striving to actualize its unimaginable potential?

As we look back over the landscape, we can see no clear boundary between the land of non-living systems and the kingdom of life. The country changes gradually, with only a few remnants of old fences to mark the border. And as we journeyed through the heart of the kingdom, we found few qualities that we had not met already in inanimate systems. Rather, there seemed to be a deepening and strengthening of universal qualities, and an indefinable sense of aliveness that was not present before. Organisms are more complex, more autonomous, more creative, more purposeful, more cooperative ... and more alive.

Life is a deeper level of the great holarchy of nature, but it emerged almost imperceptibly by self-organization from what went before. And far from being a delicate creation in need of careful nurturing, our growing knowledge of extremophiles shows it to be robust and adaptable. Hence, for many scientists today, the origin of life has lost its mystery. They expect life to appear wherever conditions are suitable. And yet there is still a mystery. How did life buck the second law of thermodynamics to assemble its library of genetic information? Perhaps organisms are quantum computers hooked into the universal internet, and drawing inspiration from the implicate order or psi field? Or perhaps scientists will soon discover a new law of complexity which drives the cosmos from energy to matter to life to mind to ... what?

In either case, the universe has produced minds capable of understanding the laws of nature that gave them birth. And it takes only small steps further to see the process of evolution as the universe coming to know itself, and to reveal the universe as purposeful and meaningful.

As we observed the development of organisms from their 'seed' cell, we came face to face once again with the power of self-organization. Contrary to what most biologists believe, development is not dictated by a genetic blueprint, but results from a cascade of self-organized processes involving the genes and their expression, the structure and biochemistry of the emerging organism, environmental influences, and, possibly, morphic fields. A small change in any of these factors has the potential to carry the system down an alternative pathway to another stable form.

A similar picture emerged from our observation of evolution. Far from being dependent on chance genetic mutations and the necessity of survival, we discovered that organisms play an active role in the process. They experiment with their genes, construct their own environmental niches, and learn new behaviors from experience – all of which they may pass on to their offspring. We saw that the basic forms of life are created by self-organizing dynamics, and that natural selection, far from being the fount of creativity, can do no more than choose from a menu of available forms, and fine-tune the structures and processes it is given.

Organisms are self-determining, autonomous, creative beings that continually strive to go beyond their current forms, processes and limitations; always seeking to transcend and include. But organisms are not rugged individuals. They depend totally on their relationships with the physical environment and other organisms in their ecosystem. These interactions bond them into a cooperative whole that forms the living planet, Gaia.

This unity extends outwards to embrace the solar system, without which Earth would be dead, and the cosmos which forged matter and the stars, including Sol. And there is a yet deeper aspect to this unity. As far as we know, all the diversity and complexity of matter and life emerged from the 'seed' energy of the big bang and whatever came before. It is truly the many from the One.

Earlier in my life, as I traveled the journey we have just made together, I experienced a growing sense that there is something more to reality than the glib explanations of objective, mechanistic science. The way living things develop and evolve can be explained by complex self-organizing networks of cause and effect. But is this sufficient to explain the wholeness, the 'Beingness' of an organism? For me, the answer is 'No'. I find the mechanisms of self-organization fascinating, inspiring and beautiful as means to an end. But when I think of them as ultimate causes, I find them frighteningly sterile. They don't reflect the natural world as I encounter it; a world of Beings not objects and mindless processes.

My encounters with myself, other people and other Beings tell me there is something more, whether we call it the life force, vital energy, soul, spirit, deva ... For me, the whole is more than integrated processes and parts; more than the one from the many. There is a wholeness that expresses itself in the multiplicity of nature; the many from the One. There is a deeper reality, a quality of Beingness that is reflected in the dynamics of development, life, reproduction and evolution.

I cannot rationally justify this perception of 'something more.' It arises from my direct intuitive experience; from an I-Thou relationship with people, plants, animals, landscapes and the planet. But neither can I see any rational reason for rejecting its reality, apart perhaps from the scientists' ultimate weapon of Occam's Razor – the belief that the simplest explanation is always the true one. On this basis, most scientists would tell me they have no need of my hypothesis, and hence that there is no life force, there is nothing more.

For me, self-organizing processes are the means adopted by Beings to achieve their ends; ends that reflect the purposes of the cosmos. The fact that the beauty of *Acetabularia* arises from natural processes comes as no surprise, but this does not mean the organism is 'nothing but' this integrated process. The life force makes use of whatever means are available to achieve its ends. The acorn feels the pull of the oak it is to become and harnesses natural processes to pursue that end. And perhaps the universe likewise feels the pull of its ultimate destiny?

I encounter this life force as an energy emanating from each organism. This is not just an inner, subjective experience, but one that has physical manifestations; and one that I share with others who are similarly attuned

– often far more acutely than I am. I can detect the fields around living things with divining rods that amplify my body's reactions. And often I can detect the 'presence' of a tree, for instance, through subtle sensations that at times may literally stop me in my tracks to pay attention.

This Beingness is an enfolding wholeness that wraps itself around and permeates the physical processes and structures of the material organism. It is the particular 'treeness' of a tree; the 'dogness' of a dog. It is who I am as an integrated mind-body-spirit miraculously bringing itself into existence for a time. Surely, this is more than an integrated mechanism without purpose, meaning or spiritual wholeness?

Dismiss it as illusory nature mysticism if you will, but those who directly encounter nature – poets, artists, mystics, indigenous peoples, naturalists, deep ecologists and many scientists – find there a something, a presence, a Beingness that is more than an integrated system. A great cathedral is perhaps as close as a human-made object comes to this, imbued as it is with the spirit of its creators and centuries of devotion – as well, often, as an energy of place that was identified as sacred millennia before the cathedral was built.

There is a spiritual dimension to life.

Reflections on Part V

These Reflections focus on connecting with life, ourselves, and the Earth. First come several quotations that prompt us to ask 'What is life?', followed by a guided meditation to connect with the miracle of life through germination and growth of a seed. Then come more quotations linked by the question 'Who am I?', and reflections on the place of humanity in the cosmos. Finally, come Reflections on our relationship with Mother Earth, the goddess Gaia.

Choose those sections and quotations that speak to you. And take time to explore their deep significance for your worldview and life.

What is Life?

What is life?

Clearly, a molecule of carbon dioxide that crosses a cell boundary into a leaf does not suddenly 'come alive' nor does a molecule of oxygen suddenly 'die' when it is released to the atmosphere. Rather, life itself has to be regarded as belonging in some sense to a totality, including plant and environment.

David Bohm[1]

What is life?

When our pet dog dies we do not mourn the irreversible loss of its power of self-reproduction, growth, metabolism, respiration, and the like – we mourn the irreversible collapse of the organized and goal-minded activities that made the dog the companion he was.

G. Sommerhoff[2]

What is life?

Life is that property of matter whereby it can remember – matter which can remember is living. Matter which cannot remember is dead.

Samuel Butler[3]

What is life?

Life is occupied both in perpetuating itself and in surpassing itself; if all it does is maintain itself, then living is only not dying.

Simone de Beauvoir[4]

Connecting with the Miracle of Life

If you've never watched a seed germinate, you might like to put some large ones, such as peas or beans, on a layer of wet cotton wool, and watch what happens over several days. When the seeds have germinated, put them with a pot plant or vase of flowers where you can see them while you meditate. Or you might prefer to meditate outdoors if you have a garden or park nearby.

Settle yourself comfortably, and spend a few minutes just looking at your seeds and plant; or, if you are outdoors, at the plants around you. Let your eyes relax, not straining to see details, but gently taking in the scene.

Become aware of your breath moving slowly in and out as you look. And as you breathe, let your body and mind relax. As you breathe in, calm your mind by letting go of thoughts. And as you breathe out, release any tensions you notice in your body.

In ... Out ... In ... Out ...

And now picture a seed in your mind.

This seed is lying where it fell on the surface of the soil, gently protected from foraging ants and birds by a few leaves. It's summer. It's hot and the ground is dry. The seed waits, unmoving, conserving its unrealized potential.

Allow yourself to be that seed in imagination as the seasons turn. Autumn and winter pass, and with the warmth of spring come gentle showers.

The seed stirs; you stir. Something tells you the time is right. Softly, slowly you drink in the life-giving moisture from the soil beneath and the dead leaves around you. Softly, slowly, you swell until your skin splits.

Tiny buds peep out, and start to grow. One shoot heads upwards, straining to reach the light and air, pushing aside its covering of leaves and twigs. As this shoot gets longer, tiny leaves unfurl, turning their delicate green faces to the light of the sun, and taking their first breaths of carbon dioxide.

The other shoot heads down, forcing aside the grains of soil, wriggling through cracks, seeking the moisture and nutrients it craves. And as this root grows, so it divides, exploring, meeting other denizens of the dark. Here you make friends with a bacterium that gives you nitrogen, and there you fight off a hostile fungus.

Your stem lengthens and thickens. More leaves appear, straining to capture the sunlight filtering through an overhanging bush.

And the network of roots extends ever downwards and outwards to find new sources of water and nourishment. Listen carefully and you may hear the gurgle of your sap rising, the rustle of leaves unfurling, and the chomping of a hungry caterpillar. You may even hear the gentle grating as your roots push through the soil.

You are now mature. Not, perhaps, quite perfectly shaped, but twisted by straining towards the light, and tattered by hungry caterpillars. But alive and ready for the next stage of your journey.

Where leaf joins stem a new swelling appears. This bud grows and its skin stretches until lines appear where the sepals join. More swelling, and now the sepals start to tear apart, and in the gaps the bright colors of the flowers peep through.

Slowly, gently these petals push open the green doors of their womb. Slowly, gently, they unfold like new–born butterfly wings. There, within their bright–painted cup, style and stigma grow. And a heady scent wafts out, drawing myriad bees to pay their pollen–laden homage. And also drawing animals which love sweet flowers to eat.

The surviving petals fade and droop and drop. But in the center of some lie the swelling ovaries, the treasure-houses of seeds to be. These growing seeds attract hungry animals, but as the season draws to a close, and you begin to brown and shrivel, some treasuries remain, bursting open, to scatter the wealth of seasons yet to come. Your children.

Stay a while and let this vision flow through your soul; a vision of co–creation by plant and environment working in harmony.

Now, when you're ready, and in your own time, return to the present. Look around you. Gently stretch and move.

Before returning to every–day life, you may like to spend some time expressing what you experienced in this meditation. Draw, paint, write, dance, play music ... Talk to a friend if you are in company. However you choose.

Who Am I?

Most of the cells in my body are free-living micro-organisms, cooperating with me to digest my food and keep me healthy.

So who am I?

Mitochondria power every cell of my body. But:

> They are much less closely related to me than to each other and to the free-living bacteria out under the hill. They feel like strangers, but the thought comes that the same creatures, precisely the same, are out there in the cells of seagulls, and whales, and dune grass, and seaweed, and hermit crabs, and further inland in the leaves of the beech in my backyard, and in the family of skunks beneath the back fence, and even in that fly on the window. Through them, I am connected: I have close relatives, once removed, all over the place.
>
> Lewis Thomas[5]

So who am I?

> There's a wonderful saying in one of the Upanishads: "Oh wonderful, oh wonderful, oh wonderful, I am food, I am food, I am food! I am an eater of food, I am an eater of food, I am an eater of food." We don't think that way today about ourselves. But holding on to yourself and not letting yourself become food is the primary life-denying negative act. You're stopping the flow! And a yielding to

the flow is the great mystery experience that goes with thanking an animal that is about to be eaten for having given of itself. You, too, will be given in time.

Joseph Campbell[6]

So who am I?

(W)e are the earth, we are the consciousness of the earth. These are the eyes of the earth. And this is the voice of the earth.

Joseph Campbell[7]

Who Are We?

The evolutionary tree of life is usually shown with humanity on the tallest branch. We see ourselves at the pinnacle of evolution, the highest stage yet reached. And often we see ourselves as the end point, no longer subject to evolutionary change. But how justified are these images?

Survival and reproduction are the hallmarks of evolutionary success. The origin of our species goes back no more than a few million years. But bacteria emerged soon after the formation of the earth, over 3 billion years ago. Some have changed little over time. And disease-causing bacteria have survived everything our modern technology has thrown at them, and are poised for a comeback.

So are bacteria the true lords of evolution?

> The earliest life forms were by far the best adapted. If the meaning of evolution was in adaptation and increasing the chances for survival, as is so often claimed, the development of more complex organisms would have been meaningless or even a mistake.
>
> **Erich Jantsch (1980)**

We pride ourselves on our uniqueness, on our difference from the lower orders of life. And yet every attempt to define that difference has ultimately failed. Whales have equally complex brains; many animals use tools; some apes and monkeys use rudimentary language and abstract thought; many animals have complex social structures; and chimpanzees share over 99% of our genes. In what respect can we claim to be at the pinnacle of evolution?

Our technologies far surpass the tools of other species, enabling us to expand our habitat to all corners of the globe. But we are genetically far less adaptable than bacteria. If we destroy the life-support systems of the planet on which we depend, we will become extinct. Bacteria and cockroaches will inherit the Earth. So who is superior?

We can now modify our own genes at will, and are taking control of our own evolutionary destiny. But do we have the wisdom to choose where we should go? Can we be sure we won't turn the present planetary plague of humanity into a galactic disaster? Or that we won't simply exterminate the miraculous experiment that is humanity?

Can we be sure that nature won't bypass us as an evolutionary dead-end? Perhaps we are already sowing the seeds of our own destruction by creating the first forms of silicon-based life in our electronic technologies?[8]

Perhaps the truth is that we are inescapably part of the interactive, co-evolving whole that is Earth? Perhaps whatever we do to ourselves, we do to the whole, and whatever we do to the whole we do to ourselves? Perhaps our destiny is inseparable from the destiny of the planet which spawned us?

Connecting with the Earth

> A condor is 5 percent feathers, flesh, blood, and bone. All the rest is place. Condors are soaring manifestations of the place that built them and coded their genes. That place requires space to nest in, to teach fledglings, to roost in unmolested, to bathe and drink in, to find other condors in and not too many biologists, and to fly over wild and free.
>
> David Brower[9]

This we know. All things are connected like the blood which unites one family. All things are connected. Whatever befalls the earth befalls the sons of earth. Man did not weave the web of life; he is merely a strand of it. Whatever he does to the web, he does to himself.

Chief Seattle[10]

Shall I take a knife and tear my mother's bosom? Then when I die she will not take me to her bosom to rest. You ask me to dig for stone! Shall I dig under her skin for her bones? Then when I die I cannot enter her body to be born again. You ask me to cut grass and make hay and sell it, and be rich like white men! But how dare I cut off my mother's hair?

Chief of the Wanapum People of North America[11]

Love all God's creation, the whole and every grain of sand in it. Love every leaf, every ray of God's light. Love the animals, love the plants, love everything. If you love everything, you will perceive the divine mystery in things. Once you perceive it, you will begin to comprehend it better every day ... for all is like an ocean, all is flowing and blending; a touch in one place sets up movement at the other end of the earth

Fyodor Dostoevsky[12]

When I went to the moon, I was as pragmatic a test pilot, engineer, and scientist as any of my colleagues ... But there was another aspect to my experience during Apollo 14, and it ... began with the breathtaking experience of seeing planet Earth floating in the vastness of space.

The first thing that came to mind as I looked at Earth was its incredible beauty ... a splendid blue and white jewel suspended against a velvet black sky. How peacefully, how harmoniously, how marvelously it seemed to fit into the evolutionary pattern by which the universe is maintained. In a peak experience, the presence of divinity became almost palpable and I knew that life in the universe was not just an accident based on random processes. ... It was knowledge gained through private subjective awareness, but it was – and still is – every bit as real as the objective data upon which, say, the navigational program or the communications system were based. Clearly, the universe had meaning and direction ... an unseen dimension behind the visible creation that gives it an intelligent design and that gives life purpose.

Edgar Mitchell[13]

Gaia is no mere formula – it is our own body, our flesh and our blood, the wind blowing past our ears and the hawks wheeling overhead. Understood thus with the senses, recognized from within, Gaia is far vaster, far more mysterious and eternal than anything we may ever hope to fathom...

Just as, in breathing, we contribute to the ongoing life of the atmosphere, so also in seeing, in listening, in real touching and tasting we participate in the evolution of the living textures and colors that surround us, and thus lend our imaginations to the tasting and shaping of the Earth. Of course, the spiders are doing this just as well ...

David Abram[14]

I feel the rock breathing, and I know that this hard, still, mute, inanimate thing is in fact alive, moving at a different pace than I am, alive in a slow way, in a way that survives a long time. I feel the love of human being for rock being and rock being for human being as we snuggle, breathing together, sun heating rock body, sun heating flesh, bone and blood body, serenaded by the pounding sea. The Earth is alive! the rhythm of the waves is singing, the Earth is alive!

Nina Wise[15]

The Earth does not need to change in order to survive – she will survive with or without us. If we are to continue, it is our values that need to change.

Allan Badiner[16]

As we discover our ecological self we will joyfully defend and interact with that with which we identify; and instead of imposing environmental ethics on people, we will naturally respect, love, honor and protect that which is of our self ...

Bill Devall[17]

I believe that the universe is one being, all its parts are different expressions of the same energy, and they are all in communication with each other, therefore parts of one organic whole. ... The parts change and pass, or die, people and races and rocks and stars; none of them seems to me important in itself, but only the whole. This whole is in all its parts so beautiful, and is felt by me to be so intensely in earnest, that I am compelled to love it ...

Robinson Jeffers[18]

Honor the earth as our mother.

Love all creation as our family.

Honor the spirits of the rocks, plants and animals that are sacrificed to meet our needs.

Go with the flow of Gaia's cycles of renewal and her transforming processes of evolution.

Rejoice in the mystery and unpredictability of the evolving cosmos.

BE where you are.

Part VI:
Brain, Mind and Consciousness

We have reached one of the most difficult, mysterious and exciting stages of our journey as we climb from life to mind and consciousness. What are they? Where do they come from? How are they related to our bodies and brains, and to the material world around us? Are other organisms conscious, and perhaps even non-living things? Such questions have taxed philosophers, mystics and scientists since the dawn of civilization, and are amongst the deepest we face. And today, they are also amongst the most actively researched. Answers are sought, and offered, by many disciplines: philosophy, psychology, anthropology, neurosciences, evolutionary biology, physics and more. Theologians, mythologists and mystics are involved as well, but their contributions will be considered mainly in Part VII on spirit and spirituality.

Before setting off again, I need to define what I mean by brain, mind and consciousness. The brain is an organ of the body, part of the nervous system, whose structure and function may be studied scientifically like any other organ. To most scientists, the brain is the site and source of mind and consciousness – a view that we will challenge.

By contrast, the mind is not a physical part of the body, but encompasses the non-physical phenomena of thoughts, emotions, perceptions, memories, choices, and so on. The presence of mind is revealed by the ability of an entity to take account of its environment and respond to it, even if it is not consciously aware of what it is doing. In other words, things with minds have a degree of freedom, an element of self-determination.

Many people use the words 'mind' and 'consciousness' interchangeably, but I want to draw a distinction. Consciousness refers to those mental activities of which the entity is aware. These include feelings, ideas and deliberations; reflection and introspection; the imaginative anticipation of possible actions and their consequences; deliberate acts of free will; and direct awareness without thought such as occurs in meditation. Conscious experience gives us the impression of an observer inside our heads who watches all we do, and who is somehow independent of our minds and bodies. It is this 'inner I' who seems to feel sensations and emotions, to think and plan, and who is free to make choices. Consciousness is about how it feels to be me; about my sense of identity and autobiography; and about the meaning, purpose and value of life.

I have sought to do three things in this Part. First, in Chapter 17, we revisit two issues raised in Part I: why does science have such difficulty in coming to grips with mind and consciousness? And how can we test the validity of knowledge in this area?

Second, in the following two chapters we examine the relationship between mind and matter. Chapter 18 looks at this issue in a general sense, and then Chapter 19 focuses on the connection between our minds and our brains and bodies.

And third is an in-depth exploration of consciousness. Chapter 20 describes all the dimensions of human consciousness – the only type of which we have direct knowledge and experience. This is followed in Chapter 21 by a review of some of the diverse theories of the nature and evolution of consciousness.

Finally, as usual, this Part closes with Review and Reflections sections.

17 Science, Mind and Consciousness

The scientific approach to knowledge, as described in Chapter 1, is based on objective observation of measurable phenomena. But the key feature of mind and consciousness is their subjectivity. They are inherently private.[1] When I say "I have a pain in my back", that is objectively true in the sense that its existence is a fact (at least to me), and is not dependent on my beliefs, values or attitudes. However, my pain is subjective in the sense that it is me, and only me, who experiences it. A scientist can observe my behavior and monitor events in my spinal column and brain. And from these she may be able to infer that I am in pain. But she cannot observe my pain directly, or be sure that I am actually in pain. Indeed, there often are protracted medical and legal wrangles about the reality of back pain when a person claims to be unfit for work but there is no obvious physical damage. All sensations, mental activities and feelings present similar challenges.

It has sometimes been suggested that these difficulties can be overcome by observing our own conscious states. However, my introspective awareness of my inner self is an aspect of my subjective consciousness, and, despite what I may feel, cannot provide independent objective observations. Nor can it reliably probe our subconscious minds, particularly unwanted characteristics that we suppress.

Free will presents another challenge for science. Most scientists believe that every event in the material world is the result of a network of physical causes and effects. From this perspective, there is no way that a purely mental decision or desire can possibly influence the material world. As a result, free will and autonomy must be illusions, and are excluded from many theories of consciousness.

We are thus left with a dilemma. The objective, deterministic approach of science leaves no space for the subject who is doing the observing. Since mind and consciousness are irreducibly subjective, we are forced to conclude either that they are not real, or that the scientific method cannot reveal all aspects of reality. I believe the former is unacceptable because it denies the reality of the deepest aspects of human experience. Hence, science must be combined with other approaches in the study of consciousness.

This situation is made clearer by Ken Wilber's four-quadrant model of reality.[2] He argues that every holon has four intimately related aspects as shown in Figure 13. These are:

- The external, observable aspects of the individual holon;
- The interior aspects of the individual holon;
- The external, observable aspects of the environment of the individual holon; and
- The interior aspects of the environment.

In order to know any holon completely, we must understand all four aspects as an interactive whole. To illustrate this, consider a human being as a holon. The individual, external aspects are those properties and processes which can be objectively measured and studied in an isolated specimen. They include anatomy, physiology, biochemistry, neurology, and all the other sciences of the body. But a description based on these sciences would fail to capture the totality of the individual. While we might be able to infer from the data that he was afraid or thinking certain kinds of thought, we could not share his subjective experience of life – what it is like to be 'me'. To get even close to understanding his sensations, perceptions and emotions we would have to communicate with him as a person, not treat him as an object of study. This is the field of love, friendship and psychoanalysis.

But even intimate knowledge of a person's external and internal states does not provide complete understanding because no-one is an isolated individual. Everybody belongs to a society on which they depend for their existence. The external aspect of this society consists of the objectively observable interactions between the member holons. It includes the economic, legal and education systems, a variety of technologies, norms of behavior, and so on. The internal aspect goes beyond these observable structures and patterns to the cultural meanings which they reflect. These include the culture's image of reality, and its shared beliefs, values, ethics, myths and rituals.

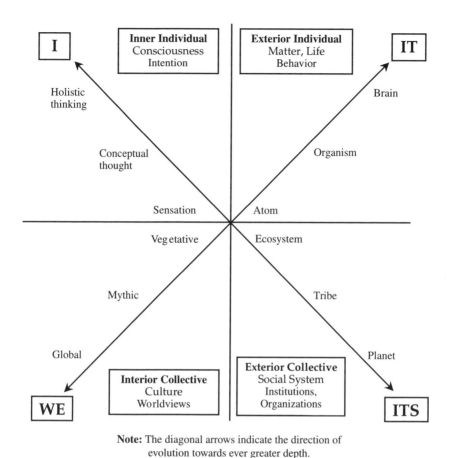

Figure 13 *Wilber's Four Quadrant Model of Consciousness*
Adapted from Wilber (1995, 2001)

A thought provides another illustration of how every holon has four aspects.[3] When I have a thought, measurable physical and chemical changes occur in my brain. This is the external individual aspect. But even if you knew what every atom of my brain was doing, you would not know the particular meaning that my thought has for me unless I shared it with you. This is the internal individual aspect. (This seems likely to remain true despite rapid advances in brain scanning technology, discussed in Chapter 19.) I would be able to communicate my thought to you only if we shared the same language and image of reality – the internal social aspect of the holon. Finally, these shared beliefs and meanings are reflected in the technologies, legal codes, written records, and so on which form the physical context for our communication. This is the external social aspect.

Classical science was concerned primarily with the exterior of individual objects, ranging from atoms to organisms. Systems and social sciences have extended the field of interest to the exterior of social holons ranging from galaxies (the environment of stars), to ecosystems (the environment of organisms) and human societies. These constitute the objective scientific reality that can be expressed in 'it' language, and which has been described in depth in Parts II-V.

Few sciences probe beyond these external aspects, and yet they are only half the story. By excluding the internal aspects, science denies the reality of experiences expressible only in 'I' and 'we' language. It denies the existence of value, meaning, beauty, motivation, understanding, intention, consciousness ... Thus, modern science reduces existence from "What does my life mean?" to "How can I get it to work better?"

The inner aspects of higher animals are readily apparent. But how can we tell if shallow holons such as atoms and subatomic particles have them? Might not these inner aspects be emergent properties which only appear at deeper levels of the holarchy? There is no way to be sure, but there is suggestive evidence that shallow holons do have internal aspects. For instance, an electron is not the same in all situations. It 'senses' the presence of other electrons orbiting the atomic nucleus, and adapts itself to them in subtle ways. Similarly, individual cells are 'irritable', sensing the world around them and responding to it. Philosopher A. N. Whitehead argued that such 'prehension' exists at all levels of the material world.[4] We will explore this idea further in later chapters.

Given that these interior aspects cannot be understood by traditional scientific methods, how can we reveal their true nature? How can we escape the prison of subjectivity and discover the underlying truth? These are the same challenges that we addressed in Chapter 4 where I outlined ways in which we can test the reliability of experiential, intuitive and inner knowledge. Let's recapitulate briefly in the current context.

The only way to understand the inner world of another person is by communicating with them; by trying to share their feelings, thoughts, meanings and values through empathy, intuition and 'resonance'; by interpreting what they say and by dialoguing about our interpretations. It is useless to seek objective truth, but we can assess the truthfulness and integrity of the other person, and the accuracy of our interpretations. And we can test the internal social aspect of any claim by examining how well it fits with its culture; and by assessing how right, just, ethical, appropriate, or 'in tune' it is in relation to values and beliefs.

Despite its limitations, the scientific approach can provide valuable insights into these inner worlds. For instance, systematic investigation of out-of-body, near-death and other-life experiences have confirmed that some of them are true in the sense that those involved learned verifiable facts or used knowledge or skills that they could not have gained by normal means. Similarly, an indication of possible mechanisms involved in such phenomena may be inferred from the findings of modern physics. However, in order to achieve greater depth of understanding, science must be complemented by other ways of knowing. In particular, we need to seek collective confirmation of subjective realities (See Chapter 4).

To illustrate how this can work, imagine I have a vision whilst meditating in a certain way. I cannot be sure if this vision is a valid reflection of reality, or simply a quirk of my brain or psyche. But if many other people report similar experiences whilst practicing the same meditation technique, I might tentatively conclude that the vision was true. However, if all the meditators had been taught by the same teacher or were from my own culture, the vision might be due to learned expectations, or shared beliefs and values. For instance, Christians tend to see visions of Jesus or Mary, Buddhists of the Buddha, and Hindus of Krishna. Including practitioners from other cultures and teachers would enable a deeper assessment of the vision's truth to be made. I can never be absolutely certain that it was not produced by some peculiarity of the human brain. But neither can scientists be sure that their models of external reality are truly objective and not images from the distorting mirrors of our minds.

In Conclusion

For the remainder of our journey into the science of oneness, we need to hold our minds open to the possibility that many inner phenomena rejected by mainstream science may be true reflections of reality; and we need to bring other ways of knowing to bear. At the same time, we need to subject all 'facts', theories and intuitions to critical investigation and questioning, never accepting anything on trust no matter what its source. I find it helpful to adopt a frame of mind in which I believe certain things to be true, whilst simultaneously knowing that they may be false.

18 Mind, Consciousness and Matter

Most civilizations have seen mind, consciousness and matter as different manifestations of a single supreme Being, or creative Spirit. But classical scientists rejected this view, proposing instead that mind and matter are distinct phenomena which interact with each other through the brain. This disembodied mind has been variously likened to the rider of a horse, the driver of a car, a musician playing an instrument, or the software of a computer.

Today, most scientists believe that mind and consciousness emerge from matter, thus completing the long process of replacing Spirit with matter as the ultimate reality. There are three versions of this theory. According to materialists, the mind is merely a side-effect of brain activity, and hence has no independent reality. This has led many theorists to deny that consciousness is a real phenomenon, and to view it as illusory.[1] To most of us, this is obvious nonsense, since consciousness is a fundamental, if not the fundamental, characteristic of human life.

The second view is that mind emerges from mindless matter when systems become sufficiently complex. Hence, mind is different in nature to matter, although not consisting of a distinct substance. This theory raises a question which occupies a lot of research effort: How complex does a system have to be before mind and consciousness appear?

A minority of scientists hold a third view, which is that mind is an inherent property of all matter, simply becoming more apparent in more complex systems.

We will discuss these alternative theories of the relationship between mind, consciousness and matter in this chapter. But before we start, recall that mind is the ability of an entity to take account of its environment and respond to it in adaptive ways. It includes thoughts, feelings, perceptions, memories, choices and so on. Consciousness is similar except that it includes only those aspects of mind of which the entity is aware.

The Materialist View

Materialists believe that mind is simply a different way of viewing the electrical and chemical activity of the brain; an epiphenomenon that arises when the nervous system reaches a certain level of complexity. And consciousness is an unavoidable but useless by-product of brain activity; an illusion which, being non-physical, cannot influence the brain or the wider material world in any way. Hence, our freedom of action is illusory, and we can achieve nothing through an exercise of will. From this perspective, the whole of human culture, civilization and history, including science itself, is reduced to a mechanical sequence of events uninfluenced by our conscious purpose.

Materialist theories reached a peak in the 1950's and '60's when behavioral psychology claimed that all our perceptions, thoughts and actions are completely determined by our genetic inheritance, and 'programming' by our life experiences. According to this theory, we are motivated simply by the drive to reduce stress by adapting to our circumstances. Our innermost sense of self is nothing but the outcome of physical and chemical processes in our brains. And our spiritual and moral characteristics are nothing but learned reflexes. It follows that both personal and social problems can be solved by reprogramming through carefully designed experiences. In short, we are nothing but complex robots.

Give me a bunch of kids, taken as they come – and I will make them doctors, lawyers, merchant men, beggars and thieves, solely by the power of conditioning.

John Watson
Quoted by von Bertalanffy (1968) p.190

The robot theory of mind and consciousness cannot account for all aspects of human behavior and experience, however. In affluent societies, the satisfaction of biological needs with minimal stress fails to produce well-adjusted people. Instead, it often leads to despair, apathy, depression, chronic boredom, juvenile delinquency, crime and addiction – mental states generated by the meaninglessness of a life without challenge.

Also, rather than seeking to reduce stress, we often create our own for non-utilitarian reasons. The world today is full of adventurers, record-breakers, and a plethora of thrill-seekers. And more and more people are intent on expressing their creativity – something that is inherently stressful since, by their very nature, creative acts differ from 'normal' behavior. It seems that, far from being robots seeking the easy life, we are imbued with an urge to transcend our boundaries, push our mental and physical abilities to the limit, and create new worlds through language, the arts, science and technology.

Today, behavioral psychology has faded from the spotlight, and the flame of the materialist view is now held aloft by the neurosciences. Encouraged by their rapidly increasing ability to map electrical and chemical processes in different parts of the brain, many neuroscientists believe they will soon be able to read and manipulate our thoughts, feelings, motives, desires and intentions. We will discuss these controversial claims in more depth in the next chapter.

Attractive though materialist theory may be to those who seek mechanistic explanations of all phenomena, it leaves us with a few puzzles. If consciousness cannot affect the material brain, how is it that we can remember experiences of conscious awareness? How can an illusory consciousness affect physical reality as quantum physics suggests? (See Chapter 9). And what are we to make of the evidence for the influence of mind over matter from parapsychology, transpersonal psychology, complementary medicine and other fields? (Evidence that we will discuss in detail in Chapter 20)

Mind and Consciousness as Emergent Properties of the Brain

Reaction against this mechanistic model, coupled with the rise of systems sciences, led to the alternative theory that mind and consciousness are emergent properties of the brain. According to this view, they are new holarchic levels which appear when the brain reaches the requisite level of complexity. Hence, mind is not a distinct essence, but is nevertheless different from, more than, and not reducible to the physical and chemical processes of the brain. Consciousness, including will and intention, is a deeper holon than mind which influences mind, brain and body through a reciprocal holarchic relationship as described in Chapter 6.

This emergence of mind and consciousness is sometimes likened to the formation of water from hydrogen and oxygen, but this is misleading. Although water has properties that are distinct from those of its constituents, it is still a material entity with mass, volume and temperature that obeys the laws of physics. Mind and consciousness, by contrast, are not material nor any known form of energy, and they appear to transcend the laws of physics as we understand them. Also, when water forms, the atoms of oxygen and hydrogen cease to exist as separate entities until the water is broken down again. By contrast, when mind and consciousness emerge, the neural networks supposedly responsible for them continue to exist unchanged as separate, observable physical structures. Clearly, the emergence of mind and consciousness is a different kind of process to the emergence of water or even life. Perhaps a better analogy is that consciousness is like the flame when hydrogen and oxygen combine to form water.

It follows that, in order to validate the emergent theory of mind, scientists must explain how it is possible for an objective, material reality to give birth to a subjective, non-material reality. And they must explain how this non-material consciousness can then control the physical brain. In the words of Christian de Quincey: "No amount of complex feedback loops in the brain or nervous system can make that jump because all those loops in the brain are themselves still objective."[2]

If we accept the theory of emergence, we are still left with the problem of determining how complex an entity must be before mind and consciousness appear. In the past, they were often thought to be exclusive to humans. But scientists are discovering gradually that non-human organisms are far more intelligent than previously thought, and that many probably are conscious. The rest of this section explores these thresholds in more depth.

Key factors in this debate are the criteria by which the presence of mind and consciousness are judged. As already noted, most biologists define intelligence as the ability of an organism to take account of its environment and respond to it, even if it is not consciously aware of what it is doing. But to count as intelligent, a response must occur within the lifetime of the organism rather than relying on evolution, and must be more flexible than a reflex determined by the genes. Hence, the abilities to learn and solve problems are clear signs of intelligence.

It is far harder to detect the presence of consciousness because inner awareness is private. Lacking a shared language, animals cannot tell us about their inner experiences. And so we must rely on intuition and inference from their behavior. In the end, conclusions about the presence of consciousness often boil down to a judgment whether or not an observed behavior would be possible without awareness of what the organism is doing. Commonly used indicators include:

- Actions that require an image or plan of what is wanted to be held in the mind as in solving new problems, making new kinds of tools, and manipulating the behavior of another member of the group.
- Actions that require 'what if' explorations of alternatives to solve problems in new ways.
- Complex communications about a situation.
- Physiological and behavioral responses that indicate feelings such as excitement and sadness.

With these criteria in mind, let's quickly survey some observed behaviors of organisms of varying complexity.

Chimpanzees can learn the basic signs of the language used by deaf people, and teach them to each other. The screams of a juvenile rhesus macaque tell its mother the rank and blood relationship of its opponent, as well as the nature of the interaction.[3] And primates can predict how their actions will be interpreted by others, and use this understanding to deceive and manipulate them.[4]

Every owner knows that dogs express joy, excitement, sadness, fear, pain and love in ways that clearly show they are conscious. More objectively, dogs communicate with humans by using different types of bark to indicate both their feelings and the nature of the situation.[5] And surely a blind person's guide dog must be aware of what she is doing?

Horses are well-known for their ability to open a variety of gate latches. And Prof. Donald Broom's cows became excited when they discovered how to open a gate: "It was as if the animals were saying 'Eureka! I've found out how to solve the problem.'"[6] And sheep, despite their reputation for stupidity, can recognize another sheep's face in a photograph even when it is taken from a different angle to the one with which they were trained.[7]

Betty, a New Caledonian crow, was far from 'bird-brained'. She got food out of a bottle with a wire hook, thus demonstrating that she could assess a novel problem, and design and make a tool to solve it.[8] Like birds, fish get a poor press for their IQ, but recent research has shown them to be "steeped in social intelligence, pursuing Machiavellian strategies of manipulation, punishment and reconciliation, ... exhibiting stable cultural traditions and cooperating to inspect predators and catch food."[9]

Moving to even lower levels of life, a spider spinning her web must adapt creatively to the location of branches and rocks, the strength of the wind and other factors. Is it conceivable that she can do this without awareness?[10] Ants can learn to associate different times and places at which food will be available.[11] And bees tell each other about the location of nectar sources through intricate dances which are choreographed by the genes, but incorporate present environmental information.[12]

We tend to dismiss plants as unintelligent, but this is a mistake. They have over 15 senses including light, sound, chemicals, vibration, touch, water, gravity and temperature. They combine, prioritize and weight these data, deciding how to respond to injury, lack of water and other environmental cues. For example, they adjust their growth in order to get the most light, or avoid root competition. And parasitic plants assess the value of a potential host before investing too heavily in exploiting it.[13] Even more surprisingly, plants have electrical and chemical communication pathways that act like nervous systems, using many of the same messenger molecules as the human brain.[14] They can even detect bacterial communications, and mimic them to foil attack.[15] More controversially, there is substantial evidence that plants sense human emotions and intentions, bonding with their carers and responding to love – perhaps the reason why some people have 'green fingers'.[16]

Even single-celled organisms show evidence of mental activity. The paramecium identifies food and toxins, moving towards the former and away from the latter. It swims around obstructions and learns to back more quickly out of narrow glass tubes. It also may be able to learn to negotiate a simple maze.[17] And close to the bottom of the evolutionary tree of life, bacteria communicate, enabling them to cooperate in complex ways and take on specialist social roles as described in Chapter 15.[18]

Finally in this survey, let's take a quick look at organs and cells within multi-cellular organisms. The heart sets its rhythm independently of the brain,[19] and contractions of the gut are controlled by local networks of nerves. White blood cells behave much like amoebae as they scavenge for wastes and bacteria. And every cell in our bodies produces the same communication chemicals as the brain.[20]

It seems from this overview that mind is a characteristic of all life. However, most biologists continue to argue that cells, micro-organisms, plants, insects, and even some higher forms of life do not show true intelligence, only genetically-determined automatic responses. A few mavericks, including Tony Trewavas of Edinburgh University, disagree. He is convinced that plants are genuinely intelligent, adapting flexibly to maximize their fitness in a variable environment, and displaying foresight, decisiveness, memory, and the ability to compute.[21]

Arguments about the threshold of consciousness are even more complex and inconclusive. But most scientists agree that primitive animals, plants and inanimate matter are not conscious, and at least some higher animals probably are. Certainly, there seem to be no physiological reasons why higher animals could not be conscious, since all their central nervous systems are alike. Further, animals respond with similar brainwaves to those observed in humans known to be conscious of comparable information.[22] And those of us who live with animals, have observed their behavior at first hand, and have learned to understand their body language, have no doubt that they are conscious. We will explore many of these issues in more depth in Chapter 21, when we discuss ideas about how consciousness may have evolved.

Finally, what of non-living matter? Gregory Bateson argued that any system is a mind if it satisfies criteria similar to those for self-organization.[23] This led Fritjof Capra to conclude that a form of mind is present even in systems like chemical clocks.[24] As we saw in Chapter 5, such systems respond to environmental conditions, and display elementary forms of purpose, creativity, adaptability and self-determination. At this point, the theory of emergent mind becomes almost indistinguishable from the alternative idea that mind is an inherent property of all matter which is discussed in the next section.

Mind and Consciousness as Inherent in All Matter

The idea that mind and consciousness are inherent in all matter is supported by a minority of scientists, and many spiritual traditions. Thus, Christian biologist Charles Birch claims that:[25]

> There is but one theory ... that casts any positive light on the ability of brain cells to furnish us with feelings. It is that brain cells can feel! What gives brain cells feelings? It is by the same logic that we may say – their molecules. And so on down the line to those individuals we call electrons, protons and the like. The theory is that things that feel are made of things that feel.

Approaching the issue from a very different direction, David Bohm reached a similar conclusion that consciousness is implicit in the underlying implicate order (See Chapter 9).[26]

If matter has come into being or if flesh has come into being because of the spirit, it is a wonder. But if spirit has come into being because of matter or flesh it is a wonder of wonders.
Gospel of Thomas
Quoted by Woolger (2001), p.203

The theory that mind and consciousness are inherent in matter comes in two versions. One holds that there are two parallel realities: matter and consciousness. Hence, a complete understanding of the universe must integrate the physics of matter with a new science of consciousness, possibly revealing an even deeper reality from which both have emerged.[27] The second version, already encountered in Part IV, is the idea that cosmic Consciousness is the source of everything, including matter.[28] From this perspective, materialist science has things the wrong way round. Rather than consciousness emerging from matter, matter emerges from consciousness. If true, this helps to explain why science has so much difficulty with consciousness.

In both versions, mind and consciousness are present in every speck of matter for all time. Thus, the 'decision' whether to appear as a particle or a wave demonstrates the presence of primitive mental capacities even in the most fundamental constituents of physical reality.[29] At higher levels of the holarchy of nature, the degree of consciousness increases until it becomes apparent to us when it acquires qualities similar to our own. Hence, it is impossible to draw a clear distinction between things with minds and things without, just as we cannot define precisely the difference between living and non-living.

Most theories of inherent consciousness are based on quantum physics (See Chapter 9). One of the most ambitious is due to Amit Goswami, who argues that consciousness, like energy, is a basic property of the universe.[30] Once again, I will use an upper-case C to distinguish this cosmic Consciousness from individual consciousness.

According to Goswami, the stuff of the universe remained a formless potential of probability waves until collapsed into a particular physical reality by consciousness. This extraordinary idea is supported by one interpretation of string theory which holds that all 10^{500} solutions to the equations exist simultaneously in the same way that Schrödinger's cat is both alive and dead (See Chapter 9). "We only experience the one state that we are in, but ... the many states of the universe are all equally real."[31]

If cosmic Consciousness solidified our actual universe from the sea of potential, conscious beings would have no freedom of choice. For example, human consciousness would not be able to determine the outcome of a quantum event. Hence, Goswami argues, material reality must have remained in potential until sentient beings evolved, whether on earth or elsewhere. Cosmic Consciousness was then able to work through the agency of individual consciousnesses, leaving us with freedom to choose our reality.

This implies that our brains mediate Consciousness in such a way that we can choose one possibility from the unlimited potential, and then collapse the probability wave to that particular outcome. In this process, we produce both the external objects themselves, and our subjective experience of them. Nevertheless, this reality remains an expression of the underlying cosmic Consciousness which acts through us rather than being a product of our individual egos.

According to this theory, it was not until conscious beings emerged that the space and time of the particular universe we inhabit was materialized. This means that much of the 'history' of the cosmos has been created outside the normal bounds of space and time, and retrospectively from our perspective. It also means that the material world does not exist independently of conscious beings, simply waiting to be discovered, but that we collectively create our own reality. Perhaps conscious beings have been progressively materializing the universe as their senses and understanding grew? And perhaps we continue to co-create the physical universe today?

This freedom to create the world would result in chaos if each one of us was free to manifest a different possibility in each moment. How is it, then, that we experience a stable world? The answer lies in the development of habits as discussed in Chapter 9. When a particular reality has been materialized once, memory of that event biases the choice next time. And when a particular reality has been materialized many times by many individuals, it becomes almost a certainty that the same reality will be manifested in future, and will be experienced the same by all people. In this sense, there is an external, objective reality. But there is always the possibility that a powerful, creative mind will break through the shell of habit; and there are always frontiers of experience and knowledge at which creation of a new reality is still possible.

According to Amit Goswami, only those entities that are aware of themselves and what they are doing can collapse the infinite potential of the wave function. And so once again we are faced with the issue of which entities are conscious. Leaving this question open for the moment, Goswami's theory implies that we humans, together with other conscious beings from earth and elsewhere, are collectively responsible for the future evolution of both matter and consciousness. Our every thought and action affects the whole cosmos in some way. Mostly, these effects are trivial and fade without trace. But sometimes we may trigger a cascade of self-organizing events that change the whole future of humanity, Gaia and the cosmos. We are not simply passive inhabitants of a given cosmos, but co-creators of its future evolution.

This view is reflected in ancient beliefs and customs still evident in many places in Europe. And indigenous peoples around the world often believe that they literally create and maintain the world through their rituals, which are thus of vital importance. This is illustrated well by this account of the Australian Aboriginal Dreaming:[32]

> We start with nothing – a total emptiness – a void. Then we have some singing and dancing. We start by forming – the singing creates the sound and the vibration forms a shape, and the dancing helps solidify it. The dancing is making the form stand out as a tree, a bird, as land. The process, the Dreaming itself, becomes a reality, something we can work with and see.

Individual and Aggregate Mind and Consciousness

Before leaving this chapter, I want briefly to explore the nature and extent of mind and consciousness beyond the individual. Charles Birch drew a clear distinction between 'individual entities' and 'aggregates of individuals'.[33] An individual entity, such as an organism, can respond to its environment as an integrated whole, and hence has a mind. By contrast, an aggregate is an unintegrated collection of individuals such as a rock or a chair which is a collection of molecules. Birch believed aggregates as such are devoid of mind because they are unable to do anything of their own volition, lack the freedom to behave creatively, are not unified wholes, and do not have self-defined values and purposes. However, every aggregate is composed of individual entities which do have minds, albeit often primitive ones.

Birch's concept of individuals and aggregates corresponds to Ken Wilber's idea of individual and social holons, as discussed in Chapter 6. The rock or chair forms the social environment of the molecules of which it is composed. Hence, Birch is suggesting that social holons do not have a higher-level mind than the individual holons of which they are composed because they do not integrate them into a deeper holon. However, at each successive level of the holarchy the minds of the shallower holons are integrated into a new and higher level of mind. Thus, increasing holarchic depth is equivalent to increasing mind and consciousness.[34]

This concept of the holarchy of mind sheds valuable light on the questions of artificial intelligence, the possibility of a planetary mind, and the nature of cosmic Consciousness.

Robots with their hardware, software and sensors such as cameras and microphones are individuals rather than aggregates. They are integrated wholes, take account of their environment, and respond flexibly to it; and neural net computers have brain-like structures that learn from experience. Clearly they have minds. Whether or not they are, or may become, conscious is not so clear.

Ecosystems do not integrate their component animal and human minds into a deeper holon, and hence are aggregates. But the situation is not so clear for Gaia. The several billion individual human brains are interlinked by increasingly dense and complex networks of communication, ranging from the spoken word and print, through broadcast radio and television, to telephones, faxes, and the internet. Thus, if each of us is analogous to a brain cell, human civilization closely resembles a developing brain.[35] But to create a truly planetary mind it would be necessary to integrate the minds of all living things, not just humans. Hence, it seems that what is emerging is a new, integrated level of the human mind, not a planetary mind. This analysis also suggests that those who experience a planetary consciousness are connecting with the collective consciousness of the organisms of which Gaia is composed, rather than a Gaian consciousness in the sense of a higher-level integration of component consciousnesses.

Holarchical analysis also helps to clarify the nature of cosmic Consciousness. The cosmos is the shallowest, most fundamental holon. Hence, its Consciousness must be of the most primitive type. But all deeper holons have sprung from the cosmos, and hence it is cosmic Consciousness, transcended and integrated into more and more complex forms, which has led to the human mind, and is leading to global mind. Ultimately, this evolution could conceivably produce cosmic mind in another, richer sense: that of a Mind which integrates and transcends all minds within the cosmos, just as the global mind is integrating and transcending individual human minds. It seems at least possible that this is the goal of cosmic evolution, as hinted in Part IV and explored further in Part VII.

In Conclusion

As so often on this journey, we are faced with a choice of theories and interpretations. Are mind and consciousness nothing more than by-products of brain activity? Or are they real phenomena that emerge from complex matter? Do they only appear above a threshold of complexity, or are they inherent in all matter? There are no absolute answers, and we are free to choose what makes most sense to us; what resonates best with our beliefs and values.

I find the materialist position untenable because it does not reflect my individual, nor our collective experience of subjective consciousness. The theory of emergence is attractive as it builds on the ideas of self-organization, the holarchy and evolution. But it leads into a quagmire of speculation about the thresholds of complexity required for mind and consciousness. And the further this track is pursued, the closer it comes to the idea that mind and consciousness are inherent in all existence: an idea that is rooted in ancient spiritual traditions. Way out though it may seem, Goswami's theory is no stranger than many aspects of quantum physics. And it is just one theory in a very young field.

In the following chapters, we go on to look at the more specific relationship between the brain, mind and consciousness, before delving into the nature and evolution of consciousness itself.

19 Brain, Mind and Consciousness

Having discussed the relationship between mind and matter in general, it is now time to turn our attention to the way our brain develops and works, how it produces sensations and perceptions, how it stores memories, and its relationship to mind and consciousness.

Brain Structure and Development

The human brain has 100 billion cells, each of which connects to an average of 3,000 other brain cells, for a total of over 100 trillion connections. If each connection is capable of 10 levels of activity, the number of potential brain states is literally mind-blowing: $10^{100,000,000,000,000}$.[1] Within each cell is a complex skeleton, the 'bones' of which are hollow cylindrical molecules.[2] These tubes are the right size to amplify quantum effects, or act as waveguides for photons of light. It is thus possible that these skeletons act as miniature computers and memories within each cell. Taking these into account increases the brain's computing power from about 10^{14} operations per second to 10^{27}.[3] If true, this means that the human brain is all but infinitely varied, capable of producing an unlimited range of behaviors and experiences.

The structure of our brains is a legacy of our evolutionary past, and has been described as the brain of a reptile, overlain by the brain of a horse, overlain by the brain of a human.[4] The reptilian brain regulates breathing and metabolism, and controls instinctive reflex reactions and movements. Around the reptilian brain evolved the *limbic system*, seat of the emotions, memory and learning. With the arrival of mammals, the limbic system was overlaid in turn with the *neocortex* – the center of perception and thought which achieves its greatest size and complexity in humans. And within this broad structure are smaller-scale features which are the location of specific functions.

Our genes specify the overall size and structure of the brain, and endow the newborn infant with certain innate abilities. Within these broad constraints, the brain grows and develops internal patterns of connections throughout our lives. The richer our experiences, the more complex the maze of connections that develops. Hence, the precise form of each brain results from the interaction of genes with internal biochemical processes and the environment. It is as if a circuit board full of electronic components was allowed to wire itself.

Brain Function

There are two competing schools of thought about the way the brain works. One believes it is modular, with lots of specialist areas in which particular functions are located. Modern imaging techniques are revealing more and more such areas associated with different senses, emotions and thoughts. A lot of publicity has been given to discoveries such as the so-called 'god spots'. These are two small knots of neurons above each ear that mediate peak religious experiences. The resulting sense of cosmic consciousness is discussed in depth in the next chapter.

The second school of thought argues that the brain works holistically, with functions widely and flexibly distributed. For instance, visual perception integrates information from 30 parts of the brain associated with different aspects such as color and movement. Also, new connections and cells develop throughout life, and the boundaries between functional modules sometimes shift. When a finger is cut off, for example, the sensory inputs from the remaining fingers gradually spread into the area of the brain previously used by the missing one, producing enhanced sensitivity.[5] And neuroscientist Vilayanur Ramachandran describes how facial functions take over parts of the brain previously associated with the arm following amputation. His work with people suffering from phantom limbs has taught him that "Perceptions emerge as a result of reverberations of signals between different levels ... (and) even across different senses. The fact that visual input can eliminate the spasm of a non-existent arm and then erase the associated memory of pain vividly illustrates how extensive and profound these interactions can be."[6]

On this basis, Ramachandran argues that both theories are true: "the real secret to understanding the brain lies not only in unraveling the structure and function of each module but in discovering how they interact with each other to generate ... human nature."[7] But in order to understand the mind, we have to look beyond the brain. It is clear that our minds do not simply emerge from brain activity, but are parts of an interactive, integrated whole that also includes the nervous system, the remainder of the body, and influences from our environment. We will explore this whole in the remainder of this chapter.

Mind-body Interactions

The close and complex interaction of mind, brain, body and environment is revealed in many ways. A common experience is of ingesting a chemical that alters our mood and senses, and hence changes our experience of reality. We drink caffeine to stimulate us, and alcohol to release inhibitions; we take aspirin for pain and valium for stress. And we don't even have to take such drugs to alter our minds, since the brain produces its own mood molecules which can be released by vigorous exercise, positive thinking, or meditation.

Connections in our brains also are modified by every experience, action and thought. Many studies have shown that the brains of young animals and children need stimulation to develop normally, and this process continues throughout our lives. Hence, thinking changes the thinker, who then recreates the thought. And a conversation is a dance between two brains as each responds to the other.

Our minds and bodies interact in many ways. Louise Hay[8] described in great detail how various physical symptoms are connected with our emotions, and many alternative therapies are based on the belief that emotional traumas become locked up in physical tensions. For those who can read them, therefore, our bodies are living autobiographies. Even mainstream medicine is increasingly recognizing the role of the mind in physical healing through such things as the placebo effect, prayer and meditation.

Specific examples of mind-body interactions include the spontaneous remission of cancer, the use of hypnosis to remove warts caused by a virus, and the triggering of an allergy to rose pollen by the sight of a plastic rose. More dramatic is false pregnancy in which a woman who desperately wants a child experiences all the symptoms of being pregnant. Menstruation ceases, she may suffer morning sickness, her abdomen swells, her nipples become pigmented, lactation starts, and after 9 months she goes into labor – all without there being a fetus. How, asks Vilayanur Ramachandran, can a wish or delusion produce such effects?[9]

When tested in the laboratory, yogi Swami Rama produced a temperature difference of 11° F between the left and right sides of his palm within a few minutes. He slowed his heart beat from 93 to the low 60's within seconds, and stopped the blood flowing through his heart for 16 seconds. He quickly produced cysts in the large muscles of his body, one of which was medically removed and examined, and then dissolved them again. He made the body's energy centers, known as the chakras, visible to an audience and appear on Polaroid film. And he repeatedly moved a compass-like object.[10]

Even more remarkable than the exploits of yogis are sufferers from Multiple Personality Disorder who have more than one identity, each unaware of the others. When one personality switches to another, the voice, mannerisms and other physical characteristics may change too: one can be diabetic and another not; one can suffer from allergies that another does not; one can by short-sighted and another have 20-20 vision; and their vital signs, immune systems and hormone profiles may be different. It is almost as if they have two bodies.[11]

Perhaps, muses Ramachandran, "it's time to recognize that the division between mind and body may be no more than a pedagogic device for instructing medical students – and not a useful construct for understanding human health, disease and behavior."[12]

More personally, my physical body supports my conscious mind thus enabling me to experience, think, feel and act. I cannot separate my conscious self from my body and brain. My sadness is accompanied by a heaviness in my limbs, and my delight is revealed by my jaunty step. My facial expressions, gestures and sighs embody my feelings, moods and desires even before I am aware of them. In the words of David Abram:[13] "To acknowledge that 'I am this body' is not to reduce the mystery of my yearnings and fluid thoughts to a set of mechanisms, or my 'self' to a determinate robot." Rather, it is my body, through its intimate connection with my environment, that enables my conscious self to engage with others and the world. If I were an immaterial mind, I would by unable to see, hear or touch, and would lack experience of anything but my inner state.

> (W)hen you say 'I have a gut feeling about such and such,' you're not speaking metaphorically, you're speaking quite literally, because your gut makes the same chemicals as your brain makes when it thinks. In fact, your gut feelings are possibly more accurate because gut cells haven't yet evolved to the state of self-doubt.
>
> **Deepak Chopra (1993)**

Mind and body form an integral whole which create and sustain each other. Our knowledge of external reality depends on our sensory nerves. Our perceptions are colored by hormones secreted by our glands and chemicals ingested from without. Our bodies mirror our emotions, reflecting them back onto our minds which are distributed throughout our bodies and not restricted to the brain.

Rupert Sheldrake goes further, arguing persuasively that the mind extends way beyond the body. He claims that the images we see are not inside the brain, as modern science believes, but outside where they appear to be. Thus, the image of the page you are reading is formed by a mental field about half a meter away, and when you look at a distant mountain the image is actually 10 kilometers away. These mental fields are rooted in the brain, just as magnetic fields are rooted in magnets, but extend way beyond it to the limits of space.[14]

What evidence is there for this hypothesis? 90% of people surveyed by Rupert Sheldrake have experienced the sense of being stared at, and over 100,000 trials of a simple experiment have proved overwhelmingly that this is a real phenomenon. This shouldn't happen if our minds are confined to our heads. Similarly, telepathy should be impossible, but evidence gathered by Sheldrake indicates strongly that it is real. Hundreds of experiments have shown that dogs react to their owners' intention to come home, even when they are many miles away, return at randomly selected times, and travel in a strange taxi. 50% of dog owners in Britain and the USA believe their animals anticipate the arrival of family members. Similarly, over 800 trials have shown clearly that people often can correctly 'guess' the identity of a telephone caller when that person is known to them, no matter how far away they are.

Memory

Rupert Sheldrake also believes that memories are held in fields outside the brain. But in order to understand his reasoning, we need to look first at standard theories of how the brain stores the vast amounts of information contained in our memories. For many years, the memory was thought to be like a computer database with billions of bits of information stored in localized brain circuits that could be retrieved when required. This idea is supported by evidence that memories are associated with the same areas of the brain as related perceptions. Visual images, for example, occur in the visual cortex. It is also known that the part of the brain called the hippocampus is essential for forming memories.

A difficulty with this theory is that memories would survive only as long as the local pattern of connections between brain cells. But the brain is dynamic. Individual molecules are replaced within days or months, brain cells die progressively throughout our lives, and connections are continually being made and remade. Hence, we might expect memory to decay with time. We do, indeed, forget many things, but this process does not seem to be a gradual decay since elderly people often remember the distant past more clearly than recent events. Also, no-one has yet located specific memories in the brain, and research shows that memories often survive removal of parts, or even most, of the brain.

These weaknesses have led to a modified theory in which multiple 'backup' copies of each memory are held in different locations. However, even this modified 'database' theory is logically flawed.[15] In order to find information in a database, we must know what we want, and be able to recognize it when we find it. This requires some kind of prior memory. But if the retrieval system itself includes summary memories, how do we access these? Through another retrieval system with yet another memory? And so on ad infinitum. These considerations led to an alternative theory in which memories are stored holographically within the patterns and rhythms of the whole network of brain cells. This implies that every part of the brain records all memories, albeit with decreasing detail as the scale is reduced.[16] Hence, every time we learn a new skill or store a new memory the whole mental landscape is altered in subtle ways. Memories stored like this could survive changes to the brain because they are not dependent on individual cells or connections. And we could recall memories by recognition of patterns and association of experiences and ideas, rather than by defined search questions. This theory is supported by evidence that recall works better when the context is similar to that in which the memory was laid down, and by the many methods for improving memory which rely on association.

Rupert Sheldrake argues that even this theory cannot explain all observed phenomena, and that memories actually may be held outside the body. Recall that in Chapters 9 and 14 we met the idea of morphic fields that guide the development of material forms. Sheldrake believes that there are also morphic fields for behavior and memory, and that we remember because of resonance with ourselves, or similar people, in the past.[17] Going one step further, Ervin Laszlo argued that the brain can access information stored in the psi field which, if he is right, contains the whole history of the universe.[18]

These ideas have not been verified, but Rupert Sheldrake provides some suggestive evidence. Theory predicts that behavioral and mental morphic fields should get stronger as more people adopt a behavior, or gain some knowledge. Hence, if morphic resonance is at work, traditional skills should be easier to learn than new ones, and new skills should become progressively easier to learn over time. It has been observed that successive generations of experimental animals learn particular tasks more and more easily, even when their parents have not taken part in the earlier experiments. Similarly, new skills developed by animals in one place often spread rapidly to other populations. An example is the way blue tits throughout England quickly learned to pierce foil milk-bottle tops to drink the cream once the skill had been discovered in one place.

With regard to humans, standardized intelligence tests have been used for many years, and the theory of morphic resonance suggests that the average performance should have improved with time. This appears to be the case, with scores increasing almost 14 points in the USA between 1932 and 1978. Sheldrake claims that no satisfactory social explanation of this phenomenon has been suggested. More recently, experiments have been specifically designed to test the idea of memory fields. In one, people found it easier to remember rhymes in an unknown foreign language if they were traditional rather than newly composed. In others, the meanings of words in a foreign language were guessed more accurately if they were in common usage; and imitation Morse codes proved harder to learn than the real one.[19]

Brain and Consciousness

It is commonly assumed that the brain creates consciousness, but there is substantial evidence to the contrary. This includes the extension of the mind beyond the brain and body as discussed above, and some aspects of human consciousness that are discussed in the next chapter, including past life experiences and memories formed when brain activity has ceased. It is true that brain activity changes as the state of consciousness changes between, say, normal waking, dreaming, meditation and anesthesia. But this proves only that the two are correlated, and not that brain activity *causes* consciousness. The brain may be like a TV set which can be tuned to different channels. The program we see depends on the activity of the electronic circuits, but is not produced by them.

In practice, the brain actually restricts consciousness. Our senses detect only a tiny proportion of what is happening, and most of what they do pick up is screened from conscious awareness to prevent information overload. Our brains filter out irrelevant information, and channel most of the rest to automatic systems of which we are unaware. For instance, as you read this sentence, you are conscious of the meanings of the words, but unaware of the pattern of ink on the page, the texture of the paper and the feel of the binding – at least most of the time. Thus our brains, far from being the source of our consciousness, are censors which limit its scope. It is when these filters fail that novel forms of consciousness may break through as we will see in the next chapter.

In Conclusion

Far from being a rigidly programmed computer, the brain is highly flexible and adaptable, wiring and rewiring itself throughout our lives. Many functions are located in specific areas, but the brain is also holistic, and possibly holographic, with complex networks of interactions across many levels and regions. Our minds are intricately integrated with our bodies, and yet may extend well beyond our physical boundaries to interact with other minds across the cosmos. Hence, it seems that the brain is not the source of mind and consciousness, but mediates their expression in the material world.

20 The Experience of Human Consciousness

As we have seen, consciousness is one of the most mysterious phenomena known to science, and cannot be studied by objective methods alone. The challenge is made doubly difficult by the fact that the only consciousness of which I have experience and that I can observe is my own. You may tell me about your consciousness, but I cannot join you in your experience, or directly observe it. In approaching the subject, therefore, we will tackle it in two stages. First, in this chapter we will simply map the terrain of human consciousness, defining the range of phenomena that a satisfactory theory must explain. And then in the next chapter we will see what light alternative theories shed on the mystery.

Normal Waking Consciousness

We are all familiar with waking consciousness. From the inside, it seems obvious and pervasive; it defines normality and what is real, who I am and what it is like to be me. But what is it? There is no agreed definition, but it is possible to describe its essential characteristics.

Perhaps its most significant feature is the presence of a witness or observer who has privileged access to my inner self. Events are experienced by an experiencer, thoughts are thought by a thinker, pain is felt by a feeler, imaginings are created by an imaginer, and choices are made by a chooser. I experience my witness as a distinct and independent being that talks to me, but is nevertheless intimately connected to my mind and body. It observes my actions, thoughts and feelings; analyzing, judging and interpreting them. It distinguishes important from trivial, good from bad, beautiful from ugly, right from wrong. My witness is always present except when I am unconscious, in deep sleep or under anesthesia. Through this witness,

I am aware of:

- Sensations including light, sound, heat, touch, taste, smell, pain, hunger, and thirst;
- An image of my body, including its position and activity. According to Ramachandran and Blakeslee (p.62) this image is an "entirely transitory internal construct that can be profoundly modified with just a few simple tricks. It is merely a shell that you've temporarily created ..."
- Perceptions – the interpretations of sensations based on past experiences and mental models of the world;
- Emotions and feelings such as pleasure, awe, wonder, fear, love, anger, joy, pride, anxiety, and compassion;
- Moods such as depression or elation which color my perceptions, thoughts, feelings and experiences;
- Rational thoughts, including ideas, logical arguments, plans, and deliberations;
- Imaginings, including visions, waking dreams, fantasies, anticipations and possibilities;
- Memories of past events, moods, emotions, thoughts, sensations and so on;
- Curiosity – the desire to explore, know and understand myself, others and the world I inhabit;
- Creativity – the ability and desire to innovate, producing new ideas, knowledge, technologies, music, poetry, or things of beauty;
- Intention and will to undertake some activity, or pursue some goal;
- The ability to focus attention on one particular aspect of consciousness at a time;
- Decisions or choices, apparently made freely in accordance with my desires, goals and will;
- Relationships with people, animals, objects and places;
- Chains of association between present experience and memories of other perceptions, ideas, feelings, and past events;
- A sense of identity and unity arising from the apparent integration of all these factors into a coherent being or persona;
- A sense of the numinous, spiritual dimension of reality.

We would be overwhelmed if we were conscious of everything at once, and so our consciousness is always focused on what is important at the moment. Less important things may remain in awareness but become unfocused and blurred, whilst the unimportant may move right to the edge of awareness, almost into unconsciousness. I often drive on a kind of auto-pilot, not consciously aware of what is going on around me, and yet when anything unusual happens it snaps immediately into the center of attention.

The Subconscious Mind

Our conscious minds are underlain by an ocean of complex mental activity, much of it to do with processing information from sense organs and memories. Except when learning a skill or performing a delicate task, the movement and functions of our bodies are controlled automatically. We are similarly unaware when we 'sleep on a problem', but this does not prevent the mind from producing solutions like magic in the morning. There are also unconscious senses, such as the 'blindsight' by which some blind people can perform actions that depend on visual cues without any awareness that they are 'seeing'.

It used to be believed that we can become conscious of all the mind's processes by introspection. It came as a considerable shock, therefore, when Freud suggested that we harbor many thoughts, feelings and desires of which we are unaware. Today, it is widely accepted that we build walls in our minds, refusing to acknowledge certain parts as belonging to us. We repress unacceptable characteristics such as aggression, eroticism or joy so that they become parts of an unconscious 'shadow'. And we bury traumatic memories of the birth process, childhood abuse, war and personal tragedy. The unconscious influences to a greater or lesser extent our perceptions, interests, 'rational' thoughts, willingness to consider alternative ideas, openness to new evidence, and judgments of truth. Hence our conscious selves are incomplete, and one aspect of personal development is to expand consciousness into the dark areas, and integrate the shadow with our persona.[1]

The unconscious is also a major factor in determining cultural worldviews.[2] According to Carl Jung, evolution has bestowed on us a psychological inheritance, or *collective unconscious*. Evidence for this is the appearance in dreams and myths from around the world of universal themes and symbols that he called *archetypes*. Archetypes are not precise images, but rather motifs that vary a great deal in detail without losing their basic pattern and significance.[3] Common archetypes include hero myths, and the symbol of the cross. Jung recounted the case of a professor "who had had a sudden vision and thought he was insane. He came to see me in a state of complete panic. I simply took a 400-year-old book from the shelf and showed him an old woodcut depicting his very vision."[4]

Transpersonal Consciousness

Around the turn of the nineteenth century, psychologist William James concluded:[5]

> Our normal waking consciousness, rational consciousness as we call it, is but one special type of consciousness, whilst all about it, parted from it by the filmiest of screens, there lie potential forms of consciousness entirely different. ... No account of the universe in its totality can be final which leaves these other forms of consciousness quite disregarded.

More recently, this thought was echoed by Ken Wilber when he wrote:[6]

> It is as if our everyday awareness were but an insignificant island, surrounded by a vast ocean of unsuspected and uncharted consciousness, whose waves beat continuously upon the barrier reefs of our normal awareness, until, quite spontaneously, they may break through, flooding our island awareness with knowledge of a vast, largely unexplored but intensely real domain of new-world consciousness.

Psychiatrist Stanislav Grof began his distinguished career as a convinced atheist and materialist. But his observations of thousands of people in non-ordinary states of consciousness induced by LSD and 'holotropic' breathing convinced him that William James and Ken Wilber were right. As he confessed: "Frankly, there were times that I felt deep discomfort and fear when confronted with facts for which I had no rational explanation and that were undermining my belief system and scientific worldview."[7]

Grof now believes that our individual consciousness connects us not only with our immediate environment and our own past, but also with other times and places, other species, and the very cosmos itself. He believes we can relive experiences in the womb, and the pain and emotion of our birth; witness events from the future as well as from the lives of our human and animal ancestors; explore mythological realities we never knew existed; and share the consciousness of other people, animals, plants, and even of inanimate matter. In rare cases we can experience unity with cosmic Consciousness, whether as awed witness or as the creative principle of the universe itself.

Such experiences are often referred to by scientists as *altered states of consciousness*. I prefer the term *transpersonal consciousness*, which reflects the fact that such experiences transcend the purely personal. To participants, transpersonal experiences are as real and authentic as anything in normal consciousness, and are often so powerful that they transform the person's beliefs and life. They may occur spontaneously, or be induced in a variety of ways including drugs, abnormal breathing, religious ritual, fasting, sensory deprivation, loss of sleep, pain, music, dance, intense physical activity, or creative arts. It seems that such activities weaken the mental filters that restrict our consciousness, thus letting through more information than normal and allowing awareness of phenomena that are usually hidden from us.

> It's a natural part of consciousness to change one's consciousness.
> **Rick Doblin**
> **Quoted by Phillips and Lawton (2004)**

Transpersonal experiences are generally dismissed as illusions. But there are many well-researched cases in which there is solid evidence for their validity. Other cases are too improbable to have happened by chance, or are shared by so many people that they may be regarded as normal. I have described several such cases in the following sections to demonstrate their reality.

> Any unbiased study of the transpersonal domain of the psyche has to come to the conclusion that the observations represent a critical challenge not only for psychiatry and psychology, but for the entire philosophy of Western science.
> **Stanislav Grof (2001), p.160**

Before Birth and Near Death

Many people in therapy recall experiences that happened before or during their birth, often accurately describing events of which they previously were unaware. Amongst these is psychiatrist Stanislav Grof.[8] Once he let go of his scientific prejudices, he relaxed into a profound experience in which he had the sensation of shrinking in size, with his head larger than the rest of his body, and floating in liquid. He could taste harmful substances entering his body through the umbilical cord – spices or other inappropriate foods, elements of cigarette smoke or alcohol. He also became aware of his mother's emotions – a chemical essence of anxiety, anger, or feelings about the pregnancy, and even sexual arousal. He vacillated between the experience of a distressed, sick fetus and a state of blissful and serene intrauterine existence.

Later he reflected that the experience would have been easy to dismiss if it had only included intellectual understanding that could have come from books or films. But the experience was sensual, including extraordinary physical sensations, and the feeling of strange textures. He had felt sick from the toxins and then recovered. There was no way he could have learnt the *feelings* of a fetus, and he concluded that his consciousness was providing him with detailed information that he had never dreamed possible.

At the time, Grof was a skeptical scientist, atheist and materialist. Together with the stories of thousands of his clients, this experience transformed his beliefs. Many people have experienced the threat of abortion, the danger of miscarriage, loud noises, toxins, and the diseases of their mothers. And others have felt their mothers' anxiety, aggression, depression, relaxation, satisfaction, happiness and love. Not a few of these memories have been checked against the accounts of friends and relatives, and the medical records. There are also accounts of the pain and trauma of birth which match medical understanding of the process. It is highly unlikely that all these are hallucinations or imaginings woven around stories told by their mothers.

At the other end of life are the experiences of those who have almost died. A 1982 Gallup poll revealed that about 1 in 20 Americans had had one or more Near Death Experience or NDE.[9] Many elements of these experiences cannot be verified objectively, and no two people have identical experiences. Nevertheless, there is a common core which includes:[10]

- A feeling of peace and well-being, without pain, bodily sensations or fear;
- An experience of being outside the body, able to view events from a distance;

- Entering darkness, floating in dimensionless space, and then moving rapidly through a tunnel;
- Being drawn towards a pinpoint of light at the end of the tunnel, which becomes blindingly bright and incredibly beautiful;
- Meeting an omniscient, compassionate and loving presence with which the person communicates telepathically;
- Re-experiencing everything the person has ever done or said or thought;
- Seeing a world of great beauty, with heavenly music, and a feeling of oneness. A very few report terrifying hellish visions.
- Reaching a boundary beyond which the person cannot go and where they meet and talk with dead friends and relatives, before deciding or being told to return to this life.

Medical science dismisses these experiences as hallucinations brought on by drastic changes in the dying brain. But quite frequently relatives have their own transformative experiences as they empathically share the dying process, directly contradicting the hallucination theory.[11] So powerful are these NDEs that in case after case the lives of those who experience them are transformed. They may still fear pain and the process of dying, but they lose their fear of death itself, becoming more tolerant, loving and interested in spiritual values.[11] Frequently, when out of their bodies, they see and hear events in other rooms or distant places that in some cases have been independently confirmed. And some people who have been blind since birth have visual experiences.[12]

During cardiac arrest, the breathing and heartbeat cease, followed within seconds by cessation of brain activity. Recent studies have shown that in the period after the brain stops working and before resuscitation, around 10% of patients experience and remember NDEs, including awareness of events such as the removal of their dentures. This suggests that the mind continues to function without the brain, and may even survive death.[13]

Such research lends some support to the claims of those who believe they have contacted loved-ones after death. There is no proof that these are objective experiences, but there is enough evidence at least to keep an open mind. For example, Stanislav Grof recounts how a close friend was drowned and later appeared to his wife. He asked her to return a book he had borrowed, telling her precisely where to find it on the shelf. Another psychologist had a vivid encounter with his dead grandmother during a session with Grof. He asked her for confirmation that the experience was genuine, and she said: "Go to aunt Anna and look for cut roses." Still skeptical, he visited his aunt the next weekend, and found her in the garden surrounded by cut roses. It just happened to be the one day in the year that she pruned them.[14]

Finally, three leading researchers of psychic phenomena died near the turn of the nineteenth century. In the following years, psychics around the world received several nonsensical communications from them. These found their way to the Society of Psychical Research where connections were discovered that revealed coherent messages. These claimed that the researchers had devised this way to prove their survival beyond death. They had chosen to communicate meaningless fragments to avoid distortions due to subjective interpretation.[15]

Past Lives, Other Lives

Thousands of past life memories have been recorded, many of which are far more than fantasies. An outstanding example is related by Sogyal Rinpoche.[16] From the age of twelve, Arthur Flowerdew experienced vivid mental images of a great city surrounded by desert. Over the years, he saw more buildings, the layout of streets, soldiers, and the approach to the city through a narrow canyon. One of the clearest pictures was of a temple carved out of a cliff. Much later in life he saw a television documentary on the ancient city of Petra in Jordan, and was astonished to see the city of his mind.

The Jordanian government flew him to Petra with a TV crew. Before he left, he was interviewed by a world authority on Petra who was baffled by his knowledge. Also before leaving, he selected three places in his vision of Petra. One was a strange-shaped rock unknown to archaeologists which was found where he predicted. The second was a well-known structure whose purpose was unknown, and which he identified as a guard room in which he had served as a solder two thousand years before. And the third was a temple, whose precise location he was able to point out on a photograph of the city. When he arrived, Flowerdew needed no map to find his way around, and was able to explain the purpose of other unexcavated structures. The expert concluded: "I don't think he has the capacity to be a fraud on this scale."

A more typical experience happened to eminent psychiatrist, Andrew Powell, in a guided meditation.[17]

(I) suddenly find myself scrambling up a rocky outcrop. At once I know exactly what is going on. This is Arizona, the year is eighteen forty-eight, my name is Tom McCann and I am being hunted down by a raiding party of Apache Indians. I heave myself up onto the flat top of the rock. I can hear the Indian braves a short way below and I know they will get to me in a couple of minutes and have my scalp. I pull out of my pocket a worn leather wallet and gaze for the last time with sadness and longing on the picture of my wife and two young daughters. Then I take out my gun, put the muzzle to my head and pull the trigger. There is no sound and no impact. I simply find myself floating peacefully up and away from the body lying on the top of the rock.

Roger Woolger is a Jungian therapist specializing in past-life regression therapy who has heard the stories of thousands of clients, and also experienced his own past lives. He notes that a very wide range of psychological problems have responded to therapy that releases past traumas. Contrary to media hype about past lives as famous people, most of these experiences are of ordinary lives from many different times and places including African tribesmen, nomadic hunters, nameless slaves, and medieval peasants. Such lives are impossible to verify historically, but often convey a strong sense of validity. Woolger's clients don't just tell a story, but relive intense experiences that may be accompanied by convulsions and contortions, sweating, hot and cold flushes, cramping, temporary paralysis, sharp pains, erotic sensations, numbness or trembling. Frequently, their faces and mannerisms change; they may speak in a strange accent or language; novel words and emotions may erupt into awareness; and they may find themselves taken over by another personality whilst retaining a strong sense of familiarity. Most people simply don't have such acting ability.[18]

Stanislav Grof similarly tells of clients who have assumed yogic postures; or performed the !Kung trance dance or Sufi whirling dervish dance; or spoken in languages, sometimes archaic, that they don't know.[19] There are many cases of young children who remember people, events and places in lives which took place only a few years before their birth, and not far from where their families live. Sometimes they have birthmarks reflecting the injuries from which they died in their previous life. Careful investigations have verified some of these beyond all reasonable doubt.[20]

What are we to make of all this? Roger Woolger identifies three possible explanations. First, these 'memories' may be unconscious imaginings fed by long-forgotten stories, TV programs and family gossip. On the basis of his experience, he finds this unconvincing. Second, it is possible that we connect with stories in the collective unconscious. But this fails to explain why some people repeatedly access memories of a specific lifetime that seems uncannily familiar. And the third possibility is reincarnation – a subject we will revisit in Part VII.

Rather than experiencing past lives, some people feel that they temporarily become other living people. Stanislav Grof recounts how one morning his wife had an overwhelming feeling that she was becoming their friend Gregory Bateson. "She had his giant body and his enormous hands, his thoughts, and his staunch British humor. She felt connected to the pain of his cancer and somehow knew with every cell of her body that he was dying. This surprised her because it did not reflect her conscious assessment of the situation." Later, they discovered that Bateson had decided at that very time that he wanted to die rather than go on fighting.[21] In another case, a woman described accurately her father's wartime experiences 14 years before her conception that he had never told to anyone.[22]

Finally, we come to the many people who have shared the feelings, sensations and consciousness of other species. Some of these stories have been studied by experts, revealing that the insights gained were scientifically accurate. One such was a woman who 'became' a whale cow giving birth, her account of which is included in the Reflections section at the end of this Part.

Roger Woolger concludes that "*we are multiple beings*, ... we have many personalities within us. Some of these other selves are surprisingly close to consciousness and can be awakened quite easily ... these secondary personalities turn out to be every bit as complex and rounded as our ego personality, even to the point of having detailed histories stretching from birth to death."[23] This conclusion from a past-life therapist is particularly interesting in the context of Multiple Personality Disorder in which each personality seems to have a different body (See Chapter 19). It is but a small step from here to acceptance of the reality of past lives.

Consciousness and Material Reality

Despite the skepticism of most scientists, there are many phenomena that demonstrate the power of mind and consciousness over matter. Some were described in Chapter 19, and I have included a few more here to highlight this dimension of human consciousness.

During peak performances, top athletes often feel calm and euphoric, with boundless energy and ample time. And they may experience drastic changes in body image which are sometimes seen by onlookers. A dramatic example is described by Stanislav Grof.[24] Morehei Uyeshiba, the inventor of Aikido, appeared to transcend the laws of physics in front of many reliable witnesses. Facing several martial arts experts armed with knives he seemed to change shape and size, disappear for an instant and reappear somewhere else. These events were not only witnessed, but also recorded in a documentary movie that his followers swear was not edited.

Less common today are the physical manifestations of religious experiences. There are many historical stories of how the bodies of saints or gurus became luminous or extremely hot, levitated, or displayed wounds. Some dematerialized after death, and others did not rot. We tend to dismiss such accounts as mass hallucinations, deliberate fabrications, or highly embroidered facts at best. But there are some that have a greater degree of scientific authenticity. Sogyal Rinpoche describes the death of a Tibetan Lama in a US hospital in 1981 whose body did not follow the usual process of rigor mortis and decay. The chief surgeon who attended him is quoted as saying: "They brought me into the room about thirty-six hours after he died. I felt the area right over his heart, and it was warmer than the surrounding area. It's something for which there is no medical explanation."[25]

Rainmaking ceremonies seem nothing but superstitious nonsense to the scientific mind, and yet they often seem to work. My son told me of an old-timer he met in a drought-stricken Australian country town who said they were afraid to pray for rain because the last time they did so they were flooded out! And mythologist Joseph Campbell tells how he attended a Native American rain ceremony. He felt amused and cynical when they began under a clear blue sky. But, to his amazement, heavy clouds covered the sky during the ceremony, and the day ended with a cloudburst.[26] Stanislav Grof tells a similar story of a Mexican shaman who attended a seminar at the Esalen Institute during a two-year drought. He agreed to conduct a rain ceremony, and the day ended with a 6 hour downpour. This could have been coincidence, but Grof notes that it is highly unlikely that so many cultures would have continued to use rain ceremonies if they never worked, and that shamans would soon lose their credibility if they often failed.[27]

Distant healing and prayer provide another demonstration of the power of consciousness. Many respected scientists have conducted controlled clinical trials with positive results that cannot be explained by psychological factors or the placebo effect. Even more remarkably, it appears that consciousness may be able to influence events retrospectively. Prof Leonard Leibovici from Israel is an outspoken critic of alternative medicine who studied the effectiveness of retrospective prayer with the apparent intent of ridiculing healing prayer. But he found statistically significant reductions in the length of stay in hospital and the duration of fever for patients who were prayed for 4-10 years after suffering septic infections.[28]

The interaction of mind and matter at a deep level is also revealed by what Carl Jung called *synchronicities*. A synchronicity is two or more events which occur at about the same time and which have a meaningful relationship to each other despite being unconnected by cause and effect. Jung collected many examples of events which he regarded as so improbable as to be impossible by chance. As Jung himself expressed it, "If an aircraft crashes before my eyes as I am blowing my nose, this is a coincidence of events that has no meaning. ... But if I bought a blue frock and, by mistake, the shop delivered a black one on the day one of my near relatives died, this would be a meaningful coincidence. The two events are not causally related, but they are connected by the symbolic meaning that our society gives to the color black."[29]

Other examples of synchronicity include the many tales of clocks which stopped, mirrors which cracked and pictures which fell at the moment of their owner's death; of people who met an old friend the day they were talking about them; or of finding that the number on your bus ticket is the same as the telephone number you just dialed. Synchronicity also may be observed when isolated scientists make simultaneous discoveries, or similar developments occur in different fields. As Stanislav Grof noted, "Jung's ideas ... show that consciousness and matter are in constant interplay, informing and shaping each other ..."[30]

Parapsychology

Beyond uncontrolled synchronicities, research has confirmed the existence of psychokinesis – manipulation of the physical world using only the mind. Robert Jahn, Dean of Engineering and Applied Sciences at Princeton University, asked participants to influence a random number generator (RNG) – the electronic equivalent of coin tossing. The results of 14 million tests showed a tiny but statistically significant result.[31] Many other scientists have conducted similar research, and an analysis of data from 832 studies by 68 investigators found that people can indeed influence RNGs. The overall odds against chance were over a trillion to one.[32]

The Global Consciousness Project used RNGs in a different way. 50 machines around the world were linked via the internet, and their output analyzed for statistically significant deviations from random. Rather than asking people to influence the results, the group focused on occasions when large numbers of people were paying attention to the same event – such as the World Cup Soccer Final, or major news stories. They found that the deviations from random were greater on 'newsworthy' days than on quiet ones, and that they increased with the amount of news.

The most remarkable result occurred on 11 September 2001 – the day the World Trade Center in New York was destroyed by terrorists. The deviation was larger than at any other time that year, with the odds against chance being close to 10,000 to 1. Even more extraordinary, the effect peaked just before the hijacked jet hit Tower No.1. Following rigorous scrutiny, the investigators were convinced that the results were valid.[33] When I checked with Dean Radin by email, he confirmed that there did indeed appear to be precognition; 'global consciousness' seems to have anticipated the events. In conclusion, Radin commented:[34] "By analogy, these observations mean that little bells perched on dozens of buoys scattered around the world's oceans somehow magically ring in coherent harmony."

Psychiatrist Andrew Powell gives a more personal experience of psycho-kinesis:[35]

> I was feeling both apprehensive and excited because it was the start of my first day as a hospital consultant. ... I tracked down the professor's secretary who took the Yale key(to my room) off a large ring, gave it to me, and I put it in my jacket pocket. ... I went back down the corridor and pulled the key out of my pocket. To my astonishment, the shank of the key had bent itself through ninety degrees. I could not enter my sanctum until I had clamped the key in a doorframe and straightened it out with brute force.

Powell also recounts a precognitive dream of a car accident that his wife had the night before they were due to make a journey. This made such a powerful impression on him that he pulled back instead of overtaking a tractor on a narrow country road. The next moment the exact green car of his wife's dream, the first car they had seen for many miles, hurtled round a bend in the road.[36]

The Stanford Research Institute, amongst many other organizations, has conducted research on extrasensory perception (ESP). In one experiment, subjects were given two numbers chosen at random representing the longitude and latitude of a particular location. They were then asked to try to 'see' what was there and sketch what they saw. In a variation of this procedure, a second person went to a randomly selected site and gazed intently at the target at the same time that the laboratory subject closed her eyes and sketched whatever image formed in her mind. In both cases, the sketches were scored by a panel of judges for their similarity to the location. Willis Harman and Howard Rheingold concluded: "If the SRI results are accurate, the ability to know what is happening at a place one has never visited is not a rare talent but a trainable skill latent within all of us."[37]

More recently, Bernard Haisch reported that the US Government sponsored such remote viewing research for 24 years. Many of the results are still classified, but from the material that is now available Prof. Jessica Utts concluded: "Using the standards applied in any other area of science ... psychic function has been well established ... It is recommended that future experiments focus on understanding how this phenomenon works, and on how to make it as useful as possible."[38]

What are we to make of all this? The majority of scientists dismiss the evidence, and there are legitimate concerns about the difficulties of replication and some contradictory results.[39] But, in the words of Stanislav Grof:[40] "There exists hardly any other realm where the expert testimony of so many witnesses of the highest caliber has been discounted as stupidity and gullibility and thus written off. We have to realize that among serious researchers were many people with outstanding credentials."

One of the most enduring issues is the 'experimenter effect'. In short, believers tend to get positive results and skeptics don't. No-one yet knows why.[41] Nevertheless, the evidence for parapsychology is strong. In the words of Robert Matthews:[42]

For years, well-designed studies carried out by researchers at respected institutions have produced evidence for the reality of ESP. The results are often more impressive than the outcome of clinical drug trials because they show a more pronounced effect and have greater statistical significance. What's more, ESP experiments have been replicated and their results are as consistent as many medical trials ... In short, by all the normal rules for assessing scientific evidence, the case for ESP has been made.

Matthews goes on to argue that no improvement to the methods, and no amount of new evidence will resolve the controversy. Mainstream scientists reject the findings because they do not believe parapsychological effects are possible. "Both camps can look at precisely the same raw data and legitimately reach utterly different conclusions, because they have radically different models for the cause of the data."

Unity Consciousness

Unity consciousness (sometimes called cosmic consciousness) is the sense of awareness of all that is. Unity consciousness, like black holes and the vacuum field, is far beyond the experience, knowledge or imagination of most of us. Such is the authority of science that we are willing to believe the physicists' strange tales whilst questioning the validity of spiritual experiences. But we should not dismiss lightly a realm of consciousness experienced by many atheists as well as religious believers, including the great spiritual teachers whose lives are shining beacons to humanity.

> It is very difficult to elucidate this (cosmic religious) feeling to anyone who is entirely without it. ... The religious geniuses of all ages have been distinguished by this kind of religious feeling, which knows no dogma. ... In my view, it is the most important function of art and science to awaken this feeling and keep it alive in those who are receptive to it.
>
> **Albert Einstein**
> **Quoted by Ramachandran and Blakeslee (1998), p.174**

The fundamental difference between unity consciousness and other levels is the apparent dissolution of the boundary between self and not-self. In normal and transpersonal consciousness there is a boundary between our conscious selves and everything else. In unity consciousness, however, the boundary dissolves, and the person ceases to be aware of subject and object, self and other, seer and seen. This does not mean that their individuality disappears into a featureless fog – after all, they return and recount their experiences. Rather, their sense of self expands to encompass all that is, and the differences between self and other become internalized. All things and events remain perfectly separate and distinct, and yet become One. They feel, at least for a while, that they are one with Reality, the Ground of Being from which they sprang.

Ultimate Reality cannot be understood intellectually. The understanding gained in unity consciousness does not compete with scientific knowledge; it is a totally different and complementary experience. Because Reality is all that exists, there is no other object or idea with which to compare it; no category to which it belongs; and no law which it obeys. It cannot be seen or heard because it is the very processes of seeing and hearing; and it cannot be remembered because it just is. It cannot be analyzed or explained, it can only be experienced directly as a whole. It is the integration of all other levels of consciousness. And since normal thought and language depend on drawing distinctions, the experience is impossible to describe adequately to others. In consequence, descriptions of unity consciousness are usually couched in metaphor, parable and allegory, and often seem paradoxical.

So what is the experience of unity consciousness like? Only having experienced it for fleeting instants, I must rely on descriptions from others. Here, I have included some attempts to describe it intellectually. The Reflections at the end of this Part include more poetic and metaphorical efforts, and spiritual aspects are addressed in Part VII.

Scientific investigations of yogis and mystics show unity consciousness to be a state in which the brain is quiet, still, peaceful, without active conceptual thought, and yet awake, alert and clear.[43] Sogyal Rinpoche calls it: "a primordial, pure, pristine awareness that is at once intelligent, cognizant, radiant, and always awake. It could be said to be the knowledge of knowledge itself. ... (It) is to realize the nature of all things."[44]

The experience is highly emotional, but the emotions aroused are of a different order to everyday feelings and hence are hard to describe. The deepest feeling is often one of serenity and lightness flowing from the perception of the whole and our place within it, accompanied by a sense that, at a deep level, everything is as it should be; all is very, very well. There also may be a sense of the holy, and feelings of awe, love, compassion, joy, timelessness, and complete understanding.

In the words of a medical researcher who experienced unity consciousness: "This knowingness is a deep understanding that occurs without words. I am certain that the universe is one whole and that it is benign and loving at its ground. ... The world seemed benign and 'right' with everything as it was 'supposed to be'. There was a great sense of inner peace."[45] And as Abraham Maslow wrote: "To have a clear perception... that the universe is all of a piece and that one ... belongs in it – can be so profound and shaking an experience that it can change the person's character and his (worldview) forever after."[46] In consequence, in the words of Willis Harman and Howard Rheingold, "one finds that the deepest motivation is to participate fully, with conscious awareness, in the evolutionary process and the fulfillment of humankind. ... one becomes aware that what appeared to be driving motivations were mainly illusory ego needs."[47] (The following note gives some recommended sources if you would like to pursue this topic further[48])

We can expand normal consciousness through practices such as meditation, biofeedback, drumming and dance. But these cannot lead directly to unity consciousness because they are actions of our separate selves, and hence sustain the barrier between self and not-self. Indeed, Ken Wilber argues that anything we do in an effort to enter unity consciousness will have the opposite effect.[49] We become open to the experience of unity only when we recognize at the deepest intuitive level that we do not exist as separate selves. This recognition may be facilitated by meditation and other spiritual practices, but they are neither necessary nor sufficient.

The world's great religious leaders have all experienced a deep Reality underlying the superficial reality of our senses. Inexpressible in words, they have given this Reality many names: God, Allah, the Tao, Atman, Brahman, Universal Spirit, Pure Light of the Void, Pure Consciousness, the Ground of Being, and more. And down through the ages, untold millions of ordinary people also have experienced unity consciousness. It comes to some unbidden and unsought. Others may spend their lives seeking it without success. It is not confined to particular races, cultures or religions, nor to the followers of particular practices. It is a gift, not a learned skill, which comes whole and immediate. To those who experience it, unity consciousness is the crowning moment of human life. Thus, the experience itself is a fact of human life. What is not so clear is what this experience means. Is it a creation of our own minds, or is it a true insight revealing that human consciousness is an expression of a single cosmic Consciousness?

Applying the tests of reliable knowledge outlined in Chapter 4 enables us to draw some tentative conclusions. The sources of information about unity consciousness include the most trustworthy in human history – the Buddha, Jesus, Muhammad and other spiritual Teachers. The understanding gained is consistent with many interpretations of modern physics and many spiritual traditions, and forms the basis for a coherent world view. Unity consciousness has been experienced by many people from all cultures, and from secular as well as religious life. Science has confirmed that it is associated with a particular brain state. It brings a deep sense of truth, and leaves people feeling peaceful, loving and joyful. And those who experience it engage joyously with life, and the problems of the world. What more could one ask for?

In Conclusion

This chapter has demonstrated that the realms of human consciousness are wide, deep and rich. And all of them are well enough validated that we cannot ignore them, or brush them aside as illusions. Any acceptable theory of consciousness must be able to encompass them all. We will survey the possibilities in the next chapter.

21 Theories of Consciousness

Consciousness is a complex experience that varies greatly between people, is inherently private, and is influenced by many factors. These include genetics; childhood nutrition and mental stimulation; the culture in which we grow up and live; our lifestyle, including spiritual practices such as meditation; and the ingestion of mind-altering substances from coffee to hallucinogens. Our theories of consciousness are deeply rooted in our beliefs about the nature of the universe as well as our knowledge of the way the brain works, which is growing explosively but is still rudimentary. It is not surprising, therefore, that researchers cannot agree on a definition of consciousness nor how to detect its presence. Given this situation, it is inevitable that current theories are partial, fragmentary and often confused as well as confusing.

An adequate theory must encompass the broad sweep of human consciousness as described in the last chapter, and answer many questions, including: What is consciousness? Where does it come from? How did it evolve? How does it work? Who or what is the 'witness'? Are animals conscious? Could computers be conscious? Is there a threshold of complexity below which consciousness does not occur? How can non-material consciousness cause events in the material world?

The self-imposed blinkers of western culture mean that it has little to contribute to many of these questions. Materialist science contributes nothing, because it holds that consciousness is no more than an epiphenomenon of brain activity. Those who believe in the emergence of consciousness from matter mostly reject the reality of transpersonal and unity consciousness, and concentrate on elucidating the nature and origins of normal consciousness. The only researchers who contribute significantly to understanding of the wider realms are the small minority who believe consciousness is inherent in all matter. As a result, the bulk of this chapter is concerned with the nature, development and evolution of normal consciousness on the assumption that it has evolved from and is created by the brain. The chapter opens with brief descriptions of the so-called 'hard problem' of consciousness, and the stages of development of human consciousness from birth to maturity. It ends with an exploration of more radical ideas that seek to explain transpersonal and unity consciousness.

The Hard Problem of Consciousness

The subjective qualities of consciousness such as the redness of a rose or the painfulness of a thorn are often referred to as *qualia* – a term coined as a counterpart to the measurable *quanta* of physics. Materialists dismiss qualia as illusions. Brain function is all there is, period. They disparagingly refer to 'qualia freaks', who in turn accuse materialists of being insensitive to the difference between a conscious person and a zombie that functions effectively but has no awareness.

A generation ago, most scientists were materialists. Now many are concerned with the 'hard problem' of qualia. The key challenge is to explain how the electrical and chemical activity of a mass of neurons gives rise to subjective experiences and thoughts, and to awareness of ourselves and the world around us. As neuroscientist Vilayanur Ramachandran put it: "By what magic is matter transmuted into the invisible fabric of feelings and sensations?"[1] Neatly concealed within this definition of the hard problem, however, is the assumption that consciousness is actually generated by the brain, and not simply mediated or transmitted by it as theories of cosmic Consciousness suggest.

As described in the last chapter, the essence of consciousness is the sense of an inner witness or observer. Despite the strength of our felt experience, however, most researchers argue that there is no witness; that the mind's 'I', the soul, or whatever we choose to call it does not exist. Rather, they believe that awareness arises from the sea of unconscious mental activity throughout the brain. Thus, a thought or feeling becomes conscious when it triggers different brain responses than events which remain unconscious.[2] But no-one has yet found a consciousness spot in the brain, nor a threshold of activity or complexity above which it occurs, nor a pattern of activity that unambiguously corresponds to consciousness. The mechanisms of consciousness will undoubtedly be unraveled one day, but there are substantial challenges still to be overcome.

Imagine that you are color-blind but completely understand the workings of the brain. You measure the wavelength of the light impinging on my eye and trace the pathway of perception in my brain, but you still cannot know what it is like for me to 'see red'. Similarly, there is no way I can understand what it is like to be a bat or whale 'seeing' with sound, or an Amazonian fish sensing electric fields.

Vilayanur Ramachandran argues that this difficulty arises because translating the patterns of neural activity into words loses the essence of the experience. He claims that the problem could be overcome by direct transfer of nerve impulses between brains. However, as noted earlier, perception integrates information from many areas of the brain. This means that the resultant experience depends on complex patterns of interconnection which, at least in detail, are unique to each brain. Hence, a direct cable connection could not provide an unambiguous sharing of experience even between two humans. Direct communication becomes even less plausible when we imagine being connected with the brain of an electric fish. Technology, it seems, is unlikely to resolve the hard problem.

The Development of Normal Consciousness

We saw in the last chapter that some young children have clear memories from their time in the womb, or of the birth process, or even of past lives, some of which have been verified beyond reasonable doubt. It is also quite common for young children to experience spirit beings, or intimate communion with animals. Such experiences may be more frequent than records indicate because infants cannot tell their parents about them, and adults tend to brush their childrens' tales aside as childish fantasies. Hence, the unformed infant brain appears to be able to tune in to the collective unconscious, the consciousness of other beings, and memories held beyond the brain.

Most theories of consciousness ignore these abilities, claiming we are born unconscious. According to this view, normal consciousness emerges as we learn, and as connections are created in the brain. This process is clearly evident in the life of Helen Keller, who was blind and deaf from birth: "Before my teacher came to me, I did not know that I am. I lived in a world that was a no-world. I cannot hope to describe adequately that unconscious, yet conscious time of nothingness."[3]

In most cases, transpersonal memories and abilities fade as normal consciousness develops. It seems that young infants are open to the full potential of cosmic Consciousness, but that as we learn and form habits, some of these windows onto reality close, and we become identified with an ego based on our personal memories and experiences. The rest of this section outlines the stages through which this normal consciousness develops.

Building on the work of Jean Piaget, Ken Wilber claims that there are 11 stages, or levels, of consciousness. He argues that we each have to climb this holarchic ladder rung by rung, transcending and including each stage in turn; we cannot jump steps, or move back down.[4] John Heron disagrees. From his experience, he believes we often jump between levels, for example developing higher spiritual abilities before circling back to ones that were missed, such as emotional development or relationship skills. He sees the process as an unpredictable, organic unfolding unique to each individual.[5] The transpersonal experiences of young children outlined above imply that they sometimes leap to spiritual levels (8 and above) before development of earlier stages is complete, thus lending support to Heron's case. Nevertheless, the typical sequence outlined below is instructive.

The first level is from birth to 4 months, during which time food and physical comfort are all that matter. The infant is selfish, narcissistic, unaware of the needs of others, and without language or logic. At level 2, the physical self hatches. The infant starts to differentiate its body from its surroundings, discovering that biting its thumb hurts, but biting the blanket does not. However, it still sees itself as the center of a magical world, filled with things that are alive, and which respond to its desires. Psychological birth as a separate emotional, feeling self occurs during the third level, from 15-24 months. This is the time of the 'terrible two's'.

From 2-7 years, level 4 sees the emergence of the 'conceptual self'. Between 2 and 4 the infant mind moves from working with images to using symbols. Then language skills and the ability to classify objects develop from 4-7 years, whilst belief in magic fades as the child realizes that the outer world does not respond to his desires. Often this belief is transferred to Daddy or God, and here, Piaget suggested, lies the source of the gods and supernatural beings of mythology, including Jung's archetypes. Also during this stage, the child starts to control bodily desires and discharges, and thus to differentiate bodily impulses from mental will.

The fifth level – the ability to form mental rules – develops between 6 and 14 years. This is when the child learns about social roles, and how to fit in with family and peers. She learns not only to classify things, but also to apply rules to whole classes. The sense of self becomes fully developed and clearly distinguished from the external world and its social roles. The world ceases to be an extension of her body, ceases to feel what she feels or want what she wants, and ceases to be there to meet her every desire. She no longer treats others as extensions of herself.

The mature ego develops from about 11 to 15. At level 5, the adolescent can think about the concrete world, but at level 6 he moves on to abstract concepts, and learns to think about thought itself. Introspection becomes possible, along with the ability to theorize, plan and design. Conventional rules can be judged, and 'what if?' scenarios explored. It is a time of new feelings, dreams, passions and idealisms.

At level 7 comes integration of body and mind into what Wilber calls 'vision-logic'. It is at this stage that ecological and systems thinking becomes possible through understanding relationships and holding different points of view in mind. Thought now moves beyond nationalism or anthropocentrism to what Wilber calls a 'worldcentric' view which grasps the universal principles of justice, mercy, compassion and equality based on mutual respect, rights and responsibilities. It is also at this stage that the witness reaches its full development as an integrated awareness of the body-mind in the world. This is the last of the conventional development stages, before moving into the realms of higher consciousness or spirituality.

Level 8, the psychic level, encompasses the shift from the rational, existential worldview to the transpersonal in which we may experience unity with nature or the 'world soul'. Wilber calls level 9 'The Subtle'. It is here that people encounter 'energies' such as inner lights, archetypes, angels and other spirit beings; and experience feelings of bliss, universal love and compassion, and unity with a god or goddess.

Level 10 is what I have called unity consciousness, and what Wilber calls 'The Causal'. This is the experience of emptiness, the void, the ultimate cause or source of all that is. It is a place of mental stillness, nothingness and timelessness. Finally comes the 'Non-dual' – the realization that there is no separation between matter, consciousness and Spirit; the awareness that every thing and every event is Spirit. Such awareness usually leads to engagement with the world rather than spiritual withdrawal.

We will revisit these spiritual levels of consciousness in Part VII.

The Evolution of Normal Consciousness

If consciousness is not an inherent quality of existence, then it must confer some survival advantage in order to have evolved through natural selection. But it seems that most of what we do could be carried out equally well by an unconscious zombie – and often is! This point is illustrated well by the phenomenon of blindsight in which brain damage renders the sufferer blind, but leaves them able unconsciously to perform tests that require vision, such as posting a letter in a slot. It seems that there is an old, unconscious visual pathway in the brain that remains undamaged. So why do we need conscious vision?[6] As yet, science has no definitive answer regarding the value of consciousness. But the theories discussed in this section offer some clues.

Igor Aleksander, who works on artificial intelligence, concluded that consciousness requires at least five mental characteristics:[7]

- A sense of place that locates the organism in the external world;
- A memory of past events that can be compared with the present situation;
- The ability to focus conscious attention on what is important at the moment;
- The ability to predict and plan by imagining alternative scenarios; and
- The ability to feel emotions that guide the choice of plans.

As we will see, this list neatly encapsulates several theories of consciousness. However, other researchers stress the importance of language, social cooperation, brain structure and other factors.

Even the most primitive organisms need to distinguish their own bodies from their environment. The cell membrane physically separates 'me' from 'not-me', excluding harmful substances while allowing the intake of food, energy and information. But any organism that can sense what is happening outside its membrane and react accordingly has a clear advantage in finding food and escaping predators. Hence natural selection favored the evolution of sensitivity (to light, touch, pressure, and so on) and mobility.[8]

At first, these abilities were limited. An amoeba's membrane, for instance, simply retracts from harm and engulfs food. Gradually, however, organisms became able to distinguish different stimuli. And by relaying information from one part of the membrane to another, they developed more complex responses such as swimming away rather than simply recoiling.

Most signals from sense organs remain unconscious, triggering automatic responses. But, argued Nicholas Humphrey, a signal may become conscious if it is prolonged internally, thus giving the organism time to consider and choose its response. In other words, the essence of consciousness is the combination of sensation and short-term memory.[9] In his view, other mental activities, such as thoughts, become conscious only when accompanied by 'reminders' of sensations as when we 'see' thoughts as pictures in our heads, or 'hear' them as voices.

Unfortunately, this theory is little help in deciding which organisms are conscious. We don't know enough about sensory systems to tell which organisms have the requisite time delays. Also, this mechanism implies that consciousness may develop in stages, starting in one response circuit and then spreading to others without a sharply defined threshold of full consciousness.

The importance of memory was emphasized by artificial intelligence researcher Marvin Minsky who claimed: "When somebody says they are conscious, what they are saying is 'I remember a little bit about the state of my mind a few moments ago.'"[10] Hence, the next step may have been the evolution of memories that last longer than the response to a particular stimulus. Long-term memory makes it possible to learn by comparing a fresh stimulus and possible responses to it with the results of similar events in the past. Such behaviors are slower than instinctual responses because of the information processing required, and hence may be less effective in dealing with routine events or emergencies. But they enable creative, flexible responses to environmental changes within the lifetime of the organism. Long-term memory also brings the organism a sense of continuity as the subject of its own experiences which is essential for self-awareness.

An alternative theory is that the earliest form of consciousness was the ability to feel emotions. Pleasure and pain may be ancient inventions that enable an animal to evaluate sensory data and choose between different desires and goals. They cause an animal to stop, consider its actions, and choose the one that offers the greatest potential for pleasure, or minimizes the possibility of pain.[11] Over time, these two primary emotions have differentiated into fear, anger, surprise, disgust, happiness, sadness and the gamut of human emotions. Once again, this theory requires short-term memory to give time for evaluation and decision. And long-term memory enables an organism to learn by interpreting a stimulus and the resulting emotions in the light of past experience.

This theory is supported by growing evidence that emotions are essential for effective decision-making. People whose ability to feel emotions has been destroyed lack judgment and are unable to make decisions or plans despite the fact that their memory and rational faculties are unimpaired. They 'know' but cannot 'feel'. And it is these 'gut' feelings which enable us to prioritize goals and evaluate actions. When Michel Cabanac gave his lizards the choice between warmth and food, their behavior indicated that they balance different objectives to maximize pleasure.[12] Hence, according to this theory, lizards are conscious. Recognizing the importance of emotions, artificial intelligence researchers have begun to seek ways to enable computers and robots to feel.[13]

This theory gives an interesting insight into human psychology. If pleasure was permanent, we would become addicted to pleasurable activities like lying in the sun, and fail to pay attention to threats such as lurking predators. By contrast, pain lasts as long as the cause remains potentially harmful. Hence, pleasure is always transient, and cannot be the source of enduring happiness. Pleasure brings joy but happiness springs from indifference to the outcome, or what Buddhists call non-attachment.

Antonio Damasio distinguishes between emotions and feelings.[14] In his view, emotions are physiological processes that guide behavior, whereas feelings arise when the brain consciously processes and reflects on emotions. He argues that even a fly can be happy or angry, but questions if it has feelings. He believes emotions trigger feelings only when the brain is complex enough to create an inner representation of the organism's body. Unfortunately, we don't yet know what level of complexity is required for this. Most scientists are confident that other primates create such 'maps', and some argue that all higher animals feel pleasure, fear, pain, joy, anger, embarrassment, love and grief, and hence are conscious.[15]

Feelings season the stew of life, letting us know we are alive, and making it sweet or bitter, exciting or dull. As Charles Birch put it: "The most important thing about us is that we have feelings. That is how I know that I am. What we feel is what gives value to our lives. To be alive is to feel. To be alive intensely is to feel intensely."[16] Similarly, Nicholas Humphrey claimed that: "A person may be conscious without thinking anything. But a person simply cannot be conscious without feeling. I feel, therefore I am."[17] Nevertheless, experiments have shown that human behavior can be influenced by emotions that are not associated with conscious feelings. For instance, drugs known to induce pleasure can change behavior without the recipient necessarily feeling pleasure; and subliminal exposure to pictures of happy or angry faces affects subsequent actions.[18]

Following sensation, memory, emotion and feelings, the next stage in the evolution of consciousness may have been when passive monitoring of the environment gave way to the active search for useful information. Mammals, including humans, can see things only when the image on their retina changes. Hence, they are blind unless something moves, and, in order to see continuously, their eyes flicker rapidly. Such active monitoring requires a large brain to cope with the flood of information and its interpretation, but enhances the ability to make predictions.

An organism which can predict events, no matter how crudely, has a great advantage. But in order to predict correctly, the brain must understand the nature of its environment. For instance, an object with a vertical axis of symmetry could be an animal facing you, raising the urgent question 'Is it prey, predator, mate or something else?' When such an alarm bell rings, an animal stops what it is doing, and uses all its senses and brainpower to work out what to do. This can happen unconsciously if the nervous system is programmed to respond to cues of size, shape and color. But conscious thought enables the animal to try out possible interpretations and actions in its head, and thus leads to a deeper assessment of the situation, and a more effective response.[19]

Prediction requires the answer to two questions: 'What is happening to me?' and 'What is happening out there?' Nicholas Humphrey argues that these two ways of representing events may be the origin of the split between subjective feelings and the objective world of perception that creates such difficulties for consciousness research.

In daily life, humans depend mostly on unconscious assessments of the environment based on past experience and learned skills. In this way, we avoid colliding with other pedestrians, return serve at tennis, and carry out our routine activities without conscious thought. But at the first hint of something going wrong we snap into conscious attention. Hence, effective prediction and response are not unambiguous indicators of consciousness, and cannot enable us to tell which organisms are conscious.

Marian Stamp Dawkins argues that any animal that tries to change the world to suit itself is likely to have subjective experiences like ours. On this basis, she concluded that chickens, the subjects of much of her research, are conscious and have feelings.[20] Others suggest that consciousness is indicated by versatile behavior and creative problem-solving, including tool use. My family used to live in a house with lever door handles which had to be pulled down to unlatch the door. With great determination, one of our cats discovered how to leap up, grab the handle with his front paws, and pull it down as he fell back to the floor. And we saw in Chapter 18 how Betty the New Caledonian crow designed and made a tool to solve a novel problem. It is hard to see how such innovations could be possible without awareness of the situation and the ability to imagine possibilities; or without intention, will and memory.

Some theorists argue that full human consciousness had to await the development of language. As described in Chapter 18, many primates share information by calling to each other, thus helping both individuals and the group to survive. Daniel Dennett suggested that perhaps a hominid called for help one day and no-one heard. But, rather than being in vain, that call stimulated a helpful vocal response from the animal itself. This is possible if the information was available in its brain, but there was no internal link to make it accessible. The audible question and response could create an external channel via the ears. Such a valuable trick would spread quickly. The advantage of not drawing attention to themselves by calling aloud could then lead to the evolution of an internal link and silent questioning.

John McCrone argued that the evolution of this inner voice was the critical event in the emergence of human consciousness.[21] He believes that animal minds cannot recall things without a stimulus from their environment to trigger an associated memory. However, internalized speech enables humans to use words like handles to pull out particular memories. In this way we can live and relive any moment at will. A slightly different hypothesis links the emergence of consciousness to the ability language gives us to describe ourselves and our situation, and to communicate it to others. This, it is suggested, leads to the awareness of ourselves as separate beings which is the hallmark of consciousness.[22]

Yet another theory suggests that it was the development of social groups, not language, that drove the evolution of self-awareness. Each individual in a group is forced to adjust to others. This leads to an awareness of the distinction between self and others, and hence to both individual and group consciousness.[23] Another way of expressing this is that consciousness enables us to put ourselves in someone else's position so that we can empathize with them, and predict or manipulate their behavior. A similar theory suggests that the way the group reacts to us provides us with an image of our personalities like the reflection in a mirror. We then use these reflections to form our sense of identity and self-awareness.[24] Such theories imply that all animals which *learn* social behavior, rather than adopting it by instinct, must be conscious.

Robin Dunbar claims that primates differ from other animals in the complexity of their social relationships. One of the key differences is that apes and humans deliberately mislead or deceive others by anticipating how they will respond. They do this through inner reflection, or what Dunbar calls *intentionality*. Most mammals and birds probably have first-order intentionality which is indicated by statements such as 'I suppose ...', 'I think ...', 'I wonder ...' or 'I believe...'. Second-order intentionality involves reflection on someone else's mind: 'I suppose that you think ...' Mature great apes reach this level, as do human children at about 5 years old. Adult humans, however, can handle fifth-order intentionality: 'I suppose that you believe that I want you to think that I intend ...'!

It turns out that this ability depends on the size of the frontal lobe of the brain, which was the last to evolve and is generally associated with consciousness. However, Dunbar argues that no new abilities are involved beyond those possessed by all mammals and birds. Rather it is the level of ability that has increased in primates. They can reason better, and predict further into the future. And he suggests that high-level intentionality is simply an emergent property of these abilities.[25]

From this discussion, high-level intentionality appears to be a unique feature of human consciousness. But many earlier attempts to define the uniqueness of humanity have foundered as research has revealed the ability of animals to learn, make tools, plan, solve problems creatively, and communicate complex messages. Whether or not this one will survive remains to be seen.

Yet another claim for human uniqueness comes from science fiction writer Brian Aldiss. He argues that, unlike animals, humans have an extended consciousness that enables us to foresee our own deaths, and reflect on our births and the birth of the universe. He suggests that this ability evolved for our pleasure. It "adds to the pure biological enjoyment of being alive. It assures us we are alive, while the squirrel is alive but cannot realize the fact."[26] But how can we be sure that squirrels don't know they are alive?

So far in this section, our focus has been on mental attributes that might have led to, or require, consciousness, rather than on the brain structures and processes that would make consciousness possible. Neuroscientist Vilayanur Ramachandran concluded that consciousness may result from a summary overview of unconscious processes in the brain. Such an overview, he suggests, could be held in a separate area of the brain, be distributed in a network, or result from particular brain processes. He likens it to the long-discredited idea of a little man sitting in our heads and watching a screen on which qualia are projected.[27]

Mathematician Roger Penrose took a different approach. He claims that conscious human thought is often 'non-computable' in the sense that it cannot be simulated using logical rules. Taking mathematics as an illustration, he interprets Gödel's theorem (see Chapter 1) as stating that no set of rules can ever prove basic propositions of arithmetic whose truth is intuitively obvious to humans.[28] In other words, "mathematical understanding is something different from computation and cannot be completely supplanted by it."[29] Other examples of non-computable thought include aesthetics, morals, strategic evaluation of situations in chess, and interpretation of ambiguous words. The difficulty of developing computers capable of playing world-class chess or interpreting natural language lends support to this idea.

However, all the known laws of nature are computable, leaving Penrose searching for an undiscovered, non-computable process. His candidate is a quantum computer hidden within the microscopic tubular skeleton of each cell. On the basis of rough calculations, he concluded that individual cells probably are not conscious, but that any large collection of cells, such as a reasonable sized brain, could be.[30] According to Penrose, these computers may not be the source of consciousness, and may simply amplify quantum processes to a level at which they can influence the body. In other words, consciousness may originate outside the brain. Penrose rejects the possibility that it is a separate, non-material substance, but nevertheless his ideas are compatible with those of Amit Goswami regarding cosmic Consciousness and quantum processes as described in Chapter 18 and the next section.

Theories of Transpersonal and Unity Consciousness

We saw in the last chapter that normal consciousness is no more than a sample of our potential. Our minds reflect what our culture tells us is real, neglecting or rejecting the rest, and becoming shrunken and distorted in the process. It is as if we are hypnotized, unable to lift our arms because we have been told we cannot. In the words of Alan Watts, we experience ourselves as 'skin-encapsulated egos', limited by our senses and environment.

We are unable to see through walls or beyond the horizon; are blind to all but a tiny fragment of the electromagnetic spectrum; and are deaf to the thoughts of friends across the ocean, the ultrasound of bats or the infrasound of whales. But we transcend all these limits and more when we enter transpersonal and unity consciousness. How can this possibly be?

The only theories that make sense of this question hold that consciousness is inherent in all matter rather than emerging when matter achieves sufficient complexity. In the words of Christian de Quincey, nature is "sentient to its roots, composed of matter that *feels* something of the nature of wholeness and love all the way down (to whatever lies at the root of physical reality) ... (A)ll matter has its own interiority, an ability to feel, to have a point of view, and the ability to move itself from within. ... matter is (and always was) sentient, 'alive.'"[31]

Many mystical traditions believe there is a cosmic Consciousness which is the ground and source of everything: space and time, matter and energy, mind and consciousness.[32] This idea, introduced already in Chapters 11 and 18, is sometimes likened to the light from a movie projector. Our perceptions, sensations, dreams, memories, thoughts and feelings are like the images in a film that are projected onto the screens of our minds. Watching the movie, we get caught up in the images and are unaware of the light itself which makes the experience possible. The images are the contents of consciousness, not Consciousness itself which is the light from the projector without a film – luminous but intangible, insubstantial, and empty of content.[33]

Many spiritual practices seek to empty and still the mind until there is pure awareness with no content or object. Hindu and Buddhist contemplatives claim that the resulting experience is the light of the underlying Consciousness that is the source of existence. They believe that human consciousness emerges from this substrate, and not from the brain.

Once it is associated with a physical form, Consciousness becomes constrained and shaped by it, whether it be a rock, tree or human. The nature of awareness becomes defined by the 'rockness', 'treeness' or humanity of its material shell. And the very existence of that particular consciousness becomes dependent on its material form. But correspondingly the rock, tree or person is plugged into cosmic Consciousness, participating in the creation of reality through the collapse of quantum possibilities as described in Chapter 18.

Using different imagery, Stanislav Grof suggested that our consciousness may be a field, encapsulating and reflecting the whole universe, just as a fragment of a holograph contains the whole picture.[34]

> (I)n a mysterious and as yet unexplained way, each of us contains the information about the entire universe and all of existence, has potential experiential access to all of its parts, and in a sense is the whole cosmic network, as much as he or she is just an infinitesimal part of it, a separate and insignificant biological entity. The ... individual human psyche (is) essentially commensurate with the entire cosmos and the totality of existence.

In similar vein, David Peat suggested that:[35]

> Just as the elementary particle unfolds out of the quantum field, the soliton appears in the nonlinear field, and the vortex emerges from the river, so may an individual consciousness emerge out of the ... consciousness that extends into the whole universe. The individual mind is therefore a sort of localization or concentration of consciousness that unfolds into the brain and body of the individual.

Whichever imagery we prefer, this approach to consciousness has no difficulty dealing with the otherwise intractable issues of qualia, free will, and transpersonal consciousness. The hard problem of qualia simply fades from view because we are no longer trying to explain how consciousness could have arisen from unconscious matter: subjective experience is potential in everything. Likewise, the question of how a non-material thought or desire can influence the material world disappears if we accept that mind and consciousness are the root causes of all material objects and processes.

In Chapter 9, we explored the idea of an implicate order or psi field that retains a memory of every event in the cosmos. It seems reasonable to assume that this field is at least an aspect of cosmic Consciousness. Hence, every action of ours is imprinted in cosmic Consciousness, from where it can instantly affect events remote in space and time, possibly through the agency of quantum entanglement or the intricacies of spacetime. Access to, or resonance with, this memory is all we need to connect with the collective unconscious, past lives or other lives.[36] Here, too, lies a possible mechanism for telepathy and prayer, and effects beyond time such as retroactive prayer and precognition.

There is not space to explore these theories more thoroughly here. And a lot of work is needed to clarify how Consciousness can give rise to matter and physical processes. However, this emerging field of study offers the best hope of understanding the full scope of human consciousness.

In Conclusion

Mainstream theories of consciousness provide valuable insights into its nature and evolution. And neurosciences are making rapid strides in revealing how the brain works. Hence, it seems likely that we will have a far better understanding of the mechanics of mind and consciousness within a few years. But most theories are based on the premise that consciousness is created by complex brains, thus forcing scientists to deny the reality of transpersonal experiences, many of which are well verified.

It seems that any approach that seeks a comprehensive theory of all aspects of consciousness must be based on the premise that the brain mediates consciousness rather than creating it. This means that consciousness arises outside the brain, and is inherent in all matter, either as a parallel aspect of reality or as the fundamental reality from which all material existence springs.

Review of Part VI

Humanity is split by a deep philosophical chasm. On one bank are the forces of materialist science and philosophy holding banners proclaiming that energy and matter are the source of everything including mind and consciousness. And facing them are the hosts of believers in the primacy of cosmic Mind and Consciousness, of Spirit, God, Tao, the Source of all existence. There is no objective, rational proof that one or the other belief is true, and there can be none. Once again, we must weigh the evidence, consider the arguments, listen to our intuition, and choose the belief that makes most sense to us; the belief that resonates most strongly with our inner being.

For me, there is no contest. My experience as a conscious, living being tells me that I am imbued with a spirit that permeates all humanity, all life, all matter. The idea of cosmic Consciousness resonates with my innermost self, bringing meaning and purpose to my life, and a deep sense of belonging and connection. By contrast, the materialist view seems sterile, lifeless, bleak, leaving me alienated from myself and the world, without hope or meaning. It is possible that I am wrong. It is possible that consciousness is no more than a meaningless by-product of individual brains, and that transpersonal experiences are illusions arising from their structure. But we cannot know for sure, and I would rather put my money where my heart is.

This belief turns scientific reality on its head. But, in one stroke, it also dissolves the hard problem of consciousness, and resolves the puzzle of how it could have emerged from matter. The science of cosmic Consciousness and its relationship to the material universe is in its early infancy. But, given how little effort has been devoted to it so far, it shows at least as much promise as the physicists' dream of a Theory of Everything – a pursuit that is faltering, and does not even recognize the question of consciousness.[1]

Today, it is the turn of neuroscientists to be gung-ho, riding a wave of optimism whipped up by a storm of new brain imaging technologies. Their spotlight is penetrating the murky depths of the brain, and many of them are confident that soon they will have revealed the secrets of the mind and the source of consciousness; that they will be able to 'see' what we're thinking, analyze our values and ethics, and control our minds. But will they? They do well to remember the similar boasts of behavioral psychologists fifty years ago, and the more recent claims for artificial intelligence. It is interesting to note that the grand dreams of artificial intelligence quietly faded from the limelight in the same decade that neuroscience burgeoned.[2]

Undoubtedly neuroscientists will learn a lot about the nature of mind, and many of our preconceptions and pet theories may be overturned. But even when they fully understand the intricate mechanisms of the brain, it will not be certain that it creates consciousness rather than mediating it. Nor can they escape the challenges presented by the uniqueness of each brain, the interactive whole which it forms with the body and environment, and the mind's extension far beyond the brain-body in transpersonal experiences. My guess is that the hard problem of consciousness will be with us for a while yet.

The idea of cosmic Consciousness sheds light on many observations we have made on our journey, and begins to integrate them. It makes sense of the purpose and creativity displayed even by simple systems, and of the holarchic connections between all that exists. It brings a new and deeper perspective to the role of consciousness in quantum processes, and the timeless, spaceless entanglement of particles. It offers explanations for the direction of cosmic evolution, the apparent existence of a 'law of complexification', and the fine-tuning of the cosmic constants. And it makes sense of our inner awareness and spiritual hunger.

From this perspective, we are collectively co-creating reality moment by moment. At the quantum level, we help to collapse the probability wave; and at the macro-level we inevitably influence the sensitive chaotic systems around us, and the very process of evolution itself. In the words of Andrew Powell:[3] "Our task is to bring the instrument of human consciousness to bear on the quantum wave with the greatest care, for whether we do it in love or hate literally determines whether we create heaven or hell." We will explore this perception in more depth in the next Part, together with the nature of Spirit and our relationship with it.

Reflections on Part VI

This section opens with a reflective discussion of the relationship between mind and matter. This is followed by reflections on the reality and experience of transpersonal consciousness, and several evocative descriptions of unity consciousness. For those interested in accessing other states of consciousness, an introduction to mindfulness meditation is then given. Finally, the section closes with a warning and some advice about the potential power of transpersonal experiences. Enjoy your selections from the smorgasbord!

The Relationship of Mind and Matter

> We are what we think.
> All that we are arises with our thoughts.
> With our thoughts we make the world.
>
> The Buddha[1]

The Buddha's insight is mirrored by Amit Goswami's theory of cosmic Consciousness, outlined in Chapters 18 and 21. These ideas are so at odds with the accepted reality of our materialist culture that you may wish to reflect further on them.

Did the Buddha mean that we *literally* make the *material* world with our thoughts? If not, what did he mean? In what ways is our world the product of our minds?

Many eminent scientists and thinkers besides Amit Goswami have claimed that the cosmos more closely resembles a mind than a machine. Instances include:

... the stuff of the world is mind-stuff.

Sir Arthur Eddington[2]

(The evidence) makes it seem more and more likely that reality is better described as mental than material ... the universe seems to be nearer to a great thought than a great machine.

Sir James Jeans[3]

God is not the creator, but the mind of the universe.

Erich Jantsch[4]

Matter is merely mind deadened by the development of habit to the point where the breaking up of these habits is very difficult.

C. S. Peirce[5]

(I)n the absence of mind ... the world ... is formlessly present, like a song that waits for a singer to sing it. ... Each act of seeing creates a world.

Darryl Reanney[6]

But how can this be? Things do happen when there is no-one there to see. Trees do fall when there is no-one there to hear. The material universe burst into being before there was consciousness to observe it. Earth formed before life emerged. And reality appears stable, solid, not the evanescent fantasy of myriad minds.

Does this mean the Buddha and physicists are wrong? Or could it be that there is a cosmic Consciousness that has materialized our universe since time began? Could it be that this cosmic Consciousness chose not to construct a universe complete and fully formed according to its design, but to empower a creative process, and to see what would emerge? Could it be that cosmic Consciousness chose to subordinate its creative power to the free will of independent beings rather than impose its will on creation? And could it be that the stability of the material world is no more than a reflection of the habits of conscious beings?

How do you feel about these ideas?

Perhaps they strike a chord somewhere deep within you?

Or perhaps they seem like a rehashing of outworn religious mumbo jumbo? If so, you might like to ask yourself if it is conceivable that consciousness could have sprung from unconsciousness? Is it possible that raw, inanimate matter, the dust of the earth, could not only come alive, but also become aware of itself?

We create the material world with our minds in a more everyday sense as well. Scientists create atoms, molecules, materials and organisms new to nature. Engineers create structures, machines and electronic devices which have never existed before. Architects and planners create buildings and cities. Artists create paintings, sculptures, music, dance, poetry, ... We are rapidly transforming the Earth into the likeness of our visions and nightmares. Every day, each one of us uses our mind to create objects and processes that did not exist before we thought of them and told our bodies to act on the thought. You might like to list some examples from your daily life.

And there is yet another sense in which we create the world with our thoughts. As we saw in Chapter 2, what we know and experience is not reality as such, but a mental model of it. Change the model, and our world changes too. Build your model on the foundations of materialism, and consciousness appears to be an unnecessary and philosophically troublesome byproduct. But base your model on the inherent consciousness of all matter, and suddenly cosmic Consciousness appears. How can we make sense of this hall of mirrors? Where can we find solid ground on which to base our lives?

The Experience of Transpersonal Consciousness

According to psychologist William James:[7]

> (M)ost people live ... in a very restricted circle of their potential being. They make use of a very small portion of their possible consciousness ... much like a man who, out of his whole bodily organism, should get into a habit of using and moving only his little finger. ... We all have reservoirs of life to draw upon, of which we do not dream.

Mainstream western culture distrusts and fears any experience which is beyond the bounds of normal waking consciousness and our limited scientific knowledge. Those who experience transpersonal realms often are diagnosed as mentally ill, and treated with drugs to suppress their 'symptoms'. And in the past they were hunted down and burnt as witches.

In other cultures transpersonal experiences often are recognized as valued gifts, sometimes part of a process of spiritual transformation which fits the recipient for spiritual leadership.[8]

Why do we treat other realms of consciousness in this way? Is it the desire of psychiatrists and psychologists to protect their image as experts? Is it fear of the unknown and inexplicable? If so, why do we fear them? What would happen if we embraced them, supporting and valuing those who undergo such experiences?

We demonstrate the power of mind over matter every time we lift a finger or utter a word; every time we use a technology; and every time we admire a work of art. The key difference between creating a picture by talking to a computer and bending a spoon by psychokinesis is that we don't understand how the latter works. But voice-controlled computers would have seemed supernatural to a scientist of 100 years ago. Are we sometimes too quick to dismiss phenomena we can't explain?

We accept without question the existence of memories stored outside our brains in books, photographs, magnetic tapes, computer disks, CD's, DVD's, and electronic chips. We read memories of our ancestors preserved in deserts, bogs and ice; of extinct organisms immortalized in stone; of past climate recorded in sediments and ice; of earth's history encoded in the rocks. And we read cosmic memories, too: of the birth and death of stars and galaxies, and of the origins and history of the universe itself.

Why, then, does it seem far-fetched that human memories should be stored in morphic fields? Or that the psi field should encode the whole history of the cosmos? And why shouldn't the human mind be able to read these memories? Are we again in danger of dismissing possibilities too soon?

The following quotations go some way towards communicating the essence, beauty and power of transpersonal and unity consciousness. I suggest reading them slowly and meditatively.

To know that what is impenetrable to our senses really exists, manifesting itself as the most profound wisdom and most radiant beauty, which our dull faculties can comprehend only in their most primitive forms, this knowledge, this feeling, is the center of true religion.

Albert Einstein[9]

At one point, I felt cold air streaming through my head and had a taste of salty water in my mouth. A variety of (alien) feelings ... took over my consciousness. A new, gigantic body image started to form out of the primordial connection to the other large bodies around me and I realized I had become one of them. Inside my belly I sensed another life form and knew it was my baby. There was no doubt in my mind that I was a pregnant whale cow.

And then came ... the birth process. ... It had gargantuan proportions, as if the ocean were stirred from its very depth; at the same time it was surprisingly easy and natural. I experienced my genitals in the most intimate way, with all the nuances of these birthing activities associated with profound visceral understanding of how whales give birth. What I found most amazing was how they use water to expel the baby by sucking it into their genitals and working with hydraulic pressure.

Belgian woman[10]

My body actually had the shape of the Sequoia tree, it was the Sequoia. I could feel the circulation of sap through an intricate system of capillaries under my bark. My consciousness followed the flow to the finest branches and needles and witnessed the mystery of communion of life with the sun – the photosynthesis. My awareness reached all the way into the root system. Even the exchange of water and nourishment from the earth was not a mechanical but a conscious, intelligent process.

... photosynthesis was not just an amazing alchemical process, it was also direct contact with God ... The natural processes such as rain, wind, and fire had mythical dimensions ...

I had a love-hate relationship with Fire, who was an enemy as well as a helper, cracking open my seed pods for sprouting and burning out other vegetation on the forest floor that might compete with my new growth. Earth itself was a goddess ... and her soil was permeated by gnomelike beings, fairy-like creatures, and elementals. ...

The deepest level of the experience was purely spiritual. The consciousness of the Sequoia was a state of profound meditation. I felt amazing tranquility and serenity, as a quiet, unperturbed witness of the centuries.

Client of Stanislav Grof[11]

The Essence of Unity Consciousness

To see a World in a grain of sand,
And a Heaven in a wild flower,
Hold Infinity in the palm of your hand,
And Eternity in an hour.

William Blake[12]

I lay on the bowsprit, facing astern, with the water foaming into spume under me, the masts with every sail white in the moonlight, towering high above me. I became drunk with the beauty and singing rhythm of it, and for a moment I lost myself – actually lost my life. I was set free! I dissolved into the sea, became white sails and flying spray, became beauty and rhythm, became moonlight and the ship and the high dim-starred sky! I belonged without past or future, within peace and unity and wild joy, within something greater than my own life, or the Life of Man, to Life itself! To God, if you want to put it that way ... like the veil of things as they seem drawn back by an unseen hand. For a second, there is meaning.

Eugene O'Neill[13]

Crossing a bare common, in snow puddles, at twilight, under a clouded sky, without having in my thoughts any occurrence of special good fortune, I have enjoyed a perfect exhilaration. I am glad to the brink of fear. Within these plantations of God, a decorum and sanctity reign, a perennial festival is dressed, and the guest sees not how he should tire of them in a thousand years. ... Standing on the bare ground ... I become a transparent eyeball; I am nothing; I see all; the currents of Universal Being circulate through me; I am part or parcel of God.

Ralph Waldo Emerson[14]

The Lord of all,
the knower of all,
the beginning and end of all –
that Self dwells in every human heart.
Look out – it's gone.
Look in – it's gone.
Don't look – it's gone.
It cannot be remembered,
it cannot be forgotten,
it cannot be grasped by any possible means.
It is beyond all limits and bounds.
It is the pure oneness
where nothing else can exist.
To know it, you must become it!

Upanishads[15]

(T)here came upon me a sense of exultation, of immense joyousness accompanied or immediately followed by an intellectual illumination impossible to describe. Among other things, I did not merely come to believe, but I saw that the universe is not composed of dead matter, but is ... a living Presence; I became conscious in myself of eternal life. It was not a conviction that I would have eternal life, but a consciousness that I possessed eternal life then; I saw that all men are immortal; that the cosmic order is such that ... all things work together for the good of each and all; that the foundation principle of the world ... is what we call love ...

Now came a period of rapture so intense that the universe stood still, as if amazed at the unutterable majesty of the spectacle. Only one in all the infinite universe! The All-loving, the Perfect One. ... I saw with intense inward vision the atoms or molecules, of which seemingly the universe is composed ... rearranging themselves, as the cosmos (in its continuous, everlasting life) passes from order to order. What joy when I saw there was no break in the chain – not a link left out – everything in its place and time. Worlds, systems, all blended into one harmonious whole.

R M Bucke[16]

It's as if everything was there and everybody was there.
The sense was of absolute, total fulfillment.
And yet there was no sense that I was there.
That's the most extraordinary thing – John vanished at that moment.

John Wren-Lewis[17]

Profound and tranquil, free from complexity,
Uncompounded luminous clarity,
Beyond the mind of conceptual ideas;
This is the depth of the mind of the Victorious Ones.
In this there is not a thing to be removed,
Nor anything that needs to be added.
It is merely the immaculate
Looking naturally at itself.

Nyoshul Khen Rinpoche[18]

Unity consciousness is not restricted to spiritual masters. The following short quotations come from a survey of experiences of cosmic consciousness conducted by Pearl Hawkins in 2003.[19]

It was as if I had an eye inside every atom ... since whatever 'I' was seemed to look out of everything, every piece of matter, every being ...; I was all beings ... I was within galaxies exploding ... I was events unfolding ... Knowledge of many things ordinarily unknowable was perceived by whatever was perceiving.

It seemed to me that I saw and understood everything that had ever been or would be, all at the same time, and that all was well. What I saw was good and the source of itself, constantly there. No doubt. No need for anything else because it was everything.

This knowingness is a deep understanding that occurs without words. I am certain that the universe is one whole and that it is benign and loving at its ground. ... The world seemed benign and 'right' with everything as it was 'supposed to be'. There was a great sense of inner peace.

No space remained – no me, no you, no them, or they, only ONE.

For one instant ... it was as if I was one with everything that existed, every atom, every stone, every word, every star, seeing creation ... from the inside out.

You can no longer say 'I believe', you must say 'I know'.

Mindfulness Meditation

One of the simplest and most ancient ways to make contact with transpersonal states of consciousness is mindfulness meditation. This seeks to empty the mind of its ceaseless chatter, and open our awareness to other realms.

The following guided meditation may help you experience a state of calm awareness. Don't be disappointed if nothing seems to happen – most people need months or years of regular practice.

If you wish to try it, choose a time when you will not be disturbed for at least 30 minutes. Find a comfortable place to sit, with your back upright and straight, but relaxed. Ideally, meditate with a friend and read it aloud to each other in turn. Or make a recording of yourself reading it so that you can focus on the meditation. Either way, remember to read it slowly, and with pauses for contemplation. If neither of these is possible, I suggest you read the meditation through as many times as you need to get a sense of the process, and then sit and observe your breath without reading it.

As you sit, allow your body to breath naturally. Don't try to change the depth or speed of the breath. And notice where you can feel the breath touch you as it enters and leaves the nostrils.

Bring your attention to this sensation as the air flows in and out. Watch it like a guard at the door who notices who goes in and out. Become aware of the subtleties of movement and sensation.

Keep your attention at this one point and notice how it feels as each breath passes in and out of the body in its natural rhythm.

If you notice your attention has strayed, bring it gently back to the point of touch, noticing the breath as it comes and goes through the nostrils.

Simply note what is happening: "breathing in" " breathing out" without thinking about the breath or visualizing it. Just be with the sensation of the flowing air touching the nostrils.

Allow other sensations and thoughts to float in the background of awareness, arising and passing away. Sounds happen. Thoughts come and go. An itch or tingling arises in your body, and fades again. Just happening, without attention.

In the foreground, gently hold awareness of the breath coming and going. Not pushing away thoughts, sounds and other sensations. Not forcing attention to the breath. Just relaxed, gentle observation. Being mindful of breathing.

Allow sensations and thoughts to arise and disappear like bubbles in water.

Allow the distractions of the outside world, the chattering of the mind, the discomforts of the body to flow past without anger or frustration, without chasing them.

Allow each moment to arise and pass away of its own momentum. No pushing away of thoughts, no hanging on to the breath. Just gently bringing awareness back to the sensations of the breath whenever it strays. Gently but persistently returning.

The awareness of breath in the foreground. Everything else in the background; the open, soft mind not sticking to anything.

Notice that each breath is unique: sometimes deep, sometimes shallow, always changing slightly.

Feel the whole breath going in, pausing, and coming out; the whole breath experienced as sensation, as touch.

Breathing happening by itself. Awareness simply watching.

The whole body relaxed. Feel the softness of the eyes, lips, face. The looseness of the neck and shoulders. The belly softly rising and falling. The buttocks, legs and feet relaxed. No tension anywhere.

Just awareness and breathing.

Just consciousness and the object of consciousness, arising and passing away moment to moment in the vast space of mind.

When the mind drifts away, return it gently to the breath, with no judgment, no clinging.

Note the whole breath, from its beginning to its end, precisely, clearly, from sensation to sensation.

The body breathing itself. The mind thinking itself. Awareness simply watching without getting caught in the content.

Each breath unique. Each moment completely new.

When a sensation arises, let awareness recognize it as sensation. Notice it coming and notice it going, without thinking of it as body or leg, as pain or tingling. Simply note it as sensation and return to the breath. And, if necessary, gently and mindfully move to a more comfortable position or scratch the itch. Absolute stillness is not necessary.

Let go into the breath. Experience the breath fully, without trying to get anything from it. Don't think of concentration. Just allow yourself to be aware of sensations arising of themselves and by themselves.

The touch of the breath becoming more distinct, more intense with each breath.

The mind focusing on the sensation of breathing.

Noting thoughts rising like bubbles, and flowing through and out of the mind. Noticing them, and returning to mindfulness of breathing.

When thought or feeling intrudes, softly note it as "feeling" or "thinking," as "hearing," "tasting," or "smelling." Then gently return to the breath.

Don't linger with thoughts or identify their contents. Just note the experience of thought or feeling or sensation arising one moment and passing away the next.

Always return gently to the flow of the breath. Not clinging to anything, and not pushing anything away. Just a clear awareness and acceptance of all that arises.

Thoughts arise and are noted within the focused awareness of breathing.

Returning more and more deeply to the point of sensation that marks the passage of each full breath.

Becoming aware of subtler and subtler sensations.

The eyes soft. Shoulders soft. Belly soft. The awareness clear.

Watch the flow of the mind, continually switching like a kaleidoscope from object to object, breath to breath, sensation to sensation.

Moment by moment thoughts arise and pass away in the space of mind. Sensations arise and pass away in the body.

Relaxed, open awareness watching the process of arising and passing away. Awareness of whatever predominates, always returning to the sensations of the breath.

When feelings arise, name them: "sadness" "itching" without becoming absorbed by them.

When thoughts arise, name them: "planning" "judging" without getting caught in the content.

Experience their movement through the mind with awareness. Words and images arising from nothing, disappearing into nothing. Just an open space in which the mind and body are experienced as a flow of change from moment to moment.

Sound arises and passes away.

Feeling arises and passes away.

All of who and what we are, coming and going like bubbles in the mind.

Arising and passing away in the vast, open space of mind.

All things which arise also pass away.

Everything we think of as "me" disappears moment by moment.

Moment by moment, just seeing it all as it is, coming and going of itself.

A star at dawn, a bubble in a stream, a flash of lightning in a summer cloud, a flickering candle – phantoms in a dream.

Continue to sit, breath and watch if you wish.

And when you're ready, slowly and gently bring yourself back to awareness of what is happening around you, and to normal consciousness.

Take time to reflect on your experience.

You may wish to write or draw about it.

The Power of Transpersonal Consciousness

Experiences of transpersonal and unity consciousness are normally beautiful and transforming. But sometimes they can overpower normal consciousness, and become frightening and disorienting. A small minority of people experience dramatic spiritual openings (sometimes called Kundalini experiences) that are treated as psychotic episodes leading to hospitalization.

At such times, the help of a spiritual guide or therapist experienced in spiritual work is essential to complement mainstream medical expertise, to ground the experience in reality, and to complete the spiritual transformation. There are established support networks for this in many parts of the world, some of which are listed in the note.[20] Even when such rare difficulties are experienced, the long-term effects are usually beneficial, with reports of enhanced consciousness, including improved intuition, clairvoyance and clairaudience; the emergence of healing powers and prophetic perception; and increased sensitivity and love to others.[21]

Part VII:
Spirit and Spirituality

At intervals in our journey, we have caught glimpses of another reality underlying the material world we know; glimpses through the windows of self-organization, the holarchy of nature, the quantum vacuum, space-time, the implicate order, the psi field, life, evolution, cosmic Consciousness ... In this final stage of our journey, we seek a clearer, more holistic vantage point from which to view the nature of this reality and our place within it. In particular, we will explore four questions: What is the nature of ultimate reality, or Spirit as I have called it? What is our relationship with Spirit? Who am I? And why am I here?

Setting off on this stage of the journey, I feel I'm walking a narrow ridge. On one side is a slippery slide into yet another version of the shrinking "God of the gaps"; a Spirit with space to exist only because science has not yet completed its task. On the other side, is the trap of withholding some of what I believe because it is not, and cannot be, proven by scientific criteria. I want to share what I have discovered, whilst leaving the door wide open for other interpretations, other conclusions.

It is impossible in the confines of a few chapters to present a balanced survey of all scientific, mystical, religious and philosophical views of the nature of ultimate reality, even if I were competent to do so. Instead, I've adopted a particular perspective, supported as far as possible by others. Hence, this Part of the book is more personal, more partisan, perhaps more polemical than the rest. And it relies more heavily than earlier Parts on intuitive and spiritual knowing as described in Chapter 3.

I have attempted to integrate the scientific vision emerging from the earlier parts of the book with various spiritual traditions and theories. But my approach is open and non-dogmatic, leading to a way of seeing all religions and spiritual paths as potentially valid. It is a perspective that honors the freedom to choose. Naturally, it is a perspective which makes sense to me, and with which I resonate deeply. And it is one which I hope will inspire you, my readers, too. But my journey of exploration into the science of oneness is far from over, and it is possible that some of my ideas will have changed even before this book is published!

So come with me as we explore the road not taken by science – at least so far – and a way of co-creating the future.

In Chapter 22 we discuss what can be known about the Mystery of Spirit. Chapter 23 then explores the diversity of human relationships to Spirit, before continuing in Chapter 24 to present a model of spirituality for today. The next two chapters discuss the perennial human questions 'Who am I?' and 'Why am I here?'. The Part ends as usual with Review and Reflections sections.

22 The Nature of Spirit

What is Spirit?

At the heart of existence lies absolute and unfathomable Mystery. Whether we use the methods of science or philosophy, religion or mysticism, we arrive eventually at the same question: Why is there something rather than nothing? Cosmologists imagine an ocean of pulsating energy from which the universe sprang. But what is the source of these pulsations? Plato saw the material world as shadows of an Ideal realm cast upon the walls of our cave. But where did these Ideals come from? Christians believe in an uncreated Creator, an uncaused Cause. But why should such a Being exist? And mystics encounter cosmic Consciousness, the One Mind, the Source and Ground of all Being. But what is this?

No matter how far back we trace the thread of existence, we always end at a blank wall of incomprehension; awestruck and speechless in the face of the Mystery. Where did THIS come from? I feel as if I'm standing on the edge of a cliff gazing into the mists of time and space, into a mental black hole which sucks in all understanding. I feel vertigo. I daren't look too deeply lest I lose myself in the abyss. Why is there something rather than nothing? Why should there be this potential that turns itself from nothing into a universe? Of necessity, I must accept that IT JUST IS. And unavoidably, I AM THIS.

Our souls shy away from the depths of this Mystery. We draw back from the brink, inventing stories to distract ourselves – stories of a big bang, of God or the Dreamtime. We paint a mural on the blank wall of our incomprehension; and project images onto the impenetrable mist.

The Mystery is Absolute. And yet we can perhaps detect vague outlines of Reality within the mist. For at the deepest levels, we are manifestations of the Mystery, and we share Its nature. Hence our understanding of the ultimate questions of life can be enriched by probing the fog with the light of our minds and inner knowing.

Humanity has given many names to this Mystery, from amongst which I have chosen to call it 'Spirit'. 'Spirit' is the motivating principle or potential of the cosmos. Spirit just *is*, existing beyond time and space and matter. In this chapter, we gather together the fragmentary evidence for the nature of Spirit.

The Nature of Spirit

So powerful is the hold of science over the modern mind that we swallow with scarcely a blink speculative theories with 10 dimensions or a frothing foam of universes. By contrast, mystical descriptions of transcendent spiritual realms often are regarded with suspicion, and rejected as superstition. But science and mysticism are complementary approaches to knowledge of Spirit which combine to yield a richer understanding. What, then, can we deduce about the nature of Spirit?

The fount of all existence is an unknown essence imbued with creative Potential. This Potential is called by many names: the quantum vacuum, the implicate order, the psi field, Spirit, God, Goddess, cosmic Consciousness, the Tao, the Ground of Being ... But whatever we call it, the core idea remains the same. There is a mysterious something which can turn itself into something else. What is this Potential and where does it come from? We cannot know. It just IS.

Perhaps this Spirit is eternally, timelessly, ceaselessly creative. Or perhaps it has the power to rouse itself from slumber, unleashing its own Potential. Either way, Spirit has the power to dream our universe into being through the big bang, divine creation, or other means. It has the power to create time and space, energy and matter, life and consciousness, and all the processes of cosmic evolution.

Cosmologists tell us our universe may not be alone. An infinity of universes may have sprung from the Potential of Spirit. Ones that are fundamentally different to our own, as well as those embodying the myriad evolutionary paths ours did not take. Failed universes which collapse almost before they are born, as well as ones flourishing in unimaginably alien ways. Is this boundless creativity pure playfulness, or does it emanate from some mysterious purpose? Both ideas are represented in mystic intuitions.

We cannot hope to understand why Spirit creates. And yet, as manifestations of Spirit's potential, we share Its essence, and may catch glimpses of Its nature and motivations within ourselves. Perhaps our desires and intuitions reveal fragments of the Truth. Our deepest urges include a love of play; a passion to create beauty and novelty; the curiosity to explore, experiment and understand; a longing to know ourselves and realize our potential. Perhaps Spirit is moved by similar forces?

At first glance, Spirit's creativity may appear to be unbounded. But deeper thought reveals that even Spirit is constrained. A universe riven by contradictions will destroy itself – as many proto-universes may do. Hence, a universe must be self-consistent if it is to reach maturity. In its early stages, many possibilities may lie open. But as it evolves and becomes more diverse and complex, so the network of interconnections becomes denser and the demands of consistency increase. Habits develop, and possibilities freeze into laws that eventually become all-but unbreakable within the fabric of the whole. The structure, processes and direction of evolution of each universe gradually rigidify, until it can realize its potential only within a tight framework.

This constraint of consistency is illustrated well by the Anthropic Principle (see Chapter 12). Because our universe has particular physical constants, it has spawned the kinds of life and consciousness with which we are familiar. But in other universes, these constants may be quite different, giving rise to other forms of matter, life and consciousness, or perhaps to sterility and early death.

An even more fundamental constraint is that Spirit cannot create anything that is not latent in its Potential. Its only raw material is its own essence, which must form the foundation for everything it creates. Hence, Spirit must contain within itself the seeds of energy and matter, space and time, life and consciousness; of creation and destruction, autonomy and dependence, cooperation and competition, love and hate, joy and despair.

Since Spirit is the source of all existence, each and every object, each and every process, each and every being must consist of Spirit transcended and included through the many levels of the great holarchy of nature. In its deepest essence, everything *is* Spirit, fully and completely. Just as a droplet is fully and completely water, so every being is fully and completely Spirit. And everything has value in and of itself as a perfect manifestation of Spirit. Also, just as water is one substance comprised of many molecules, just as each of us is one body composed of many cells, so Spirit is one undivided whole with many aspects.

God sleeps in the rocks
Dreams in the plants
Stirs in the animals
And awakens in Man

Sufi saying

This wholeness is revealed in many ways. As we have seen, spacetime, energy and matter are one; and fundamental particles are entangled with each other across the cosmos, which is underlain by the wholeness of the implicate order and the psi field. We have seen that the boundaries between systems, between living and non-living, between mind and consciousness are fluid and permeable, little more than illusions. And we have seen how the web of life connects all things on Earth and beyond.

Creativity is one of Spirit's most obvious characteristics, and one shared by our universe. We have seen this creativity in self-organizing systems and the process of evolution, and we encounter it directly in human languages, cultures, arts and science, and in the technologies by which we prod evolution to form novel elements, chemicals, and organisms. Is creativity an inevitable quality of Spirit-filled creations, or has Spirit chosen to impart it to us and not, perhaps, to some other universes? We cannot know. But perhaps creative universes are more interesting and beautiful to Spirit than planned, controlled ones?

Creativity implies freedom and autonomy – the freedom to create whatever we can, whatever we choose within the laws of our particular universe. This freedom means the outcome of evolution is open-ended, and cannot be predicted. Hence, when Spirit creates a universe like ours, it cannot know what will happen, and can only set the process in motion. Each universe is an experiment in which Spirit limits itself to setting fundamental laws and constants, and empowering the process of evolution. Hence each universe co-creates itself with Spirit. Every field, every particle, every grain of consciousness becomes an active element in the unfolding of the whole by repeatedly transcending and including itself in deeper and deeper levels of the emerging holarchy. Perhaps, as we will see in later chapters, the nature of Spirit itself may not be fixed, but be co-created within each creation. Perhaps we are co-creating God?

If Spirit cannot bring forth anything that is not latent in its Potential, then it must contain the seeds of consciousness which have germinated and flourished in our universe. And if Spirit does no more than set each universe upon its way, then elementary consciousness must be present from the moment of creation, pervading every thing and every process like the water which flows through all life.

Is Spirit a fully Conscious Being that is watching the evolution of the cosmos? Or does Spirit simply have a potential for consciousness which can be realized only through creative evolution in a particular universe? Again, we cannot know, and both views can be found in spiritual intuitions. But either way, the essential nature of this Consciousness is likely to be too alien for us to recognize, any more than we can sense the energy of the vacuum field from which matter springs. We meet the Mind of Spirit in unity consciousness, near-death experiences and other transpersonal phenomena, but only as Spirit chooses to reveal itself to us, mediated by our body-mind and culture.

We have seen how sub-atomic particles remain mere ripples of possibility until frozen into solidity by the act of observation; an act which many physicists believe requires a conscious mind. And we asked if cosmic Consciousness could have materialized our universe, concluding that our creative autonomy would be compromised if this were so. We would live in a universe that was determined by Spirit at least until a consciousness emerged that could collapse the possibility waves into reality for itself; until conscious life, humanity or sentient aliens were ready to take on the mantel of co-creators. Perhaps, as Amit Goswami argues (see Chapter 21), matter remained as unmanifested potential before sentience evolved? And, before then, perhaps evolution happened outside normal spacetime?

However it happened, consciousness somehow condensed matter from energy like liquid from vapor. Matter then evolved until it could channel, mediate and focus consciousness through living things. From the materialist perspective, matter is the ultimate reality, whilst consciousness and spirit are by-products. But from the spiritual viewpoint, matter is a condensation of Spirit, and Spirit is what is ultimately real.

In near-death experiences, people commonly encounter a Being that radiates light, love and compassion. And in unity consciousness, there is often a sense that love lies at the heart of existence; that at some level the world is as it should be and all is very, very well. It is possible that these are direct experiences of the nature of Spirit. Or they may reflect how Spirit chooses to, or is able to, reveal itself to human consciousness. But why should Spirit be loving and compassionate? The answer is that these and other virtues form a better basis for evolution of a self-consistent, interactive whole than the divisiveness of hatred and coldness.

This brings us to the perennial questions of evil, pain and suffering. In traditional terms, how can a loving God cause, or even allow, such things to happen? Our journey so far suggests two answers. In Chapter 3, we explored the logic of yin and yang, and saw how the existence of any quality implies its opposite: good and evil, pleasure and pain, love and hatred, cooperation and competition, freedom and control. The one cannot exist without the other to define it. And Spirit cannot create the one without also creating its opposite.

The second answer is that Spirit cannot give us creative freedom without also granting us the power to choose unwisely; the power to create pain, hatred and competition as well as pleasure, love and cooperation. Hence, we share responsibility with Spirit and other conscious beings for the evil and suffering of the world.

God and Spirit

This vision of an evolving, creative, conscious Spirit of which we are integral parts and co-creators appears to have little in common with the traditional conception of God in Judaism and Christianity, or of Allah in Islam. However, many theologians and scientists are reinterpreting their faith in the light of modern science. One of the leaders of this movement within Christianity was biologist Charles Birch. In Chapter 4 of his book *On Purpose* he described his beliefs about God, creation and humanity's purpose. Summarized here, they are deeply similar to the world of Spirit described above.[1]

Before matter existed there was the possibility of its existence within the potential of cosmic mind. In Judaism and Christianity this potential, this "ordering principle at the heart of the universe" is called God. God seeks to bring all possibilities into reality through love. The creative love of God draws His creation towards ever-greater richness as each individual seeks to realize its potential. And the possibilities increase with each evolutionary step. They were tiny for the energy following the big bang, and, Birch believes, reach their peak in humanity.

When human love meets divine Love our response is "passionate and transforming." And as each entity responds to the lure of God's love, God becomes "concretely real in a way God was not concretely real before." With each step forward in cosmic evolution or in an individual life, "God becomes conscious in a way that God was not conscious before." Thus Birch suggests that God's nature is dynamic and evolving, rather than eternally perfect. Also, God is not a detached spectator, but the synthesis of the feelings of a feeling universe. He provides the purposes and values of creation, but leaves it free and self-determined.

Birch claims that, as far as we know, human society is the pinnacle of this cosmic adventure, the frontier of cosmic evolution. "Here is the great upreach toward values higher than any which have ever visited the realm of existence. Here the existing universe is groping out into the vast realm of the possibilities of God as yet unrealized on earth." And hence here is where the sufferings and joys of God and creation are most intense.

The future, Birch concludes, is open. We can change as individuals and as a society. But God's involvement does not mean that He will look after everything and all will be well. "There is a real sense in which the future is in our hands. ... For us to fail to respond to the forward call of life is not just a personal failure. It is a cosmic tragedy."

In Conclusion

Spirit represents the ultimate, absolute Mystery of existence. Nevertheless, because we originate in Spirit, we share its essential nature and can intuit some of its inherent characteristics, including wholeness, creativity and freedom, consciousness, co-creative evolution, love and compassion. In the next two chapters we go on to explore the human relationship with Spirit.

23 The Diversity of Spirituality

Having deduced what we can about the nature of Spirit, this chapter seeks to clarify the relationship between Spirit and humanity. The aim is not to review particular religious beliefs or spiritual practices, but rather to gain an overview of spirituality that encompasses the broad scope of the religious impulse and the diversity of spiritual paths. Before continuing, you may want to refer back to the short introduction to spirituality in Chapter 3.

To Be Human Is To Be Spiritual

We are manifestations of Spirit, and hence are spiritual in nature. In the words of John Hick, spirituality is "a fifth dimension of our nature which enables us to respond to a fifth dimension of the universe. In this aspect of our being we are ... either continuous with, or akin to and in tune with, the ultimate reality that underlies, interpenetrates and transcends the physical universe."[1] Similarly, a recent consultation paper on spirituality and youth development in Britain referred to spirituality as a dimension, like the third dimension of space, which cannot be removed from us.[2] And Mircea Eliade claimed that "The 'sacred' is an element in the structure of consciousness and not a stage in the history of consciousness."[3]

The spiritual impulse is common to all human cultures, even today's secular, materialist society. While people are deserting traditional religions in droves, research indicates that as many as two-thirds are interested in an alternative, more meaningful form of spirituality.[4] And fundamentalist religions are booming.

Many scientists claim that spiritual experiences are no more than artifacts of particular kinds of brain activity. As evidence, they point to experiments in which an experience of God can be induced by magnetic stimulation of certain parts of the brain; the existence of the so-called 'god spot' which becomes active when thinking about spiritual or religious topics; and the intense spiritual experiences associated with some epileptic seizures.[5] But eminent neuroscientist Vilayanur Ramachandran takes a more open-minded approach.[6]

> Why is the revealed truth of such transcendent experiences in any way 'inferior' to the more mundane truths that we scientists dabble in? Indeed … one could use exactly the same evidence – the involvement of the temporal lobes (of the brain) in religion – to argue for, rather than against, the existence of God. By way of analogy, consider the fact that most animals don't have the receptors or neural machinery for color vision. Only a privileged few do, yet would you want to conclude from this that color wasn't real? Obviously not, but if not, then why doesn't the same argument apply to God?

The way we experience spirituality, and the way we express spiritual truths are mediated by the structure of our brains, and by our culture and personalities. as noted earlier, those brought up in a Christian society tend to see visions of Mary or Jesus, and to express their visions in familiar metaphors. Similarly, Hindus tend to experience Krishna, Muslim's might encounter Muhammad, and New Age believers may see Pan. And an alien's experience of Spirit would be quite different to that of any human. But this diversity tells us nothing about the underlying truth or falsity of spiritual reality. In the rest of this chapter, we will explore the diversity of human spirituality assuming that Spirit is real.

The Varieties of Spirituality

Spiritual experiences are not confined to religious ritual or quiet contemplation. They are often triggered by the peace and beauty of nature, and the sense of connection with the whole that contact with it brings. Many people discover wells of love and compassion in themselves through caring for others. And sport, dance or lovemaking may evoke a sensual unity of mind, body, emotions and soul. At other times, spiritual experiences may come through a flash of insight, immersion in creativity, a child's smile, an image of suffering, or the scent of grass after rain. There seems to be no limit to the diversity of spiritual experiences, and their expression.

Many people have categorized religious experiences, and this is not the place for a detailed survey. Instead, I will illustrate its diversity using two simple classifications that draw on the work of several authors.[7] The first is due to Mike King, and summarizes the complex history of religion in four stages.

Shamanism was the earliest expression of spirituality, and appears to have emerged with the evolution of humanity itself. It perceives every rock, mountain, tree, river, animal and Nature herself as imbued with indwelling spirits. Shamanism aims to contact and experience the spirit realms which are the abode of totem animals, spirit guides, the spirits of tribal ancestors, and other spirit beings. The Shaman communicates and identifies deeply with the natural world, mediates between his or her tribe and the spirits, and makes the world sacred through ritual. Shamanism is mainly associated with hunter-gatherer societies, but has experienced a strong revival in modern society.

Polytheism evolved from shamanism with the emergence of farming, and the transition from tribal societies to agricultural civilizations. The nature spirits became goddesses and gods who had more human characteristics, and needed to be propitiated in different ways. The Shaman's role as self-effacing spiritual guide and healer was taken over by an institutionalized priesthood, and religion became centered in great temples rather than the natural world. The first phase of transition was from shamanism to goddess religions in which the relatively balanced influence of male and female gave way to the feminine principles of fecundity, growth, seasons and cycles. In the second phase, the masculine principle became dominant, bringing with it a pantheon of warring gods, and myths of warriors and heroes.

The third stage in the development of religions was the rise in the Middle East of patriarchal monotheism in which a single God, or supreme being, replaced the pantheon of gods. This transition is associated with the invention of the alphabet which led to the written Word taking on sacred significance. In Judaism, Christianity and Islam the existence of other gods is vehemently denied, and followers are forced into a single spiritual path rather than being able to choose from a plurality of deities and practices. By contrast, Hindus worship Brahman through many other deities who represent his various attributes, thus blending monotheism and polytheism.

The most recent stage of development was the emergence of transcendent religions. In the mystical strands of Judaism, Christianity and Islam, God ceased to be seen as a separate supreme Being, and became the 'Beingness' at the heart of existence with which the mystic seeks union. This concept was seen as heretical by religious authorities in both Christianity and Islam, often leading to persecution. By contrast, Buddhism does not include the concept of 'God', and focuses on the goal of extinguishing the separate sense of self and becoming absorbed into the oneness of everything in the state of enlightenment.

These four types of religion appear to have emerged more or less in this sequence, although it does not exactly match the development of any particular religion. The sequence has often been interpreted as evolution to higher levels of spirituality, with shamanism at the bottom of the ladder, and transcendent forms at the pinnacle. However, a more inclusive view is emerging that accepts all four as equally valid expressions of the spiritual impulse. Shamanism is the ultimate form of spirituality for those who believe that antiquity is the key to authority, or who relate to Spirit through nature. But polytheism or monotheism may be the highest form for those who are seeking a relationship with a personal God or gods; and the transcendent is the ultimate for those who believe that human consciousness is evolving 'towards the light'.

Not only do religions vary over time and with location, but also individuals have widely different spiritual temperaments which are expressed in a variety of spiritual paths: devotional, contemplative, ascetic, solitary, communal, intellectual, sensuous, creative, and so on. Drawing on the work of Mike King and others, I have classified these paths under five polarities: Devotional – Intellectual, Detached – Engaged, Solitary – Social, Exoteric – Esoteric, and Transcendent – Immanent. The first three of these need little elaboration, but the last two require more extensive discussion.

Devotional spirituality is oriented towards the Divine Being. It emphasizes love, the heart, prayer, surrender and our relationship with God. It is based on faith and devotion. By contrast, the intellectual path is based more on knowledge and reason. It is concerned with the infinite and eternal, wisdom and mind. Both paths are present in Hinduism whereas Christianity is largely devotional and Buddhism emphasizes the intellect. In western culture, the intellectual spiritual path effectively became absorbed by non-spiritual philosophy.

Detached spirituality is inward looking, tending to withdraw from the world to focus on meditation, contemplation or prayer. By contrast, engaged spirituality looks outwards, embracing nature and the material world. It sees God as manifest in everything, or, as William Blake put it, that "eternity is in love with the productions of time." This path is often favored by artists, scientists, and social activists.

The solitary path is relatively self-contained, seeking peace, stillness, and communion with nature. By contrast, the social path is more concerned with community life and collective practices such as ritual, chanting and meditation. There is often a tension between the demands of community life and practice, and the need to nurture the inner life. Hence, it is not surprising that the lives of many spiritual leaders include a period of solitude and withdrawal followed by a return to community and teaching.

None of these three dimensions is a fixed polarity. Individuals may embody a mixture of characteristics, and move between the poles over time. Thus it is possible to combine devotional and non-devotional practices, periods of detachment with others of engagement, and times of solitude with ones in community.

The Exoteric – Esoteric and Transcendent – Immanent polarities are discussed at greater length in the next two sections.

Exoteric and Esoteric Spirituality

The polarity between exoteric and esoteric spirituality appears in various guises. It is the divide between outer religious forms and inner spiritual experience, between organized religion and mysticism, or between what Timothy Freke and Peter Gandy call literalism and gnosticism.[8] It is often claimed that both strands are represented in all the world's religions, and that the esoteric strands have more in common with each other than with the exoteric branches of their own traditions. Both can take devotional or intellectual, detached or engaged, and solitary or social forms.

Exoteric religion is an institutionalized activity that takes place in particular locations with the aid of appointed mediators, or priests. In theory, it provides access to spiritual experiences for its adherents. But in practice religions often lose contact with their spiritual roots. They become hierarchical, concerned more with power and money than spirituality, and often fail to satisfy spiritual needs. When this happens, true spirituality may be confined to the esoteric or mystical branches.[9]

Exoteric religion is concrete and literal. 'Literalists' believe that their scriptures are the actual words of God, and their myths are factual history.

Thus, Moses really did part the Red Sea, and the world was literally created in six days. But, as Freke and Gandy put it: "If we become fixated with the words, as Literalists do, we mistake the message for the meaning and end up eating the menu, not the meal."[10] Literalists focus on the outward trappings of sacred symbols, scriptures, rituals, the ecclesiastical hierarchy, and divinely ordained religious customs. They believe that their particular tradition has a unique claim on Truth; and that faith in their particular creed is the only route to salvation. Literalists and fundamentalists the world over are prepared to silence dissenters in the name of God, and are the source of religious oppression, persecution and wars. In the words of Timothy Freke and Peter Gandy:[11]

> Literalist Christianity is often credited with inspiring positive social reforms in Western society. But the truth is that the driving impetus for humanitarian change has come from humanists and non-conformists. The conservative forces of the established Churches have resisted every step towards greater compassion, from the ending of slavery to the abolition of the death penalty. In recent decades, unable any longer to simply bully us into submission, Literalist Christianity has developed a gentler, more attractive face. Yet its darker side continues to be a nefarious force in the world.

Names are very deceptive because they turn the heart aside from the real to the unreal. Whoever hears the word 'God' doesn't think of the reality, but of what is unreal. Likewise with words such as 'Father', 'Son', 'Holy Spirit', 'life', 'light', 'resurrection', 'church', and so on.

Gospel of Philip

By contrast, esoteric religion, mysticism or gnosticism involves a relationship between the individual and Spirit that is based on direct inner experience. It does not usually need a special place or person to mediate contact with Spirit, although mystics often emphasize the importance of spiritual guidance by a wise Teacher and the support of a community of fellow seekers. Esoteric traditions are regarded as hidden, not because they are secret but because they involve inner, mystical experience.

J. C. Cooper noted that esoteric teachings are reserved for initiates who have 'ears to hear', and who can see beyond the world of the senses to the deeper reality. Nevertheless, the teachings are "open to anyone who is prepared to take the necessary journey, 'path', or 'way', first to self-knowledge, then to knowledge of the Self, the all-embracing Tao."[12] However, esoteric teachings are not concerned with factual or experimental proof in a scientific sense, but only with revealed truth about the spiritual realms through direct experience. (See Chapter 3 for a fuller discussion of intuitive and spiritual knowledge).

Timothy Freke and Peter Gandy used the term 'gnosticism' to describe the esoteric traditions. The aim of Gnostics is not to make converts, but to gain gnosis, or knowledge of the Truth, through a spiritual journey of personal transformation. They interpret the traditional teachings and stories as signposts pointing beyond words to a mystical experience of Spirit, and adopt the wisdom of other traditions if it enriches their own. They tend to be free spirits and idealists, unbound by tradition or their own culture, who follow their hearts not authority or the herd.

There are two views of the nature and purpose of esoteric practices. One holds that they are like personal experiments, the results of which can be validated against the claims of other trained practitioners in much the same way that scientists validate their results by reference to results obtained by other scientists.[13] The opposing view is that mystical teachings are not descriptions of reality to be confirmed or falsified, but prescriptions for life to be practiced in order to transform ourselves and the world. In support of this view, Jorge Ferrer points out that any seeker whose experience contradicts their tradition will be regarded as deluded or simply wrong, and will be told to continue meditating until she sees things correctly.[14] And John Heron concludes: "budding practitioners have the kinds of experiences they have been taught to have."

If we accept the latter argument, how can we establish the truth or falsity of esoteric traditions? A more appropriate criterion is the quality of practitioners' lives. The true mystic, according to Coventry Patmore, appears ordinary, harmless and generally inferior, with a lively sense of humor, sound common sense, imperturbable good nature, and always with time to be of service. But faced with any laxity or affront, "he will be sure to answer you with some quiet and unexpected remark ... which will make you feel you have struck rock and only shaken your own shoulder."[15] In similar vein, J. C. Cooper described the Taoist sage in these words:[16]

> (His) whole life is held in balance and harmony. He neither wants 'status', ... nor does he want more than enough of anything. He does not strain after money in excess of his needs or persuade himself that luxuries are

necessities ... There is no need for asceticism and renunciation since undesirable qualities will die naturally when the sense of values is changed. ... But there is a need for joy ... Joy is a vital quality, an important experience, balancing the sorrow in life.

Transcendent and Immanent Spirituality

Transcendent spirituality focuses on the non-material realms; on the infinite and eternal; on the void or emptiness; on the one God, Allah, Brahman, Buddha Nature, Ground of Being, or discarnate Spirit; on gods and goddesses, and lesser spiritual beings and powers. Its concern is our relationship with these realms through contemplation, meditation, salvation, faith, awakening, enlightenment, sacrifice, faith, service, ritual or other means. Its practices include worship, devotion, prayer, meditation and asceticism. Transcendent spirituality is represented in all mystical traditions, and is a strong strand in mainstream Christianity where the transcendent aspect of God is balanced by the immanent aspect of Christ.

The transcendent focus on unity consciousness, emptiness, or union with God emphasizes the virtue of contemplation and withdrawal from the world. It denigrates the validity and importance of spirituality in everyday life. By contrast, as we'll see later in this section, immanent spirituality focuses on our embodied nature, the material world, and living a spiritual life in society. It is more concerned with the depths of our inner beings, with 'the god within', or 'that of God in every man' as the Quakers expressed it, than with an outer, transcendent Being.

The Gnostic cosmology introduced in Chapter 11 is an example of a transcendent image of reality. It envisages that Spirit desires to know itself, and stirs into Consciousness. In the process it splits the pre-Conscious Oneness, and creates the duality of subject and object. The whole panoply of creation then emerges from a cascade of further splits. But Consciousness can achieve full knowledge of itself only when each sentient being awakens to its identity and unity with the whole. The incarnation of Christ then provides a redemptive path through which this liberation can be achieved.

This is an example of the 'descent and ascent' model of spirituality which appears repeatedly in religious myths. Spirit, God, or whatever name we choose for the Mystery, descends into material reality in the creative process of *involution*. In this process, the primal Oneness becomes split, and the material world is separated from its Source, thus becoming imperfect, fallen, sinful or illusory. As consciousness evolves, the separate objects of creation gradually realize their true nature as fragments of the holographic Whole. Thus, creation ascends or returns back to Spirit through a process of rebirth, salvation, awakening or enlightenment facilitated by appropriate spiritual beliefs and practices. As a result of this liberation, love and compassion flow between all beings.

From this perspective, Spirit is not only the source of all existence, but also the force that motivates existence to transcend its limits and achieve its potential, and the end towards which it moves. Spirit reaches down to inspire material creation, and then reaches up again towards the goal of its own perfection. At each cycle it achieves higher levels of integration, consciousness and manifested Spirit. This two-fold process is Divine Goodness, Love and Compassion descending into, inspiring and embracing creation. And it is the Goodness, Love and Wisdom of creation responding, stretching up and ascending towards Spirit.[17]

As Ken Wilber expressed it: "Spirit *knows itself objectively* as Nature; *knows itself subjectively* as Mind; and *knows itself absolutely* as Spirit – the Source, the Summit ... of the whole sequence."[18] Once this is achieved, in the words of Timothy Freke and Peter Gandy, there will be only "The Mystery knowing itself. Love making love with itself. Beauty delighting in itself. Truth true to itself. Being being itself. ... We are co-creating this shared dream which reflects our wisdom and foolishness back to us to help us wake up."[19]

The 'descent and ascent' model forms the core of what is known as the perennial philosophy which claims that all religions share fundamental transcendental insights.[20] One of the most influential theories of spirituality during the last century, the perennial philosophy did not arise from cross-cultural research or interfaith dialogue, but from two beliefs. First, that all religions point in their various ways to Spirit, the source of all existence. Second, that mystical practices can lead to a knowledge of Spirit that is free of linguistic, cultural and doctrinal biases. In other words, mystics from all traditions have similar experiences and insights. Doctrinal differences are superficial, and result from culturally-loaded interpretations of those experiences. Thus, the Buddhist experience of the Ground of Being is claimed to be the same as the Hindu experience of Brahman, or the Christian union with God.

Ken Wilber is one of the strongest and most influential proponents of the perennial philosophy, but a cogent critique has been launched in recent years by John Heron and Jorge Ferrer amongst others.[21] They challenge the possibility of discovering universal spiritual Truths, and the very notion of a Spirit with eternal, fixed characteristics. They also suggest alternative models of spirituality which will be discussed in the next chapter.

Opponents of the perennial philosophy have long argued that the mystical experience is no different to any other. They claim that the language, culture and religious doctrine of the mystic not only determine the interpretation of his experience, but also the very nature of the experience itself. Thus, a Hindu does not experience Spirit and then describe it in Hindu language and symbols. He has a Hindu experience. And a Christian mystic does not simply label as 'God' her experience of Ultimate Reality, but has an experience that is at least partly determined by her beliefs about God and Jesus. Hence, the mystical traditions have no special claim to Truth, and direct, unbiased knowledge of Spirit is impossible.

This contention is supported by significant differences between the insights reported by mystics that centuries of debate have not resolved. Ferrer claims that the more they get to know each other the more bewilderingly different religious traditions become. He concludes that there is a variety of possible spiritual insights which are often incompatible and cannot be reduced to a single Truth. Further, the Dalai Lama has suggested that the various traditions have different aims.[22] If you want to reach Nirvana, you must follow the Buddhist Dharma, not Christian practices; and if you want knowledge of Brahman you need Vedic study and meditation, not Tantric Buddhism, Sufi devotional dance, or psychedelic shamanism.

The perennial philosophy aims to be tolerant and inclusive of all religious traditions. Yet it establishes a single spiritual truth against which traditions are judged. Those that do not match its criteria are rejected as inauthentic, merely exoteric, or as representing lower levels of spiritual insight. For instance, according to Ken Wilber's influential ladder of consciousness development (see Chapter 21), the Eastern concept of non-duality is the highest form of spirituality, whereas the Sufi and Christian traditions of union with an impersonal One come a rung lower at the Causal level. Platonic archetypes and Christian gnosis are a further rung down at the Subtle level, whilst the mystery religions and many indigenous faiths are relegated to the Psychic.[23]

We have seen that the evolution of matter and life is open-ended and unpredictable, creating holons of ever-greater depth. It is a truly creative process, with no predetermined path or goal. But the perennial philosophy limits the evolution of consciousness and spirituality to a path defined and established by Spirit. It leaves no room for creative emergence of new forms, and assumes that the ancient spiritual Masters fully explored and defined all levels of human consciousness and spiritual potential. Only the trappings of religion can be changed; Spirit and spirituality are fixed.

In reaction, John Heron asserted that "A theory which tells us that we have surface novelty and also tells us what will deeply and serially constrain it, is taking away much more than it gives. ... True human novelty is not mere superficial flexibility; it is rooted in creative choice *at depth*."[24] And Ferrer noted that "The fundamental spiritual value and beauty of the various traditions derives precisely from their unique creative solutions to the transformation of the human condition."[25] We will explore alternative models of spirituality which overcome these objections in the next Chapter.

Exclusive focus on the transcendent path as advocated by the perennial philosophy leads to denigration of earthly life as irrelevant at best, and sinful at worst. Our aim should be to abandon material existence and climb the ladder of consciousness to the enlightened state from which we have fallen. Thus, at each rung we must choose between denying or engaging with life; between death of the self or attachment to the illusory world; between 'bad life' or 'good death' as John Heron put it.

Not surprisingly, the result is often withdrawal from the world, asceticism, abuse of the physical and emotional self, and denial of human relationships. John Heron cites Ramana Maharshi as an illustration. Often portrayed as a modern exemplar of great spiritual attainment: "He sustained this state by going off to sit in a dirty pit, attending to the One, while neglecting and abusing his life. He let his unwashed body rot, attacked by bugs and covered in sores, leaving it to others to provide some minimal care. ... While being consumed by terminal cancer, he said 'The body itself is a disease.'"[26] Heron goes on to argue that the Christian assault on sexuality similarly led to centuries of perversion.

From an outside perspective, such forms of spirituality appear distorted and unbalanced. The transcendent path needs to be complemented by an immanent spirituality that offers the choice of a 'good life'; a spirituality whose focus is the 'god within', our inherent spirituality as individuals, and the challenge of living spiritual lives in the material world. Immanent spirituality emerges from recognition that every aspect of the world is an expression of Spirit and cannot be otherwise because everything originates in and is permeated by Spirit. Just as drops from the ocean are wholly and completely water, so we and all existence are wholly and completely Spirit. Hence, there is nothing second rate about physical existence, and we don't need to return to Spirit because we are already there. We are Spirit but don't yet realize it; enlightenment is awakening to this fact.[27]

This is not to claim that everything is perfect. The material world is an experiment, an exploration of potential and possibilities, not a controlled environment. Things can and do go wrong; evolution makes mistakes. And once duality is set loose, the polarities of good and evil, beautiful and ugly become part of the way things are. But duality is the creative force that releases all potential. Without differences to work on, the processes of self-organization and evolution are impotent. Without subject and object there can be no consciousness, except as a formless mist of illumination. So duality is not something to shun. We should celebrate it and revel in it whilst being aware that beyond it lies the One primordial Spirit that is our source.

Immanent spirituality honors each person as a unique manifestation of Spirit. As such, we are not just isolated, powerless planets passively reflecting the sunlight of a great spiritual teacher around whom we orbit. Rather, we are co-creative sources of love, compassion and wisdom; stars in our own right, each of us bringing our own distinctive contribution to the brightness of the whole.

From the perspective of immanent spirituality, our physical bodies and individual selfhood are not obstacles or illusions obstructing spiritual progress. We are not perfect, but neither are we innately flawed or sinful. Our physical nature is rooted in Spirit, and can be a source of spiritual energy just as much as the transcendent realms. Similarly, the material world is not a barrier to spirituality, although we can use it to divide ourselves from Spirit if we choose to do so. The universe is not flawed, but is a rich, wonderful and creative embodiment of Spirit.

From the immanent perspective, the spiritual realms are present and accessible in every aspect of life and existence. We don't need to embark on a lifelong journey towards a distant spiritual goal: all of life is sacred. Nor are the spiritual realms higher than the physical in some great hierarchy of being. Rather, the physical and spiritual realms both contribute uniquely to the evolving Whole in an interactive relationship.

The practice of immanent spirituality is not confined to solitary, silent meditation, nor to collective religious ritual. It is the active, aware unfolding of our lives, expressed in whatever way suits us: through individual and collective movement and music, poetry or prose, artistry, service, and relationships with people, nature and the universe. Immanent spirituality recognizes that marriage, conscious parenthood, vocation, aware consumerism, good neighborliness, environmentalism or social activism may be spiritual paths every bit as much as years of solitary meditation.

According to John Heron, immanent spirituality includes development of the emotional maturity and interpersonal skills required to empower ourselves, others and our relationships. It calls for imagination and creativity. And it demands self-determination and cooperation in co-creating a spiritual lifestyle, and in social and planetary transformation.

Neither the transcendent nor the immanent path is the higher. They are dimensions of many possible paths, and we can synthesize what suits ourselves. Spirit can manifest equally well through unity and diversity, non-duality and duality, oneness and multiplicity, non-self and self, the universal and the particular, the cosmos as a whole or through my personal self. And this is not *the* Truth, but simply my perception at this moment.

In Conclusion

This brief survey of the varied expressions of human spirituality shows them to be far too diverse to fit within any of the traditional religious frameworks. We need a new model of spirituality that is broad and flexible enough to encompass revitalized shamanism, polytheism and monotheism as well as transcendentalism; a model that welcomes and nourishes those who variously follow the paths of intellect or devotion, engagement or detachment, isolation or community, exotericism or esotericism, immanence or transcendence; a model that leaves space for the synthesis and evolution of new expressions of spirituality; and a model that embraces the new vision of reality being revealed by science. This is the subject of the next chapter.

24 Towards a Spirituality for Our Age

Ancient teachings are valuable pioneering efforts, but are not definitive. They are products of their particular cultures and times which often lack frameworks for engaging with the issues of the twenty-first century. And they sometimes embody values which are no longer appropriate, such as authoritarianism, patriarchy, emotional repression, and the denigration of women and the body.

Rather than being constrained by these traditions, John Heron claims that our spiritual potential is unlimited and a matter of creative choice. We have "a profound array of dynamic options emerging from the infinite potential within. We can co-create our innovative path with our inner spiritual life-impulse and the possibilities it proffers."[1] In other words, Spirit is not a dictatorial Creator, but the instigator and sustainer of an experimental creative process with genuine freedom of choice amongst a diversity of possible spiritual paths and goals. Higher spiritual states are not predetermined but emerge through communion and cooperation with each other and Spirit.

Whilst the old religions generally sought to escape from 'this vale of tears' by dissociating from life, perhaps we are ready now to embrace its richness; to discover a spirituality of transformative action by opening to co-creative relationship with Spirit. Perhaps our way is not to return to Spirit along a pre-ordained path, but to ride the rapids of Spirit's creative outpouring in an open-ended exploration of our spiritual potential. This chapter attempts to lay some foundations for a new, inclusive and holistic model of spirituality that speaks to the spirit of our age.

The Integration of Transcendent and Immanent Spirituality

Based on his personal experience and the results of cooperative inquiry (See Chapter 4), John Heron suggested a figure-eight model of spirituality that integrates the transcendent and immanent perspectives.[2] As shown in the diagram, this image of the spiritual path starts and ends with immediate present experience as the fulcrum for development. And the path alternates between expansion of the transcendent or immanent aspect, followed by return to present experience. If each cycle expands a little further, then immediate present experience will deepen gradually, and there will be a rhythmic growth towards wholeness.

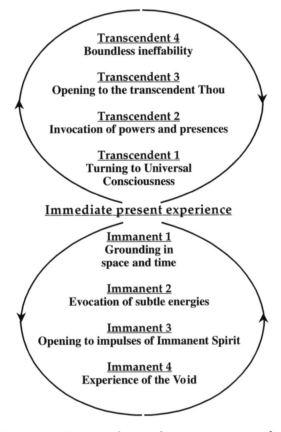

Figure 13   Transcendent and immanent spirituality

By immediate present experience, Heron means attunement to and resonance with the here and now; full, conscious awareness of our world and what is happening in it; active engagement through everyday life in interpersonal, social and planetary change and the divine unfolding of a global civilization. It is knowledge as an instrument of action as well as contemplation, and art as prayer.

The various levels of transcendent and immanent spirituality are illustrative, not definitive. They do not exclude other paths, nor form a hierarchy which must be ascended in sequence, nor represent a perennial philosophy.

The first stage in developing immanent spirituality involves intentional deepening of our physical experience of the body and everyday life through practices such as dance, yoga, tai chi, singing, drumming, conscious breathing, body work, emotional healing, diverse expressions of creativity, and conscious participation in the rhythms of daily life. By contrast, the first transcendent level is becoming aware that our ordinary consciousness is a part and local focus of cosmic Consciousness.

Immanent 2 is concerned with awakening subtle energies often associated with transpersonal consciousness as discussed in Chapter 20. These energies are recognized by many spiritual traditions under various names including kundalini, chi, prana, bioenergy, auras and the chakras. Associated practices may include rituals (including sound, movement, color and scent), and evoking transpersonal experiences. Circling back to the second level of transcendent spirituality, we encounter the archetypal principles of creation. This is the realm of Plato's ideal Forms, the Angels of Christianity, and spirit beings such as ancient Pan as encountered by the New Age movement. Here, we get a sense of intimate, loving communion with these beings, but without normal communication.

Return to immanent three opens us to the inner depths of everyday experience. At this stage we discover an indwelling spiritual companion that may prompt us to take certain actions, impart a sense of what is right or beautiful, reveal options, guide inner healing, or remind us of our role within the web of interbeing. Its promptings may come unbidden or in response to our evocation, but either way we shape and form them as much as they move us. In transcendent three comes awed communication with the creative source of all existence. In Heron's words: "It is, I find, a Thou-I communion in which my personhood is utterly transfigured within the embrace of whence it issues."[3]

The distinction between immanent and transcendent fades at the deepest levels which represent what I have called unity consciousness. Of immanent four, John Heron writes:[4] "Here I enter primordial emptiness pregnant with all existence ... Here I participate, if that is the word, in the divine." And of transcendent four:[5] "I find this area to be beyond ... words ... It is boundless beyond space. It is light beyond differentiated light. It is supreme awareness beyond all determinate name and form. ... It is ecstatic beyond bliss." However, Heron stresses that while this experience goes beyond individuality, it does not annihilate it. The fact that he can recall and recount the experience means that his individual identity does not disappear.

Summing up, Heron writes:[6]

> My spiritual development, then, cannot be measured simply in terms of hours of meditation or number of extended retreats or stabilized attainment of some inner, transcendent state of mind ... On its own this is vertical flight from full spiritual development, which I believe finds its primary consummation in the unfolding of my immanent spiritual life. And this, fully followed through, involves attention to social change and social justice through promoting participative forms of decision-making in every kind of human association with which I am involved.

Participatory and Co-creative Spirituality

In the view of John Heron and Jorge Ferrer, we co-create our own spiritual paths with each other and Spirit.[7] The resultant spiritual knowing is not objective, neutral or merely cognitive, but draws all dimensions of our nature into participation. It opens our minds, hearts, bodies and souls, and engages us in passionate activity. This process is saved from anarchy and pure relativity by fundamental spiritual values. This section elaborates these points.

We have seen how evolution of the holarchy of nature is open-ended, and can be influenced by internal disturbances. And it seems likely that the levels of consciousness have emerged in a similar way to form a complementary holarchy. If this is true, it means that the future course of human consciousness is not predetermined, and has not been mapped completely by the ancient spiritual masters such as Jesus and the Buddha. It is open-ended and open to influence. Hence we can co-create consciousness with each other and Spirit.

Even more radically, Jorge Ferrer argues that Spirit itself does not have an eternal form and characteristics. It is not just that the various religious traditions have different perceptions and ways of describing a constant and unique Spirit, but that Spirit assumes different forms in different contexts. And Spirit is not simply becoming aware of itself through the vehicle of creation. Rather, it is co-creating itself in cooperation with its creatures through open-ended evolutionary exploration of its potential. It is seeking self-awareness by experimenting with various forms of incarnation. Hence, we are not embarked on exploration into God, but on co-creation of God.

And yet behind this process there IS an empowering and enabling Power and Potential. There is the Mystery of an indeterminate Spirit that manifests in different ways over time and space in our world, and perhaps in even more diverse forms in other worlds and other universes.

> On the one hand, the human mind does not just produce concepts that "correspond" to an external reality. Yet on the other hand, neither does it simply "impose" its own order on the world. Rather, the world's truth realizes itself within and through the human mind.
>
> **Richard Tarnas**
> **Quoted by Ferrer (2002) p.155**

How do traditional religions fit within this framework? The Dalai Lama once wrote: "If we view the world's religions from the widest possible viewpoint, and examine their ultimate goal, we find that all of (them) ... are directed to the achievement of human happiness. ... To this end, (they) teach different doctrines which help transform the person. In this regard, all religions are the same, there is no conflict."[8]

Hence, all traditions can help free us from psychological patterns and self-imposed sufferings, open our hearts, and transform our self-centeredness towards a fuller, more compassionate participation in life. But nevertheless, the mystical evidence suggests that the many forms of enlightenment and innumerable ways of experiencing Spirit are not reducible to a single Truth. Some of them overlap, but often they are incompatible as, for example, the vision of a personal God and an impersonal 'Beingness', or the duality of God and creation as against the perception of a non-dual One.

Rather than searching for a single reductive Truth, or regretting this diversity as a source of conflict, we should welcome and celebrate it as the bringer of opportunities. Each person and culture can contribute their unique gifts to the evolution of the Whole. We can learn from our varied experiences, and be enriched by cross-fertilization of traditions. Indeed, it is encounters between these varied expressions of spirituality that will drive their further evolution.

From this perspective, the various religious traditions, doctrines and practices are participatory experiments in which humanity is exploring alternative approaches to our relationship with Spirit and material existence. Through our participation we individually and collectively channel and shape the creative energies of Spirit, helping bring its visions into being. In the process, we draw on old wisdom and models where these are helpful, but rely on personal and collective intuition and experience rather than authority. Our spiritual journey has no fixed goal, but is an endless exploration and manifestation of the inexhaustible possibilities of Spirit.

If our minds, hearts, bodies and souls are open to Spirit, we may become clear channels for Spirit's energy. In the process, we may be transformed and illuminated by knowledge grounded in and coherent with Spirit's unlimited creative power; a knowledge open to visions of what is possible and desirable, and attuned to the unfolding of existence; a knowledge that is never final but constantly evolving.

Such participatory knowing is not knowledge of something by someone, but lived, embodied and meaningful experience that transforms both the person and the world. It is the active bringing forth of a world co-created by individual and collective intentions and dispositions; the cultural, religious and historical context; archetypes (or morphic fields); and the creative potential of Spirit. Participatory knowing is not only intellectual, but also involves the emotional and empathic knowledge of the heart, the sensual knowledge of the body, and the visionary and intuitive knowledge of the soul. It integrates all dimensions of our being: perceptual, cognitive, sensory, emotional, sexual, imaginal, intuitive, interpersonal and more.

The teachings of different religious traditions are sometimes portrayed as rivers all leading to the same ocean of enlightenment, liberation or salvation. Adapting this metaphor, participatory spirituality is like an ocean with many different shores of emancipation, each of which can be reached on an appropriate raft. These shores are not predetermined, and waiting to be discovered. Rather they are co-created with Spirit and there is no limit to the number of shores and rafts that are possible. However, Jorge Ferrer claims that once a particular shore has been co-created, it becomes more readily accessible to individual consciousness and will attract other voyagers, in a process similar to morphic resonance and the development of morphic fields.

The danger of this open approach to spirituality is that it will descend into a relativism in which anything goes, and an anarchy in which there are no solid foundations or guidelines for action. The best safeguard against this is to judge alternative paths by the effects they have on human behavior compared to universal spiritual values. But what are these values?

As we have seen, Ken Wilber believes human consciousness develops through a series of levels until it reaches its highest attainment in non-dual consciousness. Humanity is evolving upwards, and, as our consciousness rises, so our moral and ethical codes change to reflect higher spiritual values. Thus, modern civilization has moved beyond slavery, the oppression of women and children, sacrifices to the gods, and cruelty to animals. It recognizes individual human rights, and global values beyond race, religion and the nation state. And it has traded autocracy for democracy, however flawed. Ultimately, our values will correspond to non-dual consciousness.

By contrast, John Heron believes that the evolution of consciousness is open-ended, but founded on constant spiritual values which provide absolute guidelines for the co-creative process. Jorge Ferrer similarly argues that, even though Spirit is evolving, it imposes certain limits on what we can co-create. He suggests that these limits are probably moral and ethical rather than doctrinal. But what are they?

Ken Wilber distinguishes three types of value.[9] Every holon, ie everything that exists, has equal *ground value* as a perfect manifestation of Spirit. But every holon is simultaneously a whole in its own right, and a part of a larger whole. As a whole, it has *intrinsic value* in and of itself, and the right to the conditions necessary to maintain its wholeness. The deeper a holon is, the more of existence it embodies, and hence the greater its intrinsic value. An ape, for example, has a higher intrinsic value and greater rights than an atom. However, as a part, each holon has *extrinsic value,* or value for others, and a responsibility to maintain the wholeness of every holon of which it is a part. Hence, the extrinsic value of atoms is higher than that of apes because they are vital parts of so many other holons.

From this discussion, we can conclude that the value attributed to a holon gives rise to both rights and responsibilities which it must meet in order for the whole to prosper. Hence, moral attitudes and actions are ones that support the evolution and development of everything that exists towards its potential.[10] Similarly, John Heron claimed that to live well means celebrating the fullness of life and creation, and working to release the potential of individuals in relationship with others and nature, to empower communities, to achieve social justice, and to facilitate the inner transformation of human consciousness.

These broad general principles can be summed up by the universal imperative to love your neighbor as yourself. Religious traditions and philosophers through the ages have elaborated this imperative into many more specific values some of which are listed in Table 4.[11]

Balance	Honor	Self-respect
Compassion	Humility	Service
Cooperation	Joy	Simplicity
Courage	Justice	Social harmony
Devotion	Love	Temperance
Fairness	Non-attachment	Tolerance
Freedom	Openness	Truth
Frugality	Peace	Unity
Generosity	Preservation of nature	Vocation
Happiness	Respect	Wisdom
Harmony	Responsibility	
Honesty	Reverence for life	

Table 4   Some Universal Values

Towards a New Model of Spirituality

Our spiritual dispositions vary, and we are embedded in particular material, cultural and religious contexts. As global culture becomes more complex and individuality comes increasingly to the fore, more and more people are finding that traditional religions cannot provide a spiritual path that meets their needs and inclinations. Increasingly, we are seeing genuine spiritual impulses whose emergence and expression is not supported by traditional practices. And we are realizing that all traditions reject, neglect, or are unaware of some dimensions of human spirituality. Christianity, for example, has rejected sexuality and often neglected other aspects of intimacy; and Buddhism has largely neglected our emotional nature.

Our challenge is to develop a holistic model of human spirituality that can draw upon the ancient wisdom of the shamanic, polytheistic, monotheistic and transcendent religious traditions; welcome the devotional, intellectual, detached, engaged, solitary, social, exoteric, esoteric, transcendent, immanent and other spiritual paths; and embrace a co-creative, participatory view of our relationship with Spirit.

The diagram below is an attempt to encapsulate the key features of such a model. I have deliberately presented it as a flower to represent the blossoming of human spirituality. At its heart lies immanent Spirit; our inner, present experience; our sense of identity, and the meaning and purpose of our lives. And on its outer fringe, undefined and unbounded, lies our co-created revelation of transcendent Spirit. In between lie alternative spiritual paths, examples of which are shown here as petals. These 12 are for illustration only, and are not definitive nor inclusive. There are undoubtedly other paths, and still more will be co-created in the future.

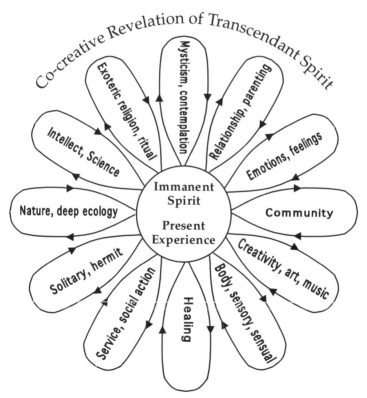

Figure 14   Conceptual model of spirituality

The arrows indicate movement from present experience along a chosen path towards transcendent Spirit and back to the center. How far we go at each cycle depends on where we are headed, and the stage we have reached on our journey. We may stay close to the core of immanent spirituality, progress a small way towards transcendent Spirit before returning and taking another path, or we may reach enlightenment along a single path.

Each of us, with our unique spiritual temperament and cultural context, will relate most strongly to one or a few of these paths at any given time. Thus, I resonate most strongly with the paths of the intellect, nature and relationship at present. But in the past, I have been deeply involved with community, service, healing, contemplation and the bodily paths. And I am open to the idea that I could be led into any of the other paths in the future. At every stage, I choose my path and my beliefs, and co-create my spiritual experience as a result of that choice. There is no single Truth for which I am searching, but at each stage, on every path, I gain a richer understanding of myself and Spirit.

At any stage of life, the 'flower' of our spiritual development is likely to have very uneven development. Indeed, those who devote a lifetime to a particular path may only have one or a few petals. It is tempting to suggest that 'balanced' spiritual growth would see each of us develop more or less equally along each path, so that we have a symmetrical flower. But I don't think that's how it works. We are all unique, and carve out our unique combinations of paths towards our unique revelation of Spirit. But the compass that ensures we remain headed in the right direction is universal spiritual values, and our community of like-minded pilgrims.

In Conclusion

An emerging view of spirituality is that all paths are valid expressions of our relationship with Spirit and that we can co-create our own that blends immanent and transcendent with a variety of practices. A relatively recent school of thought goes so far as to suggest that Spirit does not have a fixed nature that we perceive through different cultural and religious lenses, but actually experiments with different forms. Hence, we do not explore the spiritual realms, but co-create them. What keeps us from falling into anarchic subjectivity is adherence to universal spiritual values.

25 Who Am I?

Whatever worldview or spiritual path we choose, we expect it to answer the big questions of life. In this chapter and the next we see what we can glean from the science of oneness about two of the biggest questions: Who am I? And why am I here?

The Experience of the Self

Introspection reveals several identities. I am my body with its senses, and a passionate being with emotions and feelings. I manage my own life, and play many roles in society. I am the storyteller who weaves a coherent tale from past memories, present experiences and future dreams, and I am the witness, consciously observing and reflecting on my life and self.

Not only do I consist of different 'selves', but also they change with time and context. My body seems stable enough with its bones and muscles, its rhythms of breathing and heartbeat, and its unique patterns of iris, fingerprints, voice and handwriting. But even that identity is fragile. Vilayanur Ramachandran describes how easy it is to fool someone into perceiving their nose is 3 feet long, and how a stroke victim may believe that his left arm belongs to his mother![1] And recall how the bodies of those with multiple personality disorder change as their persona changes.

The social self provides another powerful illustration of the chameleon-like nature of our identity. Depending on the situation, I am father, son, partner, friend, writer, meditator, committee member, community member, car driver, photographer, gardener, carpenter and many more. My persona is different in each situation, sometimes subtly so, but sometimes quite radically. Our emotional selves are similarly fluid, shifting from moment to moment, and changing our perceptions of the world within minutes as our mood shifts from joy to sadness, or anger to compassion.

Somehow, most of us weave these 'selves' into a coherent whole. We integrate the interactions between them, our reflections on them, feedback about them, and life's experiences to gain self-understanding, self-respect, self-esteem, self-confidence, self-despair, self-hatred ... But those with schizophrenia and multiple personality disorder cannot achieve this integration, thus appearing to embody two or more distinct personalities.[2] From his observations of people with brain damage, Ramachandran concluded that the different aspects of the self are held independently in the brain, leading him to muse "there are curious parallels (with) the Hindu philosophical ... view that ... the self is an illusion"[3] – a perception shared by other mystical traditions and many philosophers.

Let's try to understand how this conclusion is reached. When I say "I hear a noise", I am distinguishing three things: the self who hears, the act of hearing, and the sound which is heard. But when I focus my attention inwards and search for the 'hearer', all I can find is the experience of hearing. In reality, it seems there is a stream of sound but no subject, no separate entity, which stands back and hears it.[4] Philosopher David Hume reached a similar conclusion in the eighteenth century:[5] "For my part, when I enter most intimately into what I call myself, I always stumble on some particular perception or other, of heat or cold, light or shade, love or hatred, pain or pleasure. I never can catch myself at any time without a perception, and never can observe anything but the perception."

This idea is hard to grasp because the distinction between subject, object and the act of perception is so deeply embedded in western culture and language. Another example may make it a little clearer. A moment ago you may have thought "I am confused." But you were not aware at the same time of a thinker who was thinking "I am confused." There was just the thought. And if you had looked for the thinker, all you would have found is another thought, perhaps the awareness that "I am thinking I am confused" or the question "Who was it who thought 'I am confused'?" Thinker and thought are one and the same.[6] In the words of Alan Watts: "No one ever found an 'I' apart from some present experience, or some experience apart from an 'I' – which is only to say that the two are the same thing."[7]

But if there is no 'I' and only present experience, how is it that we have this strong sense of identity? Recall that in Chapter 21 we likened normal consciousness to the images cast on the screen by a movie projector. Psychologist Stephen Levine claims that our sense of self arises from identification with these contents of consciousness, these present experiences.[8] In contrast, when the mind is quiet, we are able to identify with the white light of the projector, with Consciousness itself, which has no personal identity. At this point, the sense of self loosens its grip and we become aware simply of 'being', of the sense that 'I am'.[9] We mistake our identity in this way, and fail to live in union with Consciousness, because we cannot survive in the material world unless we distinguish what is us from our environment. Thus, whilst Consciousness is our ground and source, we cannot dwell continuously in unity with it.

The Self as Story

Most theories of consciousness connect the sense of identity with memory. We recall past experiences alongside present ones, thereby gaining a sense of continuity like a golden thread running through the fabric of our lives. From this thread, we weave an image of our self as the subject of our own story, adjusting it continuously in response to our experiences, social feedback, and our desire to match who we are with who we want to be.[10] Unless there are sudden, major shifts, we seldom notice how our self-image changes because it depends on our fallible, malleable memories. Hence, we normally believe we have a constant identity that is as unique as our genes, experiences and thoughts. And we tend to believe this self-image is complete and whole, remaining unaware of the ocean of unconscious thoughts and the horde of automata that control most of what we do and who we are.

Far from being a powerful inner identity, the 'self' turns out to be the 'center of narrative gravity' or the 'songline' of our lives.[11] We each tell our story with 'our selves' at the center, and the myths of our culture place our individual stories in a universal context. But, as Daniel Dennett put it, "for the most part we don't spin (our stories); they spin us. Our human consciousness, and our narrative selfhood, is their product, not their source."[12]

When the continuity of this thread breaks or it unravels into separate strands, we become demented or mentally ill. As Owen Flanagan put it:[13]

> Multiple chapters, novel twists and turns, even radical self-transformations are allowed. But these have to be part of the life of a single self. A self can change, but the changes had better make sense, or so we prefer. I need to

understand your conversion from hedonism to ascetic Buddhism in a way that locates you both before and after the conversion. In those rare cases where this is not possible, we say that some individual is no longer the same person.

The Holistic Self

The idea that we are no more than stories woven from the events of our lives cuts through our treasured perception of being unique identities in charge of our own destinies. It leaves us struggling in the quicksand of existence, adrift on a sea of experiences. But fortunately there is another side to this picture.

We are not only separate individuals, but also integral parts of larger systems. Our bodies can't exist without our natural environment, and our identities grow from our social context. Thus, our real 'self' is not an isolated individual, but a person-within-society-and-environment. And the survival of this larger holon depends in turn on each person playing their unique role within the whole. Thus we are simultaneously autonomous organisms, and dependent, but vital, components of larger systems. We are both fragments and wholes.

Many of us resist this perception because it threatens the sense of self we have constructed. But this fear is mistaken, as Joanna Macy stressed: "Do not think that to broaden the construct of self in this way involves an eclipse of one's distinctiveness. ... From the system's perspective this interaction, creating larger wholes and patterns, allows for and even requires diversity. You become more yourself."[14]

This sense of becoming more one's self strengthens when we enter transpersonal consciousness. Normally we connect most strongly with our own stories, but sometimes we resonate with the harmonies of other lives, or other beings. And then, for a while, we share their stories and their identities. Sometimes, too, we connect with the collective unconscious of humanity, weaving our personal stories into the fabric of ancient myths whose archetypal themes are perennial aspects of the human psyche.

The boundaries of our selves soften and blur still further when we enter unity consciousness. Knower and known seem to merge; the distinction between self and other dissolves; and our song blends with the eternal music of the spheres. For many of us, this prospect is terrifying because we believe the personal self is all we have, and to lose it is to die. But, to those who have experienced it, unity consciousness is an inexpressibly wonderful expansion of the limited 'self' to encompass the whole. From this vantage point, it is clear that to sin means to alienate myself from the cosmos by becoming a separate individual as the Hindu Vedas claim.[15] And what is best for 'me' becomes what will best harmonize my song with the "one song of the Uni-verse."[16]

In Jewish and Christian tradition, "I AM" is the name of God. This is echoed by the twelfth-century Sufi mystic, Ibn-Al-Arabi, who wrote "If thou knowest thine own self, thou knowest God"; and by the eighth-century Hindu saint, Shankara, who said of his own enlightenment "I used to think of (Brahman) as being separate from myself. Now I know that I am All." To many, such statements are blasphemy, claiming identity with the supreme deity. But when a mystic says "I am God" she means that she experiences her deepest self as identical with the divine. Thus, when the Bible exhorts me to "Be still, and know that I am God," it is encouraging me to quiet the mind, and come to a direct understanding that the "I am" that is my essential self is pure Consciousness, Spirit, or God.[17]

The idea that we are one with God is no longer purely mystical. As our lives unfold, they interact with the 'memories' of the implicate order or psi field. If our songs are discordant with the cosmic symphony, they will fade and die. But if they are in harmony with it, they will join the swelling chorus of creation as permanent solitons. This process may be likened to a boat on the ocean. As the great rolling swells pass under it, a small boat moves gently up and down, and tiny ripples on the surface hardly disturb it at all. But when the length of the waves matches the length of the boat, it pitches wildly in response to 'memories' of the blowing wind, passing ships, and island shores. Thus, if the boat of my life fits, it will dance to the cosmic tune, guided and empowered by the energy of the universe. And my boat's wake in turn will subtly alter the wave patterns of the cosmic ocean. But whereas waves on the sea gradually fade, those in the psi field endure till the end of time.

While we are alive, we must act as separate individuals in order to survive and fulfill our functions as parts of the One. But this separateness is an illusion, and unity is the underlying reality. We are more than our mortal minds and bodies; we are one with the infinite Whole. From this viewpoint, the 'loss' of our 'selves' no longer matters. Indeed it is to be desired.

Coming to Terms with Death

Death is the great taboo of our culture. We avoid talking about it, and delay its arrival at any cost. Why? At least part of the answer is that we believe death is the end of our story, and we can't face the prospect of extinction. Similarly, Sogyal Rinpoche suggests that this fear of death arises because we don't know who we are.[18] What solace can the science of oneness bring?

Perhaps the best place to start coming to terms with death is to understand its role in life. Without the death of outworn or damaged cells, our bodies could not live. All our cells are programmed to commit suicide when necessary for the good of the whole, and any that refuse may become cancers.[19] At a larger scale, each plant or animal in an ecosystem is like a cell in a body. And over time, each organism, each one of us, is a link in the chain of generations and the ongoing evolution of life.

The spectacle of Nature is always new, for she is always renewing the spectators. Life is her most exquisite invention; and death is her expert contrivance to get plenty of life.

Johann Wolfgang von Goethe
Quoted by Sheldrake (1990) p.53

When we see ourselves as parts of the systems of life in this way, our individual deaths become normal events in the larger scheme of things. But why should death be a necessary part of this scheme? There are at least three answers. First, organisms wear out, gradually losing the ability to maintain and repair themselves. We don't know why, but it may be because death is an essential ingredient of life. A world without death would lack carnivores and deadly diseases. Even herbivores would be limited to foods that do not kill the plant. Reproduction would have to stop once the world was fully populated, and evolution would be impossible because natural selection could not weed out those less fitted to survive. Hence, life could never adapt to meet the challenge of change as continents drifted, asteroids struck the Earth, and the climate changed. Paradoxically, therefore, without the death of individuals, life itself would become extinct and the living planet would die.

The third reason why death exists is simply that it is part of the mystery of existence. We have seen that all matter appears from the restless energy of the vacuum, and disappears into it again like dust motes in a sunbeam. We have seen that the cosmos is evolving to ever-greater complexity and organization, inevitably destroying what went before in the process of creating the new. And we have seen that systems are fluid processes rather than fixed structures. Viewed in a long enough timeframe, everything, including the universe itself, is ephemeral. We must accept the reality of impermanence, and with it the reality of death.

If you want to know the truth of life and death, you must reflect continually on this: There is only one law in the universe that never changes – that all things change, and that all things are impermanent.

The Buddha
Paraphrased by Sogyal Rinpoche (1992) p.29

But what is the point of life if the universe itself is doomed? Here once again we face the loss of ultimate meaning and the existential despair wrought by the second law of thermodynamics. The science of oneness has not repealed that law. But it has opened new avenues of hope. Modern cosmology suggests that there are myriad other universes, or else that our universe is developing through endless cycles of death and rebirth. Each new universe, each cycle of rebirth, emerges from that same Ground of all being, that same Spirit, that same psi field in which is encoded every slightest nuance of the cosmic history. Hence even the death of our universe is not the end of the evolutionary story. Our influence may reach beyond the limits of this space and this time, into spaceless, timeless eternity. We will have helped to co-create God.

These ideas also shed some light on the question of reincarnation. There seem to be only two theories that are consistent with the evidence summarized in Chapter 20. One is that we are literally reincarnated as some religious traditions believe, but that only a few people are able to recall the details of their past lives. This seems unlikely if the individual 'self' is no more than a story.

The second explanation is that some people resonate so strongly with the imprint on the psi field or morphic field of a particular past life that events from it emerge in their minds as if they were personal memories. In some instances, this resonance even extends to physical manifestations, such as birthmarks that mimic the injuries that ended the previous life. This idea is consistent with those Buddhist teachings which stress that, although actions in this life affect the quality of the next life, it is not the identical human consciousness which returns. Each new life is influenced by the previous lives with which it resonates, but is not identical with them, and is free to develop and evolve according to its own unique purposes and potential. Each life is like a child's building block, resting upon those below and supporting those above, but, unlike beads on a necklace, it is not connected by a common thread or soul.

In conclusion

So who am I? I am story and song. I am resonance and cosmic memory. I am All there is. I am the dance of Shiva. I am Spirit. Now and beyond death.

And who are you? You are free to choose your story, create your own dance. You are free to choose what makes most sense to you; to choose the story that resonates most strongly in your being.

26 Why Am I Here?

We long for life to have meaning. Without it, we suffer mental, emotional and physical disorders, and civilization becomes lost in the swamps of hedonism. But where and how can we find it?

Our lives are meaningful when we have a place and role in the world, and when we know and do what is worthwhile and important. In the past, we imbibed meaning with our mothers' milk from the structures, myths and beliefs of our culture. But the modern world has lost contact with its roots. Our culture has been diluted and fragmented by materialist science, individualism, secularism and consumerism. We are left to find our own way through a world without deep meaning or purpose. We urgently need a new mythology that reintroduces meaning and purpose to life. The science of oneness can help fill this void.

The Meaning of Life

A vision of the cosmos has emerged from our long journey together which bears many similarities to ancient spiritual traditions. At its heart is the Mystery of Spirit, a creative Potential that shattered the primal One, bringing matter and consciousness into being. But even as the unity shattered, its fragments were being reintegrated into a complex, connected whole that is infinitely richer and more beautiful than the Void from which it sprang; a whole with the capacity to know itself in all its richness and beauty.

We are creatures of this process, as both independent organisms and integral parts of the whole. And the meaning of our lives is to be found in these contradictory roles. We value ourselves. We care what happens to us. The very essence of our being urges us not only to survive, but also to perpetuate our kind, and to fulfill our potential. But those who live purely for themselves often find their lives are hollow shells, empty of joy and true meaning. Paradoxically, the pursuit of personal happiness often brings mental and emotional suffering. We cling to life, while dreading death. We pursue pleasure and success, while fearing pain and failure. We hunger for love, while feeling deeply unworthy. In our efforts to escape these negative feelings, we try to eliminate their causes by conquering death, banishing pain, and vanquishing evil. Hence was born the ideal of progress which has so singularly failed to make western society happier or more content.[1]

This desire for progress is doomed to frustration by the principle of yin and yang. Pleasure and pain, goodness and evil, life and death are like the troughs and crests of waves. No matter what our absolute quality of life, positive and negative emotions continue to alternate. Rather than trying to eliminate pain and suffering, the way forward is to enter fully into the richness of life with all its joys and woes. In doing so, we risk being overwhelmed – as much by joy as by suffering – but it is only by taking this risk that we can truly live and find the meaning of life. As Joseph Campbell expressed it: "You've got to say yes to life and see it as magnificent this way ... It is joyful just as it is. ... The ends of things are always painful. But pain is part of there being a world at all."[2]

Buddhist's teach that the way to avoid suffering is through detachment – a term that is often misunderstood. Most of us repress painful memories and emotions, pushing them away into remote recesses of our being. We build almost impregnable psychological walls around them, and pretend to ourselves that they don't exist. This is not detachment, it is *attachment*. It is clinging on to the pain and binding it to us until it is reflected in our postures, gestures, movements, illnesses, emotions and attitudes.

Detachment is not reached by hiding from the pain, but by going through the fire. It is the reward for the hero's journey when he or she returns home, having survived their ordeals, and finds peace; the peace that comes from knowing who we truly are and why we are here. It took Jesus a symbolic 40 days in the wilderness, the Buddha years on the mistaken path of asceticism, and Thich Nhat Hanh the horrors of the Viet Nam war and exile from his beloved homeland. Detachment is acceptance of the horrors and traumas and failures as parts of life and of ourselves. But detachment is not indifference. It is a place of deep love and acceptance of both ourselves and others – warts and all. A place from which we can launch forth into service, giving ourselves to the whole, not in a spirit of martyrdom but from the strength of our love and oneness.

Paradoxically, detachment is a state in which nothing ultimately matters. All is illusory, impermanent. And yet it is a state in which everything matters deeply. Every person and being matters as if they were our own child or partner, or our very self.

Detachment from our individuality brings a deeper meaning as integral parts of the whole. Our well-being depends largely on the well-being of every other individual, and hence on the extent to which our actions serve others. Selfishness, hatred, injustice and intolerance weaken the links which bind the whole, while love, compassion and cooperation strengthen them. Thus the basic value emerging from the systems view matches the fundamental moral value of all spiritual traditions: love your neighbor as yourself. But we can now see that our neighbors include not only our tribe, but all humanity, all beings, and Gaia herself.

Love, wrote Darryl Reanney, is expressed when two separate entities abandon their individuality to achieve a greater selfhood in union.[3] Alongside the evolution of the many from the One is the parallel process of uniting the many, destroying their independent selfhood in the creation of richer, integrated wholes. Love in this sense is a fundamental principle of the universe. When hydrogen atoms combine to form helium they forsake their own identity, but in doing so they ignite the stars. And when Gandhi, Mother Teresa, or Nelson Mandela forsook their own lives for service, they ignited the world.

As we come to understand this, so the boundaries of our identity move outwards, encompassing more and more until we are the Whole. As we move from love for ourselves to love of the whole, so our caring and protective attitude extends until it encompasses our community, our ecosystem and our planet. And as we draw other people and nature into our circle of love, so we lose our alienation and isolation, gaining joy and a sense of being at home. Here lies the great paradox of existence. He who would find his life must first lose it.[4]

Meaning comes from wholehearted participation in the cosmic process. It ceases to be something we create for ourselves, and is found in the purposes of the whole. When we act selfishly, we foster dis-integration; but when we act with love, we promote integration and harmony. The deepest meaning is thus to be found in our capacity to co-create our planet and the cosmos.

The Purpose of Life

Purpose and meaning are closely interwoven. If we discover the meaning of our lives, we will find our purpose. And if we find a deep purpose, our lives will have meaning. Buddhists believe that everything we are and experience is an illusion, a product of our minds like a dream. And modern science sees matter and consciousness as ephemeral manifestations of underlying fields, no more solid than the reflection of the moon on water. But if everything is an illusion, does it matter what we do? Why shouldn't we do whatever pleases us?

The Buddhist response again parallels that of science. Everything we do creates karma which influences our future like cause and effect in the material realm. In the words of the Buddha: "What you are is what you have been, what you will be is what you do now." In the Christian tradition, we are told we will reap what we sow. And if the stories of our lives that we spin are to be self-consistent, then who we were in the past largely determines who we are now, and who we are now strongly influences who we will be in the future. Or yet again, in Rupert Sheldrake's terms, we resonate with the morphic fields of our own past thoughts and actions, and are thus channeled in similar directions in the future. There is but one conclusion: we will never find future happiness and fulfillment unless our present actions are right.

The evidence from near-death experiences suggests that this influence of the past may extend beyond death. According to one account: "For me, it was a total reliving of every thought I had ever thought, every word I had ever spoken, and every deed I had ever done; plus the effect of each thought, word, and deed on everyone and anyone who had ever come within my environment or sphere of influence."[5] And those who have returned agree that they encountered absolute and unconditional love. This suggests that at the moment of death our lives are measured against the yardstick of love. And we are judged, not by an almighty 'God', but by our own inner critics.

Similar ideas can be expressed in more scientific terms. The record of our lives may persist beyond death, encoded in the implicate order, psi field or morphic fields. So, even if we are not literally reincarnated, our shadow stretches down the generations and across the cosmos. When we recall the sensitivity of chaotic and evolving systems to small disturbances, it is not too far-fetched to imagine that an individual life could be a trigger that changes the future of humanity, Gaia and the very cosmos itself for better or worse. None of us will ever know if that potential was realized, but we are nevertheless responsible for the effects that we have. We are co-creators of the cosmos and Spirit. Inevitably. Whether we like it or not. It matters how we live because this will determine the quality of our experience in the future, and influence untold lives to come.

This conclusion implies that we cannot save the world by ideology or politics, science or information on their own. What really counts in the long term is the quality of each of our individual lives; whether or not they resonate with the deeper purposes of the cosmos, and we nudge it in positive directions. The transformation of our society, our world, our universe, must start and end with the transformation of ourselves.

What does personal transformation mean? We have seen how we divide mind from body; the outer objective world from our inner subjective world; subconscious shadow from conscious ego; and individual consciousness from transpersonal and unity consciousness. Our lifelong task is to heal these splits, and integrate the various aspects of ourselves into a whole being; to integrate that individual being into the fabric of society, planet and cosmos; and to learn who we truly are. As described in Chapter 24, we each must find our own unique spiritual path towards this destiny. There are many false trails, so we must choose our way carefully, looking for evidence of its fruits in the lives of our fellow travelers as well as in our own.

Whatever our path, it should lead to a life of love, compassion and service for all creation rather than the pursuit only of our own happiness, salvation or enlightenment. Indeed, we cannot succeed in our quest for personal transformation if we ignore the wider whole. And transformation to a society of higher beings can occur only if the community, economy, technology, government and other factors are transformed too. As Joseph Campbell expressed it: "The Indian yogi, striving for release, identifies himself with the Light and never returns. But no one with a will to the service of others and of life would permit himself such an escape. The ultimate aim of the quest ... must be neither release nor ecstasy for oneself, but the wisdom and power to serve others."[6]

Our ultimate purpose is to co-create a whole person, a whole community, a whole civilization, a whole planet, and a whole universe. This is our responsibility, but we do not bear it alone. We cannot create a whole self on our own, let alone a whole planet or universe. But we can play our parts. Collectively and with Spirit, we have the power to guide the future of our planet and cosmos into the paths of love and truth, beauty and wisdom.

Setting About Our Task

In the past, our power was limited and largely unconscious, but science and spirituality have brought awareness of our potential, and all but made us omnipotent. We *can* transform ourselves, humanity, Gaia and the cosmos, whether for good or ill. And we must seek the wisdom to find the right way. Such wisdom is far more than dry ethics. Such action flows from the heart, from loving identity with the whole. As Arne Naess argued, we do not need ethics to tell us to care for ourselves.[7] And as our sense of self expands, so we do not need ethics to tell us to care for our families and friends, our communities, or our homes and neighborhoods. Ultimately, when our sense of self expands to include the whole, we care for the whole, acting beautifully rather than morally.

How can we best set about our task? The ancient Taoist concept of wu-wei provides a clue. Often translated as 'non-action', it is better rendered as 'refraining from action contrary to nature'.[8] According to Lao Tzu, everything can be done by wu-wei, by 'going with the flow'.[9]

> The Tao goes on forever
>   wu-wei – doing nothing
> And yet everything gets done.
>
> How? It does it by being,
> And by being everything it does.
>
> If people and rulers go by this
> then every living thing will be well.

In Conclusion

Our purpose is to create harmony. Harmony of mind, body and spirit within ourselves. Harmony amongst all parts of ourselves, including those of which we are ashamed, and which we repress. Harmony with family and friends. Harmony within communities and our larger societies. Harmony amongst diverse races, cultures, tribes, and religions. Harmony with nature and planet.

If we live in harmony with the song of the universe, we will help harmonize the discords of the world. And if we are in harmony with ourselves and each other, we will be supported and sustained in our purpose by the whole of which we are harmonious parts.

Harmony is multi-dimensional and dynamic; a blending of diverse notes and colors and movements into a beautiful whole. Our purpose is to seek that ever-changing mix of tones and tints and flows which creates the most beautiful symphony.

Review of Part VII

We've come a long way together, and are almost at the end of our journey. We have followed the golden thread of science from the behavior of systems, through fundamental physics, cosmology, life and consciousness until it led us to the deep mystery at the heart of existence. Why is there something rather nothing? Who am I? Why am I here?

These are the most important questions of life, but they are beyond the reach of science. No matter how deeply it probes, there will always lie a Mystery behind the known; a Mystery that I have called Spirit. And, by its very nature, science can never tell us who we truly are, nor give our lives a larger meaning and purpose. At this point, we have to leave the secure platform of scientific knowledge and embrace the rationally unknowable world of Spirit.

Spirit is Potential that unleashes its own potential. It is limited only by the need for inner consistency in its creations, and to what is latent within itself. How and why It creates we cannot know. Spirit is what is ultimately real, the source of space and time, matter and energy, life and consciousness. Everything we are, everything we perceive and experience, think and do, is Spirit made manifest.

We share Spirit's creativity. Hence, we are free to experiment and innovate, make mistakes and learn; free to co-create the universe in a process begun and empowered by Spirit. We are free to create evil as well as good, suffering as well as happiness. And as co-creators we share responsibility for our world. This is the price of freedom.

Spirituality is not an optional extra. It is part of our very nature because we come from the same essence as Spirit. But our perceptions of Spirit are mediated by our brains, cultures and personalities, and hence are dim, distorted reflections of the reality. We have no way to be sure which insights are the truest images of Spirit; nor which are genuine revelations appropriate for our time and culture. The best we can do is to examine the impact of any spiritual path upon its followers. Does it empower the realization of universal values in their lives, and communities? Does it increase love, compassion, truth, beauty, wisdom and other values which foster the evolution of a coherent, harmonious society and planet?

Within these guidelines, we are free to choose our own spiritual path, our own way of relating to the Spirit of the universe. There is no single best way, but a choice amongst shamanism, polytheism, monotheism, transcendentalism and further 'isms' yet to emerge. And within these broad highways, we can match our practices to our temperaments and the stages of our life's journey, choosing amongst devotion and intellect, detachment and engagement, solitude and community, exoteric and esoteric, transcendent and immanent.

Of particular importance for our age is the balance between transcendent and immanent; the recognition that everyday life can be a spiritual path just as much as retreat and contemplation. Conscious vocation, partnership or parenting; good neighborliness; aware consumerism; community service; action for political, social or economic reform; deep connection with nature; creativity in all its forms – all these and more can be spiritual paths that may lead us to experience of the transcendent as well as our inner depths.

When we find a spiritual path, we are well on the way to discovering who we are. But who we are comes as a surprise to many of us. There is remarkable unanimity amongst scientists, philosophers and spiritual traditions that our perceived 'self' is an illusion. We are stories woven from the multiple strands of our being and our roles in the world. And we exist as separate beings only in and through our connections and relationships with nature and our society. As our awareness of this 'interbeing' grows, so our sense of self enlarges from 'skin-encapsulated ego' to 'person-in-environment-and-society'. We integrate more and more of who we are until 'self' becomes 'Self', and we recognize ourselves as One with all that is, One with Spirit. We realize that we are God.

Death is the price of life and creativity; an inevitable part of who we are. But our personal death is not the end. The quality and energy of our existence lives on, encoded in the patterns and rhythms of the universe. Our lives potentially affect the future not only of our descendents, but also of other beings who resonate with them. Our lives may trigger events that change the course of evolution of the Earth and the very cosmos.

The meaning of our lives consists in our paradoxical roles as both separate individuals and integral parts of the whole. As independent organisms, meaning comes from the simple fact of existence, and from perpetuating our kind through reproduction. But by itself this self-centered meaning produces hollow lives. Fulfillment flows from becoming loving, harmonious components of the whole; and from entering fully into the richness of life with all its joys and woes. The deepest meaning is found in co-creating our planet, the cosmos, and Spirit itself.

Meaning leads naturally to purpose. It matters what we do because our present actions determine our future, both in this life and beyond death. We cannot save the world by ideology, politics, science or information, but only by the quality of our individual lives. The transformation of our society, world and universe must start and end with the transformation of ourselves. Our life-long task is to integrate the various aspects of ourselves into a whole being. Whether we choose a life of action or contemplation, our purpose is to become deeply loving, compassionate and truthful. And through this to co-create with others and Spirit a person, a community, a civilization, a planet, and a cosmos which are whole and harmonious.

Reflections on Part VII

In this section, I have gathered a wide range of quotations that speak to me, and I hope speak to you, about the existence and nature of Spirit; our search for truth, identity, and purpose; and what it means to live a successful life. They come from scientists, philosophers, poets, spiritual teachers and others. I suggest you read them slowly and meditatively, and allow yourself time to reflect deeply on any that resonate.

On Spirit

(T)he idea of an enfolded implicate order implies that the whole of reality is enfolded within each individual ... Enfolded within each one of us is the implicit ground which is sustained by the eternal spring which bubbles up from the unnamed source of creativity.

It is clear, therefore, that this creative source is present in each of us and its manifestation unfolds not only into consciousness and the physical body but into the whole culture, civilization, and the entire universe ... Like the pure spring that bubbles out of the rock unbidden, this creativity has no necessity and no end, and pervades all of existence.

... within each moment of a person's life, or a speck of dust on the ground, is enfolded the whole universe, which is, itself, the manifestation of an unimaginable and unnameable creativity.

Physicist David Peat[1]

The Void exists beyond form of any kind. While being a source of everything it cannot itself be derived from anything else. It is beyond space and time. ... this absolute emptiness is simultaneously pregnant with all of existence since it contains everything in a potential form. ... various forms can emerge from this Void and take on an existence ... and ... they can do so without any apparent cause or reason. ... By choosing a particular reality, that reality is created in consciousness.

Psychiatrist Stanislav Grof[2]

(C)onsciousness evolves in a direction defined by ever-increasing integration. At its limit, this process ... (leads) to the insight that All is One. Which is exactly the message of the sacred traditions of the world's great faiths. We live, I believe, at a pivot-point in evolution because the two great traditions of the human species are now converging on the same spot. On the one hand we have science which ... uses 'left-brain' reasoning to arrive at the conclusion that to explain anything you have to explain everything. And on the other hand we have the sacred tradition which ... uses 'right-brain' intuition to arrive at the conclusion that the part has no meaning outside the whole.

Microbiologist Darryl Reanney[3]

We are entering a time when the worlds of the poets and mystics will blend with the world of the scientist. We are entering a time when knowledge will come from beyond time as well as from within time; from intuitive experience as well as rationality. To paraphrase Ken Wilber:[4]

Ecologists claim that at some point in history we deviated from the Great Spirit, the Goddess of nature. But the idea that it is possible to deviate from the flow of the Kosmos is arrogant and egocentric. It overlooks the fact that if the Great Spirit is truly great it must be behind even those moves that look to us like deviations. As Zen would put it: "that which one can deviate from is not the true Tao."

I had considered the universe as a gift of God. Now it occurred to me; What if the universe were not a gift of God, but the very body of God? What if I were, even now, reposing within God? What if every slug, stone, salmon berry, each snowflake and each human embodied its creator? WOW! But it made sense: the galaxies, their myriad interwoven systems, from the atoms to the stars and beyond – this pulsing organism – God made it, yes, but God also was it!

Deep ecologist Judy Brown[5]

If we are indeed Spirit ...
If the creative source of the universe is present in each one of us ...
If each thought and action of ours can influence All that Is ...
What then?

When before the beauty of a sunset or of a mountain you pause and exclaim, 'Ah,' you are participating in divinity.

Upanishads
Quoted by Campbell (1988) p.207

The Search for Truth

Science appears to give us a rational, objective and empirical foundation for our knowledge. By contrast, spirituality relies on non-rational (but not irrational) subjective and intuitive knowledge – both our own and that of others. We have been taught to distrust such knowledge. The Buddha also taught his followers to doubt and question, but his 'noble doubt' contrasts with our modern 'mean-spirited' doubt.

Our contemporary education ... has created in fact what could almost be called a religion or theology of doubt, in which to be seen to be intelligent we have to be seen to doubt everything ...

The Buddha summons us to another kind of doubt, "like analyzing gold, scorching, cutting, and rubbing it to test its purity." For that form of doubt ... we have neither the insight, the courage, nor the training. ...

The vast truth of the mystical teachings handed down to us is not something that our endangered world can afford to dismiss. Instead of doubting them, why don't we doubt ourselves: our ignorance, our assumption that we understand everything already, our grasping and evasion, our passion for so-called explanations of reality that have about them nothing of the awe-inspiring and all-encompassing wisdom of what the masters, the messengers of Reality, have told us?

This kind of noble doubt spurs us onward, inspires us, tests us, makes us more and more authentic, empowers us, and draws us more and more within the exalting energy field of the truth. ...

Sogyal Rinpoche[6]

Fired by 'noble doubt', we can find the truth within ourselves, as Robert Browning knew.

> Truth is within ourselves; it takes no rise
> From outward things, whate'er you may believe.
> There is an inmost centre in us all,
> Where truth abides in fullness; and around,
> Wall upon wall, the gross flesh hems it in.
> This perfect, clear perception – which is truth.
>
> A baffling and perverting carnal mesh
> Binds it, and makes all error: and to know,
> Rather consists in opening out a way
> Whence the imprisoned splendour may escape,
> Than in effecting entry for a light
> Supposed to be without.

Robert Browning[7]

And Kabir reminds us that we have only this lifetime to discover the Truth.

> Friend, hope for the truth while you are alive,
> Jump into experience while you are alive!
> Think ... and think ... while you are alive.
> What you call "salvation" belongs to the time before death.

Kabir[8]

Who am I?

Man studies himself ... in order to understand the universe ... This demands a much more rigorous discipline than any other form of knowledge-seeking. Self-knowledge is concerned with the quality, not the quantity of knowledge gained, it is a total awareness, not an analytical dissection, a building up, not a breaking down. It aims at direct knowledge which transcends thought.

J C Cooper[9]

Who am I?

You are ... host to a billion or so atoms that once belonged to Jesus Christ, or Julius Caesar, or the Buddha, or the tree that the Buddha once sat beneath.

Physicist Paul Davies[10]

Who am I?

What you are is what you have been,
What you will be is what you do now.

The Buddha[11]

Who am I?

This is the message of science to each of us and it is, I submit, stranger than anything the shamans of past ages taught their children: we are a star strangely and wonderfully fashioned into a thinking creature. We are starstuff made conscious of itself. We are a cosmos awakening in self-awareness, seeking to know what it was in order to understand what it is, so that it can look forward to what it may be.

Darryl Reanney[12]

(W)e are the children of this beautiful planet ... We were not delivered into it by some god, but have come forth from it. We are its eyes and mind, its seeing and its thinking. And the earth, together with its sun, this light around which it flies like a moth, came forth, we are told, from a nebula; and that nebula, in turn, from space. So that we are the mind, ultimately, of space.

Joseph Campbell[13]

Who am I?

I have a body, but I am not my body. I have emotions and thoughts but I am not my emotions; I am not my beliefs. I have desires, but I am not my desires. I recognize and affirm that I am a Center of pure self-consciousness. I am a Center of Will, capable of mastering, directing and using all the psychological processes and my physical body.

Willis Harman and Howard Rheingold[14]

I am the wind that blows o'er the sea;
I am the wave of the deep;
I am the bull of seven battles;
I am the eagle on the rock;
I am a tear of the sun;
I am the fairest of the plants;
I am a boar for courage;
I am a salmon in the water;
I am a lake in the plain;
I am the word of knowledge;
I am the head of the battle-dealing spear;
I am the god who fashions fire in the head.

Celtic Chant[15]

Who am I?

(I)dentity is not a solitary achievement but a communal experience, always implying a relationship to others. ... When I see that my very identity is always shaped in part by those I am with, then I can glimpse my soul as fluid and multiple. ...

At the practical level this ... means that I can be most myself when I am engaged with other people. I sense myself as an individual in context, in relationship to another. I don't have to insist on my separateness, thinking that only by protecting my individuality will I be an independent person.

Thomas Moore[16]

Who am I?

When we let go of identification with the contents of the mind:
(W)e discover that all we imagine ourselves to be – all our
becoming, our memory, all the contents of mind – is just old film
running off. The projectionist has died. "Who am I?" can't be
answered. We cannot know the truth. We can only be it. ... It is
difficult to let go of the security of some imagined "I," and enter into
the not knowing of just being.

Stephen Levine[17]

But when we do transcend our individual selves and open up to Spirit,
a light and power pours through and transforms us.

(T)hose persons through whom the soul shines, through whom
the "soul has its way," are not therefore weak characters, timid
personalities, meek presences among us. They are personal plus,
not personal minus. Precisely because they are no longer exclusively
identified with the individual personality, and yet because they still
preserve the personality, then through that personality flows the
force and fire of the soul. They may be soft-spoken and often remain
in silence, but it is a thunderous silence that veritably drowns out the
egos chattering loudly all around them. Or they may be animated
and very outgoing, but their dynamism is magnetic, and people are
drawn somehow to the presence, fascinated. Make no mistake, these
are strong characters, these souls ... because their personalities are
plugged into a universal source that rumbles through their veins and
rudely rattles those around them.

Ken Wilber[18]

Coming to Terms with Death

I have come to realize that the disastrous effects of the denial of death
go far beyond the individual: They affect the whole planet. Believing

fundamentally that this life is the only one, modern people have developed no long-term vision. So there is nothing to restrain them from plundering the planet for their own immediate ends and from living in a selfish way that could prove fatal for the future.

Sogyal Rinpoche[19]

Is life the incurable disease?
The infant is born howling
& we laugh,
The dead man smiles
& we cry, resisting the passage,
always resisting the passage,
that turns life
into eternity.

Blake sang alleluias
on his deathbed.
My own grandmother
hardly a poet at all,
smiled
as we'd never seen her smile
before.
Perhaps the dress of flesh
is no more than a familiar garment
that grows looser as one diets
on death, & perhaps we discard it
or give it to the poor in spirit,
who have not learned yet
what a blessing it is
to go naked?

Erica Jong[20]

There would be no chance of getting to know death if it happened only once. But fortunately, life is nothing but a continuing dance of birth and death, a dance of change. Every time I hear the rush of a mountain stream, or the waves crashing on the shore, or my own heartbeat, I hear the sound of impermanence. These changes, these small deaths, are our living links with death. They are death's pulse, death's heartbeat, prompting us to let go of all the things we cling to.

Sogyal Rinpoche[21]

The Meaning and Purpose of Life

Meaning brings motivation. Motivation leads to action. Action leads to transformation. Transformation is possible because human life can rise above present circumstance. ...

Human life feeds on purpose. Richness of life depends upon purposes we freely choose. ...

When I fail to give myself in full commitment to that which matters most I inhibit and frustrate myself. ...

To live is to feel, to think and to act. The call to the full life is to love with all our heart and mind and strength ... I know of no greater commitment that life can make. ...

The future is not closed. It is open. The resources of God have not been exhausted. ... We do not need to go on as we are now. No man or woman need stay the way he or she is. No society need live for ever with the status quo. That God is involved and that we are involved in God does not mean that God will look after it all and all will be well. .... There is a real sense in which the future is in our hands. ... For us to fail to respond to the forward call of life is not just a personal failure. It is a cosmic tragedy.

Biologist Charles Birch[22]

I don't believe life (as such) has a purpose. ... But each incarnation ... has a potentiality, and the mission of life is to live that potentiality.

Mythologist Joseph Campbell[23]

A human being is part of a whole, called by us the "Universe," a part limited in time and space. He experiences himself, his thoughts and feelings, as something separated from the rest – a kind of optical delusion of his consciousness. This delusion is a kind of prison for us, restricting us to our personal desires and to affection for a few persons nearest us. Our task must be to free ourselves from this prison by widening our circles of compassion to embrace all living creatures and the whole of nature in its beauty.

Physicist Albert Einstein[24]

Each action, even the smallest, is pregnant with its consequences. ... As everything is impermanent, fluid, and interdependent, how we act and think inevitably change the future. There is no situation, however seemingly hopeless or terrible, such as a terminal disease, which we cannot use to evolve. And there is no crime or cruelty that sincere regret and real spiritual practice cannot purify.

Tibetan Buddhist Master Sogyal Rinpoche[25]

So each time we make a choice that puts self ahead of other, each time we withhold a word of compassion from a troubled friend, we shift the balance, albeit perhaps slightly, towards our collective extinction. By contrast, each time we smile at someone in the street, each time we extend a caring hand to a fellow creature in distress, we move – all of us – towards that light which illuminates the near-death moment with love.

Microbiologist Darryl Reanney[26]

(W)e have only the world that we bring forth with others, and only love helps us bring it forth.

Biologist Humberto Maturana and Cognitive Scientist Francisco Varela[27]

Some day, after we have mastered the winds, the waves, the tides and gravity, ... we shall harness ... the energies of love. Then, for the second time in the history of the world, man will have discovered fire.

Jesuit Priest and Paleontologist Pierre Teilhard de Chardin[28]

For as long as space exists
And sentient beings endure,
May I too remain,
To dispel the misery of the world.

Lord make me an instrument of thy peace,

Where there is hatred, let me sow love;
Where there is injury, pardon;
Where there is doubt, faith;
Where there is despair, hope;
Where there is darkness, light;
And where there is sadness, joy;
Grant that I may not so much seek
To be consoled as to console;
To be understood as to understand;
To be loved as to love;
For it is in giving that we receive,
It is in pardoning that we are pardoned,
And it is in dying that we are born to eternal life.

Prayer attributed to St Francis

The Successful Life

To laugh often and much; to win the respect of intelligent people and the affection of children; to earn the appreciation of honest critics and endure the betrayal of false friends; to appreciate beauty, to find the best in others; to leave the world a bit better, whether by a healthy child, a garden patch or a redeemed social condition; to know even one life has breathed easier because you have lived. This is to have succeeded.

Ralph Waldo Emerson[29]

The yearning for a state of total fulfillment ... could never be satisfied by even the most spectacular achievements in the external world. ... He suddenly understood the message of so many spiritual teachers that the only revolution that can work is the inner transformation of every human being.

Stanislav Grof[30]

We may detect our destiny, the seeds of our life, and our calling in the needs of the world around us. With this kind of "obedience" we move toward community in a deep way. We see our own fulfillment as entangled in that of the people around us. As we find our true calling and live it out, others prosper; and as others fulfill themselves as a community, they provide us with an irreplaceable context for our own unfolding and self-discovery.

Thomas Moore[31]

Extremes are to be avoided since they are incompatible with balance, putting undue weight on one side or the other, whether religious, political or moral. Extremes are the sphere of the essentially ignorant and immature. All militancy is a mistake; being an extreme measure, it involves 'excess of strength' in which 'there exists regret' and from which it is difficult to retreat in the event of going too far – flexibility and command of the situation are lost and over-balance and catastrophe become inevitable. All extremes ... defeat their own ends in arousing strong opposition and in giving rise to fanaticism and bigotry. Taoism, Confucianism and Buddhism all preach the Doctrine of the Mean. It is a difficult path to tread.

J C Cooper[32]

(Just as) an author who has spent years creating and refining a book that is his life's work ... saves only the finished product because this ... best expresses his message ... (so) the collective singing of the universe preserves those songlines that best harmonize with its basic themes of love and unity, abandoning without judgment those that fail to rhyme.

Darryl Reanney[33]

Part VIII:
Journey's End

What a long way we have traveled together; what vast territories we have crossed. And what a long journey the writing has been for me! A journey of discovery and revelation, full of surprises; a journey guided by an inspiring vision, but without a clear destination until I got here. In many ways, as I relax at journey's end, I feel I have found what I sought, I have done what I set out to do, and I hope you, my readers, have benefited too. But there is another part of me that is disappointed at not finding a simple Truth that is the answer to life, the universe and everything. Instead, I find myself at a temporary resting place, recuperating before I shoulder my pack again for the next stage of my life's journey.

Before we part, I want to leave you with two things. A final brief Review, containing a challenge to us all. And a final meditation which encapsulates the story of our journey together and thus forms a fitting summary and conclusion.

A Final Review
and Challenge

For now, what I have written seems to me to hold the essence of Truth. But it is a Truth without certainty and riddled with paradox; a Truth in constant flux, developing and evolving with the universe around us. And so it is possible that I am wrong! Only beyond death, on the other side of time, will I truly know, and then only if my individual consciousness survives the transition. I hold the tension of living by what I believe whilst knowing I may be deluded. And that is part of what I have discovered – that human knowledge of the great mystery of existence is inevitably fragmentary, distorted and impermanent.

Science is a wonderful, beautiful part of human culture. The way it peels back the layers of nature's onion to reveal ever-deeper secrets is little short of miraculous. The progress in every field has been amazing in the decade since I began this book. Scientific knowledge is advancing faster than ever before, and the pace is accelerating. Hence the details, and even some of the main features, of the cosmos as portrayed in this book will change in the coming years and decades. Most scientists expect these changes to blow away the last vestiges of religious superstition, and prove once and for all that reality is purely material. By contrast, I expect them to reveal more and more clearly the underlying spiritual nature of reality; to show that matter flows from consciousness and Spirit, not the reverse.

At its birth, the scientific method was seen as the savior that would free humanity from want, disease, the whims of nature, and religious bigotry. In many parts of the world, it has largely fulfilled that promise, and its full potential is only now becoming apparent. And yet the image of science is tarnished. Almost every time it solves one problem, it seems to throw up another; its power is more often harnessed for war or consumer trivia than to meet human or planetary needs; and scientists have mounted the throne of authority from which they so recently deposed the priests. The fault lies not so much in science itself, as in our expectations and uses of it. In looking to science as our savior, we fail to see that it cannot answer all questions, and fail to seek wisdom elsewhere.

Yet even now, science may bear within it the seeds of our salvation because it has brought us face to face once more with the spiritual dimension of reality. The future lies in synthesizing spiritual understanding with the rational inquiry, empiricism and logical reasoning of science – as many of the world's greatest scientists have done. Science is the most effective means yet discovered of finding out how things work and what is materially possible; and spirituality is the best way of discovering why things are the way they are, and of gaining the wisdom to know what we should do.

Science's potential for good is mind-blowing. It would be an immense tragedy if this knowledge were lost through the collapse of civilization and the planet's life-support systems, or if our arrogant meddling were to destroy the qualities that make us human. And yet these things may happen unless we learn to use science more wisely. We need to take off our blinkers, and abandon the prejudice that the world is nothing but blind, inanimate, amoral matter. Science is no longer fighting religious bigotry. That war has been won, and so we can afford to re-open the windows to the spiritual dimension. If we can learn to combine material and spiritual science, we will reach a new Enlightenment in both the Renaissance and religious senses. And we will be on the way to discovering a true Theory of Everything.

Like many others, this book is a call to action; a call based on love and affirmation of our deepest human values. It is only by love that we can transform ourselves and the world; love, first and foremost, for our selves, flowing thence to others and the whole of creation. If we lovingly pursue our deepest meanings and values, we will transcend our present way of being, and thus resolve our deepest problems as individuals and societies, and as a species.

The present plight of humanity is cause for both concern and hope. Concern, because we teeter on the brink of annihilation. Hope, because our turmoil is exactly what we should expect of a system in creative transformation. It is the birth of a new era. We are at a bifurcation point in the history of humanity, of Gaia, and perhaps of the cosmos. The future is balanced on a knife's edge, and the actions of each one of us may tip the scales one way or the other. The media are filled with hate and destruction, but, amidst the nightmares, the light of love still shines like a beacon. Let us add our candle flames.

I will leave you with a final guided meditation that encapsulates the message of this book, and so forms a fitting summary and conclusion. But first, I want to share a favorite quotation from playwright Christopher Fry.[1]

> The human heart can go to the lengths of God.
> Dark and cold we may be, but this
> Is no winter now. The frozen misery
> Of centuries breaks, cracks, begins to move;
> The thunder is the thunder of the floes,
> The thaw, the flood, the upstart Spring.
> Thank God our time is now when wrong
> Comes up to face us everywhere,
> Never to leave us till we take
> The longest stride of soul men ever took.
> Affairs are now soul size.
> The enterprise
> Is exploration into God.
> Where are you making for? It takes
> so many thousand years to wake.
> But will you wake? For pity's sake!

A Final Reflection: The Inward and the Outward Journeys

I want to take you on two imaginative, meditative journeys of exploration – first, at smaller and smaller scales into the heart of yourself, and then at larger and larger scales out into the cosmos. These two journeys encapsulate the whole message of this book.

To help you relax and allow yourself to be carried by the words, it would be best if you can get a friend to read the meditation for you, or record it first. Please read it slowly, with pauses for reflection. Allow at least 30 minutes up to an hour, and make the whole journey in one session.

Please make yourself comfortable in your favorite chair, or in meditation posture, or lie on your back on the floor.

Breath slowly and deeply. Feel the air moving down into your lungs and out again. Let go any tensions of which you are aware as you breath out.

Move your attention from part to part of your body sensing any tensions and letting them go.

Your feet and ankles; calves, knees and thighs; your buttocks; your stomach and the small of your back; up your spine and chest; your hands and arms; shoulders and neck; face and head.

Feel your body making closer and closer contact with the floor or the chair as you relax.

And now breath gently and listen as I take you on a journey to the center of yourself.

Imagine you are looking at your own body through an X-ray microscope, so you can peer inside, and examine it in any detail you choose. Start with normal size.

There are your organs: Heart and lungs, kidneys and liver, stomach and intestines; all connected by an amazing network of blood vessels and nerves; and supported by bones and muscles. Pause for a few moments to explore the marvelous complexities of your body with your mind's eye.

Now ask yourself, 'Where am *I*? Where is the unique person I recognize as myself?' Are you in your heart, where you feel love? Or in your solar plexus with the butterflies of fear? Are you in your brain, where you think? Or in your stomach where you feel hunger?

Perhaps, like most of us, you identify most strongly with your mind. Pause and watch your mind for a while. Watch the flow of thoughts wandering here and there. One moment your mind may be filled with a turmoil of anger, the next with gentle love; one moment you may be confident of what you can achieve, and the next filled with self-doubt; one moment reliving a joyous occasion, the next overcome with sad memories. Pause and reflect upon this transient inner world. And ask yourself, 'Where and what am *I* in this ever-changing mindscape?'

Notice that you are able to reflect; that you are aware of these inner workings of the mind; that you are conscious. But who is this witness, this inner 'I', who watches what you do so closely? And where is it located?

We normally associate our minds with our brains. So zoom in for a close-up view of your brain. Watch it as you think. Miniature lightning flashes zip here and there amongst the brain cells. And those cells are making 'thought' chemicals which carry messages to other cells. Zooming even further in, you can see tiny tubular frameworks within each cell that house busy quantum computers.

As you watch, different parts of your brain light up when you hear something, or think about a friend, or wonder about the nature of Spirit. Keep a careful look out for your witness, the center of your consciousness, the essential *you*. ... Where are *you*? There doesn't seem to be any special place that is *you* amongst all the activity.

Now leave your brain and look around the rest of your body. Notice that all your other cells are making the same thought chemicals as your brain. And all your other cells house microscopic computers too. Could it be that all your cells are thinking, not just your brain? Could it be that when you say you have a 'gut feeling' about something, you're speaking quite literally, because your guts are conscious parts of you? Could it be that every cell in your body is imbued with consciousness?

Now imagine that you have received a heart or kidney transplant; a pacemaker or artificial hip. Are you a different person now? ... No. You do not change just because a part of you changes. So where and what are *you*? Could it be the interconnected and interactive whole, the conscious mind-body, that is you?

Let's shift focus now to your stomach and intestines, liver and kidneys. That food you ate a while ago is being broken down into simpler chemicals. Useful compounds are being carried off around your body to where they are needed for energy or to build new cells. Useless materials are being excreted.

As you watch, you see that your body is repairing itself, replacing outworn parts. Molecules which yesterday were carrots and cows are parts of you today. Your body is never the same from one moment to the next, but is constantly changing. Remember how you grew and developed from a fertilized ovum, to an infant, a child, an adolescent, an adult, and will continue to develop and change into old age. Week by week, the cells in your body die and are replaced.

So what are *you*? You are not this too, too solid flesh, but a constantly changing mirage. You are not a solid object like a rock. You are a collection of biochemical processes given temporary physical form; a form which is changing with growth and development, damage and decay. Your body is not a fixed anatomical structure, but an ever-changing river. Pause to feel the energy, materials, information and consciousness coursing through you.

Now zoom in for a closer look at the cells of which you're made. As you look around your body you see that many of the cells aren't human at all. They're bacteria. In fact, most of your cells seem to be bacteria, living cooperatively with your body's human cells.

Peer even closer, into the very heart of your cells, at the tiny blobs called mitochondria. And inside these mitochondria you see their own, independent genes. But what are they doing here, in the very heart of you? Could it be that our human cells, and those of every other animal, evolved from the cooperation of two simpler cells, one living inside the other?

Could it be that you and me and all of us are biological cooperatives? Societies of organisms living de facto for mutual convenience?

Zoom in further still on the atoms and molecules of which your cells are made. These are the very atoms of which the dinosaurs were made. These are the atoms which formed the first primeval soup of life on earth. These are the atoms which were forged in the furnaces of the stars, and which condensed from the intergalactic gas to form the planet earth. And in a little while, these are the atoms which will make the grass, and the trees, and the birds.

Travel on, into the heart of the atoms themselves, and what do you see? Tiny subatomic particles whirling in limitless space, as empty as the intergalactic void. If you brought all the subatomic particles in your body together into one solid lump, it would be a millionth the size of a tiny grain of sand.

Look again at those tiny specks of solidity. They're not really there at all, but flicker in and out of existence like dust motes in a sunbeam. Before you look at them, they are just possibilities; tendencies to exist. And in the act of looking, you freeze them into momentary existence, and then they are gone again. You are made of nothing solid, nothing tangible. You are myriad tiny ripples in the fabric of spacetime, no different to all the other tiny ripples that make the universe.

Who are you? You are a pattern of interacting waves, like those from a handful of stones tossed into a still pond. But see how the ripples that are you interact with all the other ripples in the universe. When something in you changes, the whole universe changes in subtle ways as the ripples spread outwards. And when the universe changes, you change too.

Here at the very heart of yourself is the Void of the mystics, the Ground of Being, God, Allah, Spirit. Rest a while in that awareness.

It's time now to return to your body. Change the magnification of your X-ray microscope up through the whirling atoms to your cells, and from your cells to your organs, and finally to your whole body.

Relax a moment in this present time and place. And then prepare yourself to embark on another journey, this time to the outer limits.

*   *   *

Imagine leaving your body behind, and looking down at it here on the ground.

Look at the edge of yourself, where you end and the world around you begins. At first your skin seems to form a clear, distinct boundary between you and your environment. But as you look closer, you see that the edge is fuzzy and indistinct. There are all those hairs to consider. And the openings: are the air passages through your mouth and nose, and down into your lungs inside you or outside? That molecule of water oozing out of a sweat gland – is it part of you or part of your environment? At what point does it change from being one to the other?

And over here you can see a virus working its way through a cell wall. Once inside the cell, it starts to change your genes. Is that virus part of you or not? Is the altered cell part of you or not?

The closer you look, the harder it becomes to decide precisely where you end and your environment begins. The more you look, the less your skin seems like an impenetrable barrier dividing you from your environment, and the more it seems to be simply a place where density, temperature and chemical concentrations change rapidly.

Move a little further away, and watch the air moving in and out of your lungs; unimaginably huge numbers of molecules with every breath. Watch as you breathe in air whose perfect balance of gases has been created and maintained by bacteria, plants and other living things over billions of years. Watch as you breathe in life-giving oxygen molecules which only a little while ago were released by a tree nearby. Watch as you breathe out air enriched with carbon dioxide, the very staff of life to that same tree.

Imagine drinking water. It was distilled from the ocean by the sun's rays, and carried by the winds to fall as rain over the land from where it flowed into the reservoir that supplies your house.

Watch as you eat food; bread from the fields of wheat, milk from the cattle in the pastures, fruit from the trees. And watch as you give back your wastes, food to myriad other organisms. Without your brothers and sisters, the bacteria, the fungi, the plants and the animals of this planet, you would not exist.

Now watch yourself interacting with people around you, and ask yourself 'Who am *I*?' Who is this self with whom I identify so closely?

Describe yourself. Perhaps you are a student, grandmother, daughter, golfer, partner, therapist, academic, carpenter, engineer, artist, agnostic ... whatever words fit you. ... And notice that all these descriptions are meaningless in isolation. They have meaning only in relationship to other people. So who are *you* when you're not interacting with others?

Imagine yourself isolated from the community in which you live, isolated from any human contact. You might like to imagine you were raised from an infant by a wild animal. You cannot speak, and you imitate the animal's sounds. You move on all fours, and eat the animal's food. So who are you? Without other people you would not be recognizably human.

Now describe yourself physically. You may be a male 1.8m tall and weighing 90kg. Or a female 1.6m tall weighing 70kg. But maleness and femaleness only have meaning in a world of bisexually reproducing organisms. And size and weight only have meaning in relation to something else – the standard meter and standard kilogram. How can you describe yourself without reference to things outside yourself?

Now fly up and up until you can see the whole of your town or city, the whole of your country, the whole of the living planet, Gaia. And as you look down, see how our home is indeed the living planet. See how life has created a world suitable for life, and kept it that way for uncounted millennia.

See the tiny, floating phytoplankton in the oceans, the sweeping grasslands and the mighty forests, all pumping out oxygen. See the limestone shells of myriad sea creatures sinking down, down, carrying carbon dioxide from the air to form sediments on the sea bed. See tiny marine algae giving off sulfur compounds into the air, where they form nuclei for raindrops, helping to return moisture and sulfur to the land.

See the myriad, intertwining processes by which Gaia has maintained conditions for life as the sun has grown hotter, as asteroid collisions and volcanoes have blanketed the sky with dust, as continents have broken apart, drifted and collided again.

See humanity, swarming everywhere like a cancer on the face of Gaia. But also see how we are awakening to our oneness with all life and with Gaia herself. And see the linking of individual minds through knowledge and communications into an emerging mind of humanity. A mind with the power to take control of our destiny and the destiny of Gaia. And watch the dawn of wisdom and a new age as we are transformed by a new spiritual awareness.

Look beyond Gaia to Sol, that mighty nuclear furnace without which earth would be but a lifeless lump of rock. Beyond Sol, to the Milky Way, our galaxy whose gravity drew formless gases together into countless stars and planets. And looking beyond the Milky Way, peer into the endless reaches of space, whose light comes to us from the very moment of creation.

Look back, back in time to the beginnings of it all; to the Big Bang which sent the cosmos on its way, giving it the creative energy from which all matter, all galaxies and stars and planets, and from which all life has evolved.

And at this point your inward and your outward journeys meet, at the creation of spacetime itself, in the Oneness of the cosmos. It is here that our individual consciousnesses blend and merge in unity with each other and cosmic Consciousness. It is here that we experience the nameless, indescribable Ground of all Being, the Mystery of Spirit.

It is here that we can find our true identity as eternal parts of the cosmic whole, drops in the cosmic ocean, leaves on the cosmic tree. We are one with the Universe, unconstrained by body, space or time. We are one with the cosmic stream flowing from the unimaginable past, through this present moment and on into the limitless future.

We are one with Spirit.
That art thou.
Dwell in that awareness. ...

*   *   *

Now it is time to return to your body, to yourself, and reassert your individual identity, lest you become lost in the vastness of the cosmos.

You *are* one with Spirit, but you are also a unique individual. No-one else has ever had your exact features and characteristics, or those exact atoms arranged in that exact way. No-one else has ever had your sense of identity, your experiences or your consciousness. No-one else has ever had the exact relationships with others and the cosmos that you have.

You are a leaf on the cosmic tree. Part of a living, growing, evolving, creative Being, sensitive to the conditions and actions of every part of itself.

You are indissolubly a part of the whole, and yet you are free to choose how you will live from now on. No matter what you have been, you can change yourself if you have the will to do so. You can choose to live selfishly or selflessly. You can choose to follow a spiritual path, or turn your back on the inner world of Spirit. You can choose to serve the whole and to foster love, peace and harmony. Or you can choose to serve yourself and to foster greed, alienation and violence.

You have the power to transform yourself, and flowing outward from that transformation you have the power to change your family, your community, your nation, your planet, and the future course of the very cosmos itself.

We're not trapped by the past. We need not despair at the state of the world. Every tiny action we take has the potential to tip the balance one way or the other. We can harness the creative powers of mind and consciousness, chaos, self-organization, life and evolution to co-create a better future together with Gaia and Spirit.

With the realization of our power comes an awesome responsibility. We have the responsibility to live as creative partners with humanity, Gaia and Spirit. We have the responsibility to help guide their evolution into the ways of beauty, truth and wisdom, love, compassion and harmony. And yet, even as we assume this responsibility, we can relax in the knowledge that we do not carry it alone. We share it with all humanity, all life and with Spirit.

We are not isolated, independent beings, but integral parts of the whole – the whole which created us and which sustains us from moment to moment; the whole which will support us in whatever we do for the good of the whole.

*   *   *

Now, when you're ready, and in your own time, return to your body. Return to the present time and this room.

There's no hurry. Just slowly become conscious of what's going on around you. Listen to the sounds. Open your eyes. Look around you. Breathe more deeply. Stretch and move, one limb at a time. When you're ready, get up and walk slowly around.

You may want to spend some time now in quiet contemplation. Or you may want to express what you've just experienced in some way. You may want to write about it or draw a picture; sculpt it in clay; dance it to a favorite piece of music, or go for a quiet walk.

But whatever you do, I suggest you take some time to absorb your experiences before you jump back into your daily life.

Sources and Notes

Chapter 1

1   Quoted by Goldsmith (1988) p. 162.
2   Quoted by Goldsmith (1988) p.162.
3   McCrone (2004).
4   Matthews (2004).
5   Ziman (1978), p.30.
6   Wilber (1995) p. 503.
7   Quoted by Davies, P. (1992) pp. 101-102.
8   Reanney (1994) p. 10.
9   Quoted by Siu (1957) p. 30.
10 Goldsmith (1988) pp.162-4.
11 Thorne (1994) p.96.
12 Cuthbertson (1991).
13 Einstein and Infeld (1938) p.31.
14 Lerner (2004).
15 Siu (1957) p.137.
16 Quoted by Mathews (1991) p.37-8.
17 Mathews (1991) p. 38.
18 Russell (1957) p.107.

Chapter 2

1   Chopra (1993).
2   Phillips (2005).
3   Bateson (1985) p. 41ff.
4   Davies and Gribbin (1992) p. 23.

5   Aikhenvald (2004).
6   And others disagree, seeing mathematics as a human construct. See Davies (1992) on the mathematics as reality; and Stewart (2003) and Ananthaswamy (2003) on the view that it is a construct.
7   Abram (1996).
8   Peat (1994).
9   Peat (1994) p.224.
10 Laszlo (1993) p.174.
11 Gardner (1993).
12 Goleman (1995).
13 Zohar and Marshall (2000).

Chapter 3

1   Peat (1994).
2   Cooper (1981) p.22.
3   Cooper (1981) p.20.
4   Cooper (1981) p.26.
5   Quoted by Cooper (1981) p.89.
6   Bastick (1982).
7   Diamond (1979).
8   Bloom (2004).
9   Assagioli (1991).
10 Leonard (1995).
11 Whitmore (1990).
12 Wilber (1991), p. 17.

13 Dawes et al (2005).
14 Maclean in Walker (1994).
15 Peat (1994), pp.5-6.
16 Ibid, p.65.
17 Ibid, p.67.
18 Ibid, p.8.
19 Ibid, p.276.
20 Ibid, p.62-63.
21 Ibid, p.66.
22 Cooley (1980).
23 Müller, quoted in Mitchell (1980).
24 Bortoft (1996) gives a detailed description of Goethe's approach.
25 Walker (1994).

Chapter 4
1 The Gospel of Matthew Ch. 7, v. 15-20.
2 See Wertheim (1995) for a discussion of the logic of Aboriginal culture.
3 Wilber (1995).
4 Heron (1998), Ferrer (2002).
5 Heron (1998) p.234.
6 Heron (1998) p.17.
7 Siu (1957) p.84.
8 Cited by Csikszentmihalyi and Rathunde (1990) p. 31.
9 Maxwell (1984) p.66.
10 Arlin 1990 p.230-233.
11 Robinson (1990).
12 Orwoll and Perlmutter (1990) p.169.
13 Cooper, J. C. (1981) p.8.

Review of Part I
1 Heron (1998) p.34.

Reflections on Part I
1 Midgely 1989 p. 41.
2 Peat (1994) p.251.
3 Caddy (1977) p.85.
4 T. S. Eliot *Burnt Norton* Lines 62ff.

Chapter 5
1 For more detailed descriptions of chaos see Davies (1988), Davies and Gribbin (1992), and Gleick (1987).
2 Chown (2004), Kostelich (1995), Percival (1989), Lesurf (1990), May (1989), Savit (1990), Tritton (1986).
3 See Gleick (1987) for a description of Lorenz's work. For an interesting discussion of chaos and long-range weather forecasting see Monastersky (1990) and Chapter 2 in Casti (1991). For efforts to forecast further ahead see Reich (2003).
4 McRobie and Thompson (1990).
5 Bown (1992).
6 McRobie and Thompson (1990).
7 Chown (1995).
8 Gleick (1987) p. 21.
9 Mullins (2002).
10 Graham-Rowe (2005).
11 Popular accounts of self-organization and its implications are given by Davies, P. (1988) and Prigogine and Stengers (1984). A rather more difficult but rewarding treatment is that of Jantsch (1980).
12 Kauffman (1993).

Chapter 6

1 Dawkins (1986) p. 170.
2 Clery (1992), Pendick (1994).
3 Jordan (1981).
4 Sheldrake (1988) p.57-59.
5 Brief introductions to the ideas of hierarchical or stratified systems are given in Capra (1983) and Davies, P. (1988). More detailed and technical discussions are provided by Allen and Starr (1982) and O'Neill, DeAngelis, Waide and Allen (1986).
6 Quoted by Wilber (1995) p. 49.
7 Koestler (1976).
8 Koestler (1976).
9 Wilber (1995).
10 Wilber (1995).

Chapter 7

1 Kauffman (1993).
2 Bak et al (1994).
3 Dawkins (1986) p. 242.
4 Jantsch (1980) p. 66; Holling (1978); MacKenzie (1995).

Review of Part II

1 Wilber (2000).

Reflections on Part II

1 Five different parts, each of which can be linked, or not linked, to each of the others, can form over a million different systems. The number of possible connections among 20 parts is greater than the number of subatomic particles in the universe! (von Bertalanffy, 1968, p. 25).

2 Thich Nhat Hanh (1987) pp. 45-47.
3 Quoted by Poundstone (1985) p. 24.
4 Sheldrake (1988) p.59.
5 Handy (1989) p. 4.

Chapter 8

1 Abram (1996).
2 http://encyclopedia. thefreedictionary.com/ mahayuga.
3 Barrow (1988) p.103.
4 Hawking (1988) p. 35-6.
5 Quoted by Mathews (1991) p. 68.
6 Penrose (1994) p. 230; Mullins (2004) reported the first evidence of the curvature of spacetime due to the rotation of the earth in October 2004.
7 For example, questions are being raised about the nature of gravity, eg Schilling (2004), Battersby (2003, 2004).

Chapter 9

1 Quoted by Davies and Gribbin (1992), p. 47ff. I am indebted to their discussion for much of the material in this section.
2 Bohm (1980) p.191.
3 Davies and Gribbin (1992) p. 142-3.
4 Campbell (1995), Cartlidge (2004).
5 Sheldrake (1988) p.118.
6 Quoted by Capra (1976) p. 233.
7 Capra (1976) p. 157.
8 Davies and Gribbin (1992), p.213ff.

9   Brooks, M. (2004).
10   Bohm (1980) p. 174.
11   Stein (1996), Reich (2004),
     Brooks (2004a), Anon (2004).
12   Brooks (2004a).
13   Penrose (1994) p. 300.
14   Laszlo (1993).
15   Hecht (1995).
16   Chown (2001).
17   Chown (2001).
18   Penrose (1994) p. 337ff.
19   Davies and Gribbin (1992) p.
     226.
20   Penrose (1994) p. 330.
21   Davies and Gribbin (1992) p.
     308.
22   Brown (1994).
23   Olshansky and Dossey (2004).
24   Radin (2002).
25   The information in this section
     is mainly drawn from Greene
     (1999).
26   David Bohm's ideas are expressed
     most fully in his book (Bohm,
     1980); a summary of them is
     given by Briggs and Peat (1984).
27   Briggs and Peat (1984) p.232.
28   Sheldrake (1988).
29   Sheldrake (1988) p. 131.
30   Briggs and Peat (1984) p. 231.
31   Laszlo (1993).
32   Penrose (1994).
33   Brooks (2003).
34   Chown (2004a).

Reflections on Part III
1   Thompson cited by Birch (1990),
     and Birch himself p. 54.
2   Thich Nhat Hanh (1987) pp. 37-
     39, 43.
3   Quoted by Waddington (1977)

     p. 64.
4   Quoted by Wilber (1979) p.67.
5   Wilber (1979) p.69.
6   Frodsham (undated).
7   Quoted by Davies and Gribbin
     (1992) p. 82.
8   Reanney (1994) p. 33.
9   Davies and Gribbin (1992) p. 83.
10   Quoted by Wilber (1979) p. 65.
11   Quoted by Wilber (1979) p. 69.
12   Wilber (1993) p. 85-86, 88.
13   Campbell (1988) p. 223.
14   Thich Nhat Hanh (1987) p.110-
     11.
15   Davies (1984) p. 221.
16   Briggs and Peat (1984) p. 89.
17   Davies and Gribbin (1992) p.
     283.
18   Briggs and Peat (1984) p. 90.
19   Laszlo (1993) p. 34.
20   Thorne (1994) pp. 397-405.

Chapter 10
1   For a fuller discussion of the
     limits of knowledge, see Munitz
     (1986) p.150ff.
2   Laszlo (1993) p.200.
3   Toulmin (1982) p.41.
4   Mathews (1991).
5   Quoted by Sheldrake (1990) p.
     102.
6   Hawking (1988) p.50.
7   Hick (1999).
8   Munitz (1986) p.115 ff.
9   Reanney (1994) p. 17.
10   Lawton (2001).
11   Bohm (1980); Reanney (1994)
     p. 22ff.
12   Gribbin (2004).
13   Gribbin (1994 a).
14   Davies and Gribbin (1992);

Munitz (1986).
15 Davies (1988) p. 135.
16 Battersby (2003, 2004).
17 Minkel (2002), Susskind (2003).
18 http://map.gsfc.nasa.gov/m_
uni/uni_101matter.html.
19 Battersby (2005).
20 Ibid.
21 Chown (2005).
22 Lerner (2004).
23 Reich (2004a).
24 Kamat (2004).
25 Gefter (2004).

Chapter 11
1 Hick (1999).
2 Freke and Gandy (2001).
3 Quoted by Freke and Gandy
(2001).
4 Freke and Gandy, op cit, p.138.
5 Nowak (2004) describes the
many artificial body parts
already in use or in advanced
development, some of which are
claimed to work better than the
originals.
6 To view images of these
formations, sign up for the free
crop circle database on ccdb.
cropcircleresearch.com and look
at images for 21 August 2001,
and 15 August 2002.
7 I am indebted to Paul Davies
(1994) for this scenario, and for
many of the ideas which follow
in this section.
8 Battersby (2005).
9 Davies (1994) p.125.
10 Davies (1994) p.126.
11 Davies (1994) p.155.

Chapter 12
1 Laszlo (1993) p. 85-86; Munitz
(1986) p. 239-40; Hawking
(1988) p. 131ff; Davies and
Gribbin (1992) p. 280ff.
2 Hawking (1988) p. 130.
3 Birch (1990) p. 71.
4 Sheldrake (1988) p.9.
5 Hawking (1988) p. 133ff.
6 Gribbin (1994 a).
7 For more detailed and
fascinating descriptions of black
holes see Davies and Gribbin
(1992) and Thorne (1994).
8 Davies and Gribbin (1992),
pp.271-2.
9 See for example Thorne (1994),
and Carl Sagan's science fiction
novel *Contact*.
10 Gribbin (1994 a).
11 Laszlo (1993) p. 201ff.
12 Bohm (1980).
13 Davies (1992, 1994).
14 Battersby (2005).
15 Capra (1976) p. 316ff; Chew
(1968).

Reflections on Part IV
1 Quoted by Levine (1982) p.186.
2 From the *Tao Te Ching*, Chapters
21, 25, translated by Kwok,
Palmer and Ramsay (1993).
3 Quoted by Ken Wilber (1993) p.
52.
4 Munitz (1986) pp. 233-235.
5 Hawking (1988) p. 184-5.
6 Reanney (1994) p. 26.
7 Harman and Rheingold (1984)
p.128.
8 Sheldrake (1990) p. 101.
9 Reanney (1994) p. 4.

10   Sheldrake (1988) p. 258.
11   Harman and Rheingold (1984) p. 128.
12   Quoted by Wilber (1993) p. 17.
13   Quoted by Davies (1992) p.125.
14   Davies (1992) p. 125.
15   Brooks (2002).
16   Barrow (2003).

Chapter 13
1   Holmes (2004), Ainsworth (2004).
2   Holmes (2004).
3   Raven and Johnson (1992) p.67.
4   Maturana and Varela (1992) p. 43, 57.
5   Raven and Johnson (1992) p. 573-4.
6   Raven and Johnson (1992) p. 576.
7   Ferry (1994).
8   Sheldrake (1990) pp.77-78.
9   Bentley (2004).
10   Davies (1998).
11   Davies (1988) p. 118; Laszlo (1993) p93.
12   Sheldrake (1988) p. 123.
13   Davies (1998).
14   Davies (1998), p.62.
15   Hecht (1994).
16   Cohen (2004a).
17   Lewin (1990).
18   Kauffman (1993).
19   Davies (1998).
20   Davies (2004); Brooks (2004a).

Chapter 14
1   Becker and Selden (1985).
2   Kauffman (1993) p.418-422.
3   Dawkins (1986) p. 169-70.
4   Dawkins (1986) p.52, 112, 122,

169-70, 296.
5   Goodwin (1994) p. 40.
6   Kauffman (1993) p.614.
7   Gribbin (2004), p.126.
8   Perhaps the best known example is the Mandelbrot set, the intricate traceries of which are used as art works the world over. To find many beautiful images on the internet, simply put "Mandelbrot set" into a search engine.
9   Goodwin (1994).
10   Op cit.
11   Op cit, p.154.
12   Thompson (1917).
13   Bortoft (1996).

Chapter 15
1   Wills (2003).
2   Kauffman (1993) p. 543.
3   Raven and Johnson (1992) p.388-9.
4   Anon (2004a), Kingsland (2004).
5   Holmes (2004a).
6   Coghlan (2005).
7   Kingsland (2004).
8   Weatherall (2003).
9   Caporale (2004).
10   Sheldrake (1990) p. 115; Symonds (1991).
11   Sheldrake (1988).
12   Goodwin (1994).
13   Kauffman (1993).
14   Kauffman (1993).
15   Kauffman (1993), Gribbin (2004).
16   Stewart (2003a).
17   Singer (2004).
18   Laland and Odling-Smee (2003).

19 Brown (2004).
20 Sheldrake (1988).
21 Goldsmith (1988) p. 172.
22 Tudge (1994), Miller (1994).
23 Ridley (1993).
24 Putnam (1994).
25 Sheldrake (1988) p.225.
26 Quoted by Capra (1983) p.302.
27 Coghlan (1996), Singer (2004).
28 Buchanan (2004).
29 Buchanan (2004).
30 Margulis and Sagan (1987),
    Goodwin (1994).
31 Douglas (1994).
32 Johnson (1994).
33 Dawkins (1986) p. 123ff, 170ff.
34 Bortoft (1996).

Chapter 16
1   George (2005).
2   Laland and Odling-Smee (2003).
3   Cherfas (1994).
4   Quoted by Cherfas (1994).
5   Bentley (2004).
6   Dayton (1990).
7   Anon (1992).
8   Lockett (1995).
9   Wallace and Vogel (1994) pp.6-8.
10 Lovelock (1991) p. 25.
11 Lovelock (1991) p. 32. See
    also his earlier books for more
    detailed scientific discussion
    (1979, 1988).
12 Lovelock (1991) p. 62.
13 Lovelock (1979, 1988, 1991).
14 George (2005).
15 Nielsen (1992).

Reflections on Part V
1   Bohm (1980) p.194.
2   Sommerhoff (1981) p. 148.

3   Quoted by Sheldrake (1988) p.
    15.
4   Quoted by Elgin (1980) p. 252.
5   Quoted by Capra (1983) p. 297.
6   Campbell (1988) p. 173.
7   Campbell (1988) p. 32.
8   Kaufman (1994).
9   Quoted by Devall (1990).
10 From the speech made by Chief
    Seattle in 1854 before the Treaty
    Commission. This speech, given
    in his native Duwamish, was
    jotted down by Dr. Henry Smith
    and recreated by writer Ted Perry
    in 1970.
11 Quoted by Sheldrake (1990) p.6.
12 Wise old monk in Dostoyevsky's
    The Brothers Karamozov.
13 Edgar Mitchell quoted by
    Harman and Rheingold (1984)
    p.155-6.
14 Abram (1990) p. 90.
15 Wise (1990) p.99.
16 Badiner (1990) p. xiv.
17 Quoted by Fox (1990), p. 226.
18 Quoted by Devall and Sessions
    (1985) p.101.

Chapter 17
1   Searle (1992), Humphrey
    (1994).
2   Wilber (1995, 2000).
3   Wilber (1995) p. 133ff.
4   Wilber (1995) p. 111.

Chapter 18
1   Searle (1992).
2   De Quincey (2002).
3   Small (1994).
4   Dunbar (2004).
5   Douglas (2004), Reid (2004).

6   Kettlewell (2005).
7   Kendrick (2004).
8   Anderson (2004).
9   Brown (2004).
10   Abram (1996).
11   Franklin (1994).
12   Birch (1990) p. 22.
13   Phillips (2002).
14   Phillips (2002).
15   Smith (2003).
16   Tompkins and Bird (1973).
17   McCrone (1994a); Penrose (1994) p. 357.
18   Pennisi (1995); Buchanan (2004).
19   Report in The West Australian newspaper, 27 November 1995, p. 9 based on an article in the International Journal of Physiology.
20   Chopra (1993).
21   Phillips (2002).
22   Lewin (1994).
23   Bateson (1985) p. 101ff.
24   Capra (1983) p. 315ff.
25   Birch (1990) p. 33.
26   Briggs and Peat (1984).
27   Woodhouse (2001).
28   Goswami (1993).
29   Birch (1990) pp. 78-79.
30   Goswami (1993).
31   Battersby (2005a).
32   Maybury-Lewis quoted by Reanney (1994) p. 92.
33   Birch (1990) p. 121.
34   Wilber (1995).
35   Russell (1988), Brooks (2000).

Chapter 19
1   Flanagan (1992) p. 37.
2   McCrone (1994 a); Penrose (1994).
3   Penrose (1994) p. 366.
4   Wilber (1995) pp. 96-97.
5   Sheldrake (1988); Flanagan (1992) p. 57-58.
6   Ramachandran and Blakeslee (1998) p.55.
7   Ramachandran and Blakeslee (1998) p.11.
8   Hay (1984).
9   Ramachandran and Blakeslee (1998) p.217.
10   Grof (1990) p.194.
11   Ramachandran and Blakeslee (1998).
12   Ramachandran and Blakeslee (1998) p.221.
13   Abram (1996) p.46.
14   Sheldrake (2003).
15   Sheldrake (1988) p. 215.
16   Peat (1987) p. 173.
17   Sheldrake (1988) p. 197.
18   Laszlo (1993).
19   Sheldrake (1988) p. 186ff.

Chapter 20
1   Wilber (1979).
2   Harman (2001).
3   Jung (1964) p.41-2, 57-58; Cook (1987).
4   Jung (1964) p.58.
5   Quoted by Grof (1990) p. 12.
6   Wilber (1979) p.2.
7   Grof (1990) p. 17.
8   Grof (1990) pp. 33-36.
9   Sogyal (1992) p. 320.
10   Sogyal (1992) gives a brief account of NDEs and their relationship to the Tibetan Buddhist bardo state (pp. 319-336). For more detailed accounts

see Ring (1982), Grey (1985) and Saborn (1982).
11 Moody (2004).
12 Ring (2001).
13 Parnia (2001).
14 Grof (1990) p. 142ff.
15 Harman and Rheingold (1984) pp.139-141.
16 Sogyal (1992) pp.84-86.
17 Powell (2001), p.170.
18 Woolger (1987).
19 Grof (1990) p. 125.
20 Haraldsson (2001).
21 Grof (1990) pp. 92-3.
22 Grof (1990) pp. 117-8.
23 Woolger (1987) p.34.
24 Grof (1990) p. 181.
25 Sogyal (1992) p. 267.
26 Grof (1990) p. 190.
27 Grof (1990) p. 190.
28 Olshansky and Dossey (2004).
29 Jung (1964) pp. 226-7.
30 Grof (1990) p. 12.
31 McCrone (1994 b).
32 Olshansky and Dossey (2004).
33 Radin (2002).
34 Radin (2002).
35 Powell (2001) p.172.
36 Powell (2001) p.175.
37 Harman and Rheingold (1984) p. 141.
38 Haisch (2001).
39 Anon (2004b).
40 Grof (1990) p. 182.
41 McCrone (2004).
42 Matthews (2004).
43 Harman and Rheingold (1984) p.134ff.
44 Sogyal (1992) pp. 46-7.
45 Hawkins (2004), p.144.
46 Quoted by Harman and

Rheingold (1984) p. 136.
47 Harman and Rheingold (1984) p. 135.
48 The classic work on this subject is Bucke (1991) (First published 1901). Hawkins (2004) undertook a modern version. For a first-hand account by a successful anesthesiologist, written with the guidance of an experienced transpersonal therapist, see Tart (2001), p.128ff.
49 Wilber (1979) Ch 10.

Chapter 21
1 Ramachandran and Blakeslee (1998) p.229.
2 Dennett (1991).
3 Quoted by Dennett (1991) p. 227.
4 Wilber (2000).
5 Heron (1998).
6 Ramachandran (2003).
7 Aleksander (2003).
8 Much of this account is taken from Dennett (1991) Ch. 7 and Humphrey (1994).
9 Humphrey (1994).
10 Davidson (1993).
11 Vines (1994 a), Flanagan (1992) p. 42.
12 Phillips (2003).
13 Concar (1996).
14 Phillips (2003).
15 Phillips (2003).
16 Birch (1990) p.120.
17 Humphrey (1994).
18 Phillips (2003).
19 Lewin (1994).
20 Vines (1994 a).

21   McCrone (1994).
22   Maturana and Varela (1992) p.205ff.
23   Artigiani (1991) p. 113.
24   Peat (1987) p. 220.
25   Dunbar (2004).
26   Aldiss (2001).
27   Ramachandran (2003).
28   Penrose (1994) p. 65, 70.
29   Penrose (1994) p. 199.
30   Penrose (1994) p. 410.
31   de Quincey (2002).
32   Wallace (2001).
33   Russell (2001).
34   Grof (2001), p.162.
35   Peat (1987) pp.219-220.
36   Wallace (2002).

Review of Part VI
1   Battersby (2005a).
2   Mullins (2005).
3   Powell (2001).

Reflections on Part VI
1   Quoted by Walsh and Vaughan (1980) p. 15.
2   Quoted by Laszlo (1993) p. 34.
3   Ibid.
4   Quoted by Capra (1983) p. 317.
5   Quoted by Sheldrake (1988) p. 14.
6   Reanney (1994) p. 54-55.
7   Quoted by Walsh and Vaughan (1980) frontispiece.
8   Grof and Grof (1989).
9   Quoted by Reanney (1994) p. 165.
10   Quoted by Grof (1990) p. 99.
11   Quoted by Grof (1990) p. 101.
12   Quoted by Wilber (1993) p. 60.
13   Edmund in Eugene O'Neill's play *Long Day's Journey into Night* quoted by Grof (1990) p.90.
14   Quoted by Wilber (1995) p. 286.
15   Quoted by Reanney (1994) p. 148.
16   Bucke (1991).
17   Quoted by Reanney (1994) p. 108.
18   Quoted by Sogyal (1992) p. 49.
19   Hawkins (2004).
20   In 1980, Christina Grof established the international Spiritual Emergence Network to support individuals in spiritual crisis, and to help them understand what is happening to them. Further information can be found in Grof and Grof (1989). Links to information on support services in many countries can be found on www.spiritualemergence.net.
21   Lockley (2000).

Chapter 22
1   Birch (1990) Chapter 4.

Chapter 23
1   Hick (1999) p.2.
2   Green (2005).
3   Quoted by Hick (1999) p.3.
4   Bloom (2004).
5   Ramachandran and Blakeslee (1998).
6   Ramachandran and Blakeslee (1998) p.184-5.
7   Sources include the work of Ken Wilber, John Heron, Jorge Ferrer, and Mike King (2004 and Personal Communication).
8   Freke and Gandy (2001).

9 Grof (2001).
10 Freke and Gandy (2001) p.33.
11 Freke and Gandy (2001) p.50.
12 Cooper (1981) p.77.
13 Wilber (1991) pp.175-6.
14 Heron (1998), Ferrer (2002).
15 Cooper (1981) p.116.
16 Cooper (1981) pp.33-4.
17 Wilber (1995) p. 326ff.
18 Wilber (1995) p. 489.
19 Freke and Gandy (2001) p.138.
20 One of the best known and most influential accounts of the perennial philosophy is due to Aldous Huxley (1944); Ken Wilber is also a strong advocate of this model. The summary given here is based largely on Wilber (1991) p.79ff, and Ferrer (2002, 2003).
21 Heron (1998), Ferrer (2002).
22 Ferrer (2003).
23 Ferrer (2002) p.103.
24 Quoted by Ferrer (2002) p.101.
25 Ferrer (2002) p.92.
26 Heron (1998) p.85.
27 This discussion of immanent spirituality is based largely on John Heron (1998), and David Spangler unpublished notes for an on-line forum on the Soul of Findhorn.

Chapter 24
1 Heron (1998) p.81.
2 Heron (1998).
3 Heron (1998) p.89.
4 Heron (1998) p.94.
5 Heron (1998) p.89.
6 Heron (1998) p.100.
7 Heron (1998), Ferrer (2002,

2003).
8 Ferrer (2002) p.145.
9 Wilber (2000) p.300ff.
10 Bloom (2004).
11 This list of values was compiled from two main sources: the 15 values used in a 1996 survey of representatives from 40 countries and over 50 faith communities (http://www.globalethics.org/gvs/summary.html), and the core values adopted by the Living Values in Education project supported by UNESCO and now operating in 74 countries (http://livingvalues.net/about/index.html). To these I added four ancient virtues (courage, justice, temperance, wisdom), the Taoist qualities of balance and harmony, the Buddhist ideal of non-attachment, and frugality, joy, openness, service and vocation. The values are presented in alphabetical order to avoid imputing any ranking of importance. Undoubtedly there are others which deserve to be added. My aim is not to provide a comprehensive list, but to give the flavor of the underlying quality of the good life for which we might aim.

Chapter 25
1 Ramachandran and Blakeslee (1998), Ramachandran (2003).
2 Dennett (1991), Cohen (1995).
3 Ramachandran (2003) p.115.
4 Wilber (1979) p. 50.
5 Quoted by Wilber (1993) p. 79.

6   Wilber (1979) p.51.
7   Quoted by Wilber (1979) p. 51-2.
8   Levine (1982) p. 180-181.
9   Russell (2001).
10 Flanagan (1992).
11 Dennett (1991), Flanagan (1992) p. 189, Reanney (1994).
12 Dennett (1991) p. 418.
13 Flanagan (1992) p. 199.
14 Macy (1990) p. 60.
15 Griffiths (1992) p. 61.
16 Reanney (1994).
17 Russell (2001).
18 Sogyal (1992) p. 16.
19 Douglas (1994).

Chapter 26
1   Wilber (1979) p. 19ff.
2   Campbell (1988) p. 65.
3   Reanney (1994) p.145.
4   Mathews (1991), Naess (1989).
5   Quoted by Reanney (1994) p. 111.
6   Campbell (1972) p. 227.
7   Naess (1989).
8   Capra (1983) p. 20.
9   The Tao Te Ching, Chapter 37 translated by Kwok et al (1993).

Reflections on Part VII
1   Peat (1987) p. 189.
2   Grof (1990) p. 170.
3   Reanney (1994) p. 146-8.
4   Wilber (1995) p. 453.
5   Brown (1995).
6   Sogyal (1992) p. 123ff.

7   Robert Browning *Paracelsus.*
8   Translated by Bly (1977).
9   Cooper (1981) p.85.
10 Davies (1988) p.114.
11 Quoted by Sogyal (1992) p.92.
12 Reanney (1994) pp. 19-20, 35.
13 Campbell (1972) p. 266.
14 Harman and Rheingold (1984) p. 225.
15 Campbell (1972) p. 223.
16 Moore (1994) p. 105-6.
17 Levine (1982) p. 182.
18 Wilber (1995) pp. 281-2.
19 Sogyal (1992) p. 8.
20 Erica Jong "Is Life the Incurable Disease?" from Jong (1979). Quoted with permission.
21 Sogyal (1992) p. 33.
22 Birch (1990) p. 4, 117, 174, 176; ch. 4.
23 Campbell (1988) p. 229.
24 Quoted by Sogyal (1992) p. 98.
25 Sogyal (1992) p. 92, 95.
26 Reanney (1994) p. 171.
27 Maturana and Varela (1992) p.248.
28 Quoted by Sogyal (1992) p.364
29 Quoted by Badiner (1990) p. 145.
30 Grof (1990) p.34.
31 Moore (1994) p. 110.
32 Cooper (1981) pp.39-40.
33 Reanney (1994) p. 127.

A Final Review and Challenge
1   Fry (1951).

References

Abram, D. (1990) *The Perceptual Implications of Gaia* In Badiner (Ed), pp. 75-92.

Abram, D. (1996) *The Spell of the Sensuous* Vintage Books, New York, 326p.

Aikhenvald, A. (2004) *For Want of a Word* Interviewed by A. Barnett, **New Scientist**, 31 January, pp.44-47.

Ainsworth, C. (2004) *How Did Life Begin?* **New Scientist**, 4 September, pp.25-26.

Aldiss, B. (2001) *Like Human, Like Machine* **New Scientist**, 15 September, p.40ff.

Aleksander, I. (2003) *I, Computer* **New Scientist**, 19 July, p.40ff.

Allen, P. M. (1981) *The Evolutionary Paradigm of Dissipative Structures* In Jantsch (Ed), pp.25-72.

Allen, T.F.H. and Starr, T.B. (1982) *Hierarchy: Perspectives for Ecological Complexity* University of Chicago Press, Chicago, 310p.

Allen, W. (1980) *Side Effects* Ballantine, 1980, p.13.

Ananthaswamy, A. (2003) *Think Big* **New Scientist**, 27 September, pp.38-39.

Ananthaswamy, A. (2004) *Algae Use Sex to Beat Stress* **New Scientist**, 19 June, p.14.

Anderson, A. (2004) *Don't Call Me Bird-brain* **New Scientist**, 12 June, pp.46-47.

Anderson, I. (1994) *Gene Heretic Mounts Fresh Challenge to Darwin* **New Scientist**, 10 December, p.6.

Angyal, A. (1981) *A Logic of Systems* In Emery (Ed), Vol. 1, pp.27-40.

Anon (1992) *Will They, Won't They Save the Owl?* **New Scientist**, 23 May, p.10.

Anon (2004) *Quantum Chips back on the Menu* **New Scientist**, 13

November, p.19.

Anon (2004a) *Jumping Genes Help Choreograph Development* **New Scientist**, 23 October, p.19.

Anon (2004b) *On the Edge of the Known World* **New Scientist**, 13 March, p.32.

Arlin, P.K. (1990) *Wisdom: The Art of Problem Finding* In Sternberg, R.J. (Ed), pp. 230-243.

Artigiani, R. (1991) *Social Evolution: A Nonequilibrium Systems Model* In Laszlo (Ed), pp. 93-129.

Assagioli, R. (1991) *Transpersonal Development* Crucible Books, London.

Badiner, A.H. (Ed) (1990) *Dharma Gaia* Parallax Press, Berkeley, 265p.

Badiner, A. H. (1990) *Introduction* In Badiner (Ed), pp. xiii-xviii.

Bak, P., Flyvbjerg, H. and Sneppen, K. (1994) *Can We Model Darwin?* **New Scientist**, 12 March, pp.36-39.

Barrow, J.D. (1988) *The World within the World* Oxford University Press, Oxford, 398p.

Barrow, J.D. (2003) *Glitch!* **New Scientist**, 7 June, p.44.

Bastick, T. (1982) *Intuition: How we think and act* Wiley, Chichester, 494p.

Bateson, G. (1985) *Mind and Nature: A Necessary Unity* Fontana Paperbacks, London, 251p.

Battersby, S. (2003) *The New Dark Age* **New Scientist**, 25 Jan, p.28.

Battersby, S. (2004) *The Ghost in the Cosmos* **New Scientist**, 7 February, pp.32-35.

Battersby, S. (2005) *The Unravelling* **New Scientist**, 5 February, pp.30-37.

Battersby, S. (2005a) *Are We Nearly There Yet?* **New Scientist**, 30 April, pp.30-34.

Becker, R. O. and Selden, G. (1985) *The Body Electric: Electromagnetism and the Foundation of Life* Quill, NY, 365p.

Bentley, P. (2004) *The Garden Where Perfect Software Grows* **New Scientist**, 6 March, pp.28-31.

Birch, L. C. (1990) *On Purpose* New South Wales University Press, Sydney, 195p.

Bly, R. (1977) *The Kabir Book* Beacon Press, Boston, 71p.

Bloom, W. (2004) *Soulution: The Holistic Manifesto* Hay House, London, 272p.

Bohm, D. (1980) *Wholeness and the Implicate Order* Routledge and Kegan Paul, London, 224p.

Bortoft, H. (1996) *The Wholeness of Nature: Goethe's Way of Science* Floris Books, Edinburgh, 407p.

Briggs, J. P. and Peat, F. D. (1984) *Looking Glass Universe: The Emerging Science of Wholeness* Touchstone Books, New York, 290p.

Brooks, M. (2000) *Global Brain* New Scientist, 24 June, p.22.

Brooks, M. (2002) *Life's a Sim and then You're Deleted* New Scientist¨, 27 July, p. 48.

Brooks, M. (2003) *Curiouser and Curiouser* New Scientist, 10 May, p.28.

Brooks, M. (2004) *Worlds Apart* New Scientist, 15 May, pp.30-33.

Brooks, M. (2004a) *The Weirdest Link* New Scientist, 27 March, pp.32-35.

Brown, C. (2004) *Not Just a Pretty Face* New Scientist,12 June, pp.42-43.

Brown, J. (1994) *Martial Arts Students Influence the Past* New Scientist, 27 August, p. 16.

Brown, J. R. (1995) *If the Universe is the Body of God* Newsletter of the Institute for Deep Ecology, Winter, p. 5.

Buchanan, M. (2004) *A Billion Brains Are Better than One* New Scientist, 20 November, pp.34-37.

Buchanan, M. (2005) *Charity Begins at Homo Sapiens* New Scientist, 12 March, pp.33-37.

Bucke, R.M. (1991) *Cosmic Consciousness: A study in the evolution of the human mind* Penguin Arkana, London, 384p.

Caddy, E. (1977) *The Spirit of Findhorn* Findhorn Press, Findhorn, 141p.

Campbell, J. (1972) *Myths to Live By* Viking Press, New York, 276p.

Campbell, J. (1988) *The Power of Myth* Doubleday, New York, 233p.

Campbell, P. (1995) *The Race to Create an Antiatom* New Scientist, 13 May, pp.32-35.

Capacchione, L. (2000) *Visioning: Ten Steps to Designing the Life of Your Dreams* Jeremy P. Tarcher/Putnam, New York, 256p.

Caporale, L.H. (2004) *Genomes Don't Play Dice* New Scientist, 6 March, pp.42-45.

Capra, F. (1976) *The Tao of Physics* Flamingo, London, 412p.

Capra, F. (1983) *The Turning Point: Science, Society and the Rising Culture* Flamingo, London, 516p.

Cartlidge, E. (2004) *Half the Universe is Missing* New Scientist, 4 Sept., pp36-39.

Casti, J. L. (1991) *Searching for Certainty: What science can know about the future* Scribners, London, 496p.

Cherfas, J. (1994) *How Many Species Do We Need?* New Scientist, 6 August, pp. 36-40.

Chew, G. F. (1968) *"Bootstrap": A Scientific Idea?* Science, Vol. 161, pp. 762-765.

Chopra, D. (1993) In interview with A.M. Einspruch entitled *Unconditional Life* Simply Living, March, pp. 33-34.

Chown, M. (1995) *Fly Me Cheaply to the Moon* New Scientist, 7 October, p.19.

Chown, M. (2001) *Taming the Multiverse* New Scientist, 14 July, p.26.

Chown, M. (2004) *Chaotic Heavens* New Scientist, 28 February, pp.32-35.

Chown, M. (2004a) *Quantum Rebel* New Scientist, 24 July, pp.30-35.

Chown, M. (2005) *End of the Beginning* New Scientist, 2 July, pp.30-35.

Clery, D. (1992) *Acoustic Fridge Sounds Good for the Ozone Layer* New Scientist, 4 April, p. 21.

Coghlan, A. (1996) *Slime City* New Scientist, 31 August, pp.32-36.

Coghlan, A. (2005) *Mendel Will Be Turning in His Grave* New Scientist, 26 March, pp.8-9.

Cohen, P. (2004a) *Were Volcanoes Creation's Crucible?* New Scientist, 16 October, p.14.

Concar, D. (1996) *You're Wrong, Mr. Spock* New Scientist, 27 April, p.

Cook, D. A. G. (1987) *Jung, Carl Gustav* In Gregory (Ed), pp. 403-405.

Cooley, M. (1980) *Architect or Bee? The Human/Technology Relationship* TransNational Cooperative Ltd., Sydney, 114p.

Cooper, J. C. (1981) *Yin and Yang: The Taoist Harmony of Opposites* The Aquarian Press, Wellingborough, 128p.

Crick, F. (1981) *Life Itself: Its Nature and Origin* Simon & Schuster, NY.

Csikszentmihalyi, M. and Rathunde, K. (1990) *The Psychology of Wisdom: An Evolutionary Interpretation* In R.J. Sternberg (Ed), pp.25-51.

Cuthbertson, A. (1991) *Hooked! Life in the Lab: A Story of Joy* 21C, Issue 3, pp. 16-18.

Davidson, C. (1993) *I Process therefore I Am* New Scientist, 27 March, p.22ff.

Davies, P. (1984) *Superforce: The Search for a Grand Unified Theory of Nature* Simon & Schuster, New York.

Davies, P. (1988) *The Cosmic Blueprint* Touchstone Books, New York, 224p.

Davies, P. (1992) *The Mind of God: Science and the Search for Ultimate*

Meaning Penguin Books, London, 254p.

Davies, P. (1994) *The Last Three Minutes: Conjectures about the Ultimate Fate of the Universe* Weidenfeld & Nicolson, London, 162p.

Davies, P. (1998) *The Fifth Miracle: The Search for the Origin of Life* Allen Lane, London, 260p.

Davies, P. (2004) *The Ascent of Life* **New Scientist**, 11 December, pp.28-32.

Davies, P. and Gribbin, J. (1992) *The Matter Myth* Touchstone Books, New York, 320p.

Dawes, J., Dolley, J. and Isaksen, I. (2005) *The Quest* O Books, Hampshire, 276p.

Dawkins, R. (1986) *The Blind Watchmaker* Penguin, London, 340p.

Dayton, L. (1990) *New Life for Old Forests* **New Scientist**, 13 October, pp.21-25.

Dennett, D. (1991) *Consciousness Explained* Allen Lane, The Penguin Press, London, 511p.

de Quincey, C. (2002) *Nature Has a Mind of Its Own* **Scientific and Medical Network Review**, No.80, pp.6-9.

Devall, B. (1990) *Ecocentric Sangha* In Badiner (Ed) pp.155-164.

Devall, B. and Sessions, G. (1985) *Deep Ecology: Living as if Nature Mattered* Peregrine Smith Books, Salt Lake City, 266p.

Diamond, J. (1979) *Your Body Doesn't Lie* Warner Books, NY, 208p.

Douglas, K. (1994) *Making Friends with Death-Wish Genes* **New Scientist**, 30 July, pp. 30-34.

Douglas, K. (2004) *It's Good to Bark* **New Scientist**, 12 June, pp.52-53.

Dunbar, R. (2004) *Can You Guess What I'm Thinking?* **New Scientist**, 12 June, pp.44-45.

Einstein, A. (1954) *Ideas and Opinions* S. Bargmann (Trans.), Crown Publishers, New York.

Einstein, A. and Infeld, L. (1938) *The Evolution of Physics* Simon and Schuster, New York.

Elgin, D. (1980) *The Tao of Personal and Social Transformation* In Walsh and Vaughan (Eds), pp. 248-256.

Ferrer, J. N. (2002) *Revisioning Transpersonal Theory: A Participatory Vision of Human Spirituality* State University of New York Press, New York, 273p.

Ferrer, J. N. (2003) *Participatory Spirituality: An Introduction* **Scientific and Medical Network Review**, No.83, pp.3-7.

Ferry, G. (1994) *Mad Brains and the Prion Heresy* **New Scientist**, 28

May, pp.32-37.

Flanagan, O. (1992) *Consciousness Reconsidered* MIT Press, Cambridge, 234p.

Fox, W. (1990) *Toward a Transpersonal Ecology* Shambala Publications, Boston, 380p.

Franklin, C. (1994) *Ants Are Smart Enough to Keep Appointments* **New Scientist**, 27 August, p. 17.

Freke, T. and Gandy, P. (2001) *Jesus and the Goddess: The Secret Teachings of the Original Christians* Thorsons, London, 2001, 328p.

Frodsham, J. (undated) *Intercultural Communication* unpublished ms.

Fry, C. (1951) *A Sleep of Prisoners* Oxford University Press, Oxford, 51p.

Gardner, H. (1993) *Frames of Mind* (Revised Edition) Harper Collins, New York.

Gefter, A. (2004) *The World Turned Inside Out* **New Scientist**, 20 March, pp.34-37.

George, A. (2005) *Coral Reefs Create Clouds to Control the Climate* **New Scientist**, 5 February, p.17.

Gibran, K. (1926) *The Prophet* Heinemann, London, 113p.

Gilovich, T. (1991) *How We Know What Isn't So: The fallibility of human reason in everyday life* Free Press, NY, 216p.

Gleick, J. (1987) *Chaos: Making a New Science* Sphere Books, London, 352p.

Gleick, J. and Porter, E. (1991) *Nature's Chaos* Sphere Books, London, 125p.

Goldsmith, E. (1988) *The Way: An Ecological World-view* **The Ecologist**, Vol. 18, No. 4/5, p.160-185.

Goleman, D. (1995) *Emotional Intelligence: Why it Can Matter more than IQ* Bloomsbury, London, 352p.

Goodwin, B. (1994) *How the Leopard Changed Its Spots* Phoenix, London, 233p.

Goswami, A. (1993) *The Self-aware Universe* Putnam, New York, 319p.

Graham-Rowe, D. (2005) *Glooper Computer* **New Scientist**, 26 March, pp.32-36.

Green, M. (2005) *Spirituality and Spiritual Development in Youth Work* The National Youth Agency, Leicester, 45p.

Greene, B. (1999) *The Elegant Universe: Superstrings, Hidden Dimensions, and the Quest for the Ultimate Theory* Jonathan Cape, London, 448p.

Gregory, R. L. (Ed) (1987) *The Oxford Companion to the Mind* Oxford

University Press, Oxford, 856p.

Grey, M. (1985) *Return from Death: An Exploration of the Near-Death Experience* Arkana, Boston.

Gribbin, J. (1994 a) *Is the Universe Alive?* **New Scientist**, 15 January, pp, 38-40.

Gribbin, J. (2004) *Deep Simplicity: Chaos, Complexity and the Emergence of Life* Allen Lane, London, 251p.

Griffiths, B. (1992) *A New Vision of Reality* Indus, New Delhi, 304p.

Grof, S. (1990) *The Holotropic Mind* Harper Collins, San Francisco, 240p.

Grof, S. (2001) *Non-Ordinary States of Consciousness: Healing and Heuristic Potential* In Lorimer (Ed), pp.150-168.

Grof, S. and Grof, C. (Eds) (1989) *Spiritual Emergency: When Personal Transformation becomes a Crisis* Jeremy P. Tarcher, Los Angeles, 250p.

Haisch, B. (2001) *Freeing the Scientific Imagination from Fundamentalist Scientism* **Scientific and Medical Network Review**, No.76, pp.13-15.

Handy, C. (1989) *The Age of Unreason* Hutchinson, London, 216p.

Haraldsson, E. (2001) *Children and Memories of Previous Lives* In Lorimer (Ed), pp.81-94.

Harman, W. (2001) *Towards a Science of Consciousness: Do We Need a New Epistemology?* In Lorimer (Ed), pp.23-33.

Harman, W. and Rheingold, H. (1984) *Higher Creativity: Liberating the Unconscious for Breakthrough Insights* Jeremy P. Tarcher, Los Angeles, 237p.

Hawking, S. (1988) *A Brief History of Time* Bantam Books, London, 211p.

Hawkins, P. (2004) *Dr Bucke Revisited: Cosmic Consciousness – Quantum Consciousness* Athena Press, London, 263p.

Hay, L. (1984) *You Can Heal Your Life* Hay House, Santa Monica, 200p.

Hecht, J. (1994) *'Molecule of Life' is Found in Space* **New Scientist**, 11 June, p.4.

Hecht, J. (1995) *Letting the Quantum Cat Out of the Bag* **New Scientist**, 21 October, p. 19.

Heron, J. (1998) *Sacred Science: Person-centred Inquiry into the Spiritual and the Subtle* PCCS Books, Ross-on-Wye, 271p.

Hick, J. (1999) *The Fifth Dimension: An Exploration of the Spiritual Realm* Oneworld Publications, Oxford, 274p.

Holling, C. S. (1978) *Adaptive Environmental Assessment and Management* International Institute for Applied Systems Analysis and John Wiley, Chichester, 377p.

Holmes, B. (1993) *Evolution's Neglected Superstars* **New Scientist**, 6 November, pp. 30-33.

Holmes, B. (2004) *What Is Life?* **New Scientist**, 4 September, p.32.

Holmes, B. (2004a) *The Great Inventors* **New Scientist**, 21 February, pp.40-43.

Humphrey, N. (1994) *The Private World of Consciousness* **New Scientist**, 8 January, pp. 22-25.

Huxley, A. (1944) *The Perennial Philosophy* Perennial Library Edition, Harper and Row, New York, 312p.

Jantsch, E. (1975) *Design for Evolution: Self-organization and Planning in the Life of Human Systems* George Braziller, New York, 322p.

Jantsch, E. (1980) *The Self-organizing Universe* Pergamon Press, Oxford, 343p.

Johnson, J. (1994) *First Fold Your Protein* **New Scientist**, 10 December, pp. 26-29.

Jong, E. (1979) *At the Edge of the Body* Holt, Rinehart, Winston, NY, 96p.

Jordan, N. (1981) *Some Thinking about 'System'* In Emery (Ed), Vol 2, pp. 21-39.

Jung, C. G. (Ed) (1964) *Man and his Symbols* Picador, London, 413p.

Jung, C. G. (1964) *Approaching the Unconscious* In Jung (Ed), pp. 1-94.

Kamat, S. (2004) *Postcards from the Edge* **New Scientist**, 13 November, pp.42-45.

Kauffman, S. A. (1993) *The Origins of Order: Self-Organization and Selection in Evolution* Oxford University Press, New York, 709p.

Kaufman, H. (1994) *The Emergent Kingdom* **The Futurist**, Vol. 28, No.1, pp20- 23.

Kendrick, K. (2004) *Here's Looking at Ewe* **New Scientist**, 12 June, pp.48-49.

Kettlewell, J. (2005) *Farm Animals 'Need Emotional TLC'* BBC News website, 21 March.

King, M. (2004) *The Role of Transpersonal Psychology in a Postsecular Society* British Psychological Society, Transpersonal Psychology Review, Vol. 8, No. 1, pp.6-22.

Kingsland, J. (2004) *Wonderful Spam* **New Scientist**, 29 May, pp.42-45.

Koestler, A. (1976) *The Ghost in the Machine* Random House, New York.

Kostelich, E. (1995) *Symphony in Chaos* **New Scientist**, 8 April, pp. 36-39.

Kwok, M., Palmer, M. and Ramsay, J. (1993) *The Illustrated Tao Te Ching* Element, Shaftesbury, unpaged.

Laland, K. and Odling-Smee, J. (2003) *Life's Little Builders* **New Scientist**, 15 November, pp.42-45.

Laszlo, E. (Ed) (1991) *The New Evolutionary Paradigm* Gordon and Breach, New York, 204p.

Laszlo, E. (1993) *The Creative Cosmos* Floris Books, Edinburgh, 255p.

Lawton, G. (2001) *Monsters of the Deep* **New Scientist**, 30 June, p.28.

Leonard, A. (1995) *Telling Our Stories* Darton, Longman and Todd, UK.

Lerner, E. (2004) *Bucking the Big Bang* **New Scientist**, 22 May, p.20.

Lesurf, J. (1990) *Chaos on the Circuit Board* **New Scientist**, 30 June, pp. 37-40.

Levine, S. (1982) *Who Dies? An Investigation of Conscious Living and Conscious Dying* Anchor Books, New York, 317p.

Lewin, R. (1990) *The Universal Constructor Set* **New Scientist**, 8 December, pp. 22-25.

Lewin, R. (1994) *I Buzz Therefore I Think* **New Scientist**, 15 Jan, pp. 29-33.

Lockett, J. (1995) *Incredible Journey of the Elephant Seal* **New Scientist**, 3 June, p. 16.

Lockley, M. (2000) *A Broader Look at the Spiritual Emergence Experience* **Scientific and Medical Network Review**, No 72, pp.18-19

Lorimer, D. (Ed) (2001) *Thinking Beyond the Brain: A Wider Science of Consiousness* Floris Books, Edinburgh, 287p.

Lovelock, J. (1979) *Gaia: A New Look at Life on Earth* Oxford University Press, Oxford, 157p.

Lovelock, J. (1988) *The Ages of Gaia: A Biography of Our Living Planet* Oxford University Press, Oxford, 252p.

Lovelock, J. (1991) *Gaia: The Practical Science of Planetary Medicine* Allen & Unwin, Sydney, 192p.

Macy, J. (1990) *The Greening of the Self* In Badiner (Ed), p. 53-63.

MacKenzie, D. (1995) *The Cod that Disappeared* **New Scientist**, 16 September, pp. 24-29.

Margulis, L. and Sagan, D. (1987) *Micro-cosmos: Four Billion Years of Evolution from our Microbial Ancestors* Allen and Unwin, London, 301p.

Mathews, F. (1991) *The Ecological Self* Routledge, London, 192p.

Matthews, R. (2004) *Opposites Detract* **New Scientist**, 13 March, pp.39-41.

Maturana, H. R. and Varela, F. J. (1992) *The Tree of Knowledge* Shambala,

Boston, 269p.

Matthews, R. (2004) *Opposites Detract* **New Scientist**, 13 March, pp.39-41.

May, R. (1989) *The Chaotic Rhythms of Life* **New Scientist**, 18 Nov., pp. 21-25.

Maxwell, N. (1984) *From Knowledge to Wisdom* Blackwell, Oxford, 298p.

McCrone, J. (1994) *Inner Voices, Distant Memories* **New Scientist**, 29 January, pp.29-31.

McCrone, J. (1994 a) *Quantum States of Mind* **New Scientist**, 20 August, pp. 35-38.

McCrone, J. (1994 b) *Psychic Powers: What Are the Odds?* **New Scientist**, 26 Nov., pp. 34-38.

McCrone, J. (2004) *The Power of Belief* **New Scientist**, 13 March, pp.34-37.

McRobie, A. and Thompson, M. (1990) *Chaos, Catastrophes and Engineering* **New Scientist**, 9 June, pp. 21-26.

Midgely, M. (1989) *Wisdom, Information, and Wonder: What is Knowledge for?* Routledge, London, 275p.

Miller, S. K. (1994) *Bats Sow Seeds of Rainforest Recovery* **New Scientist**, 18 June, p.10.

Minkel, J. R. (2002) *The Top-Down Universe* **New Scientist**, 10 August, p.28.

Mitchell, R.J. (ed) (1980) *Experiences in Appropriate Technology* Canadian Hunger Foundation, Ottawa, 150p.

Monastersky, R. (1990) *Forecasting into Chaos* **Science News**, Vol. 137, May 5, pp. 280-282.

Moody, R. (2004) *Near-Death Experiences and the Question of Life after Death* **Scientific and Medical Network Review**, No.76, pp.17-18.

Moore, T. (1994) *SoulMates: Honoring the Mysteries of Love and Relationship* Harper Collins, New York.

Mullins, J. (2002) *Raising a Storm* **New Scientist**, 27 July, p.28.

Mullins, J. (2004) *Relativity Tied up on a Shoestring* **New Scientist**, 23 October, p.16.

Mullins, J. (2005) *Whatever Happened to Machines that Think?* **New Scientist**, 23 April, pp.32-37.

Munitz, M.K. (1979) *The Ways of Philosophy* Macmillan Publishing Co, New York, 372p.

Munitz, M.K. (1986) *Cosmic Understanding: Philosophy and Science of the Universe* Princeton University Press, Princeton, 287p.

Murray, C. (1989) *Is the Solar System Stable?* New Scientist, 25 November, p.45.

Naess, A. (1989) *Ecology, Community and Lifestyle: Outline of an Ecosophy* Translated and revised by David Rothenberg, Cambridge University Press, Cambridge.

Nielsen, R. H. (1992) *Blowing Hot and Cold with Climate Models* New Scientist, 26 Sept., p. 15.

Norman, S. (2003) *Transforming Learning* SEAL and Saffire Press, London, 82p.

Novak, R. (2004) *Bionic Body* New Scientist, 30 October, pp.48-49.

Olshansky, B. and Dossey, L. (2004) *Retroactive Prayer: An Outrageous Hypothesis?* Network: the Scientific and Medical Network Review, No.84, pp.3-11. (First published in the British Medical J. (2003) No.327, pp.1465-8).

O'Neill, R. V., DeAngelis, D. L., Waide, J. B. and Allen, T .F. H. (1986) *A Hierarchical Concept of Ecosystems* Princeton University Press, New Jersey, 253p.

Orwoll, L. and Perlmutter, M. (1990) *The Study of Wise Persons: Integrating a Personality Perspective* In Sternberg, R.J. (Ed), pp.160-177.

Osinski, G. (2003) *Shocked into Life* New Scientist, 13 September, pp.40-43.

Parnia, S. (2001) *Near Death Experiences in Cardiac Arrest and the Mystery of Consciousness* Scientific and Medical Network Review, No.76, pp.6-8.

Peat, F. D. (1987) *Synchronicity: The Bridge between Matter and Mind* Bantam Books, New York, 245p.

Peat, F. D. (1994) *Blackfoot Physics:A Journey into the Native American Universe* Fourth Estate, London, 322p.

Pendick, D. (1994) *Magnetism's Cool Act* New Scientist, 10 September, pp. 21-23.

Pennisi, E. (1995) *The Secret Language of Bacteria* New Scientist, 16 September, pp.30-33.

Penrose, R. (1994) *Shadows of the Mind* Oxford University Press, Oxford, 457p.

Percival, I. (1989) *Chaos: A Science for the Real World* New Scientist, 21 October, pp. 20-25.

Phillips, H. (2002) *Not Just a Pretty Face* New Scientist, 27 July, p.40.

Phillips, H. (2003) *The Pleasure Seekers* **New Scientist**, 11 October, pp.36-43.

Phillips, H. (2005) *The Feeling of Colour* **New Scientist**, 29 January, pp. 40-43.

Phillips, H. and Lawton, G. (2004) *The Intoxication Instinct* **New Scientist**, 13 November, pp.32-41.

Poundstone, W. (1985) *The Recursive Universe* Oxford University Press, Oxford, 252p.

Powell, A. (2001) *Beyond Space and Time: The Unbounded Psyche* In Lorimer (Ed), pp.169-186.

Prigogine, I. and Stengers, I. (1984) *Order Out of Chaos* Flamingo, London, 349p.

Putnam, C. (1994) *Bat Mothers Share the Birth Experience* **New Scientist**, 18 June, p.18.

Radin, D. (2002) *Seeking for Whom the Bell Tolls: Exploring Mind-Matter Interactions on a Global Scale* **Network: The Scientific and Medical Network Review**, No.79, pp.12-15.

Ramachandran, V.S. and Blakeslee, S. (1998) *Phantoms in the Brain* Fourth Estate, London, 328p.

Ramachandran, V.S. (2003) *The Emerging Mind* The BBC and Profile Books, London, 208p.

Rapoport, A. (1986) *General System Theory: Essential Concepts and Applications* Abacus Press, Tunbridge Wells, 270p.

Raven, P. H. and Johnson, G. B. (1992) *Biology* Third Edition, Mosby Year Book, St. Louis, 1217p. plus appendices.

Reanney, D. (1994) *Music of the Mind: An Adventure into Consciousness* Hill of Content, Melbourne, 186p.

Reich, E. S. (2003) *Making Waves* **New Scientist**, 29 November, p.30.

Reich, E. S. (2004) *Which Way Is Up?* **New Scientist**, 2 Oct, pp.32-35.

Reich, E. S. (2004a) *Seeds of Doubt for Theory of Inflation* **New Scientist**, 28 February, p.9.

Reid, H. (2004) *Barks in the Forest* **New Scientist**, 10 July, p.28.

Ridley, M. (1993) *Is Sex Good for Anything?* **New Scientist**, 4 December, pp. 36-40.

Ring, K. (1982) *Life at Death: A Scientific Investigation of the Near-Death Experience* Quill, New York.

Ring, K. (2001) *Mindsight: Eyeless Vision in the Blind* In Lorimer (Ed), pp.59-70.

Robinson, D.N. (1990) *Wisdom Through the Ages* In Sternberg, R.J.

(Ed), pp.13-24.

Russell, B. (1957) *Why I am not a Christian* Allen & Unwin, New York, 225p.

Russell, P. (1988) *Awakening Earth: Global Brain* Arkana, London, 240p.

Russell, P. (2001) *The Light of consciousness* **Scientific and Medical Network Review**, No.76, pp2-3.

Saborn, M. (1982) *Recollections of Death: A Medical Investigation of the Near-Death Experience* Corgi, London.

Savit, R. (1990) *Chaos on the Trading Floor* **New Scientist**, 11 August, pp. 36-39.

Schilling, G. (2004) *Shadow over Gravity* **New Scientist**, 27 November, pp.25-31.

Schmidt, K. (1994) *Proofreaders for the Code of Life* **New Scientist**, 6 August, pp. 23-26.

Searle, J. R. (1992) *The Rediscovery of the Mind* MIT Press, Cambridge, Mass., 270p.

Shea, S. and Underwood, R. (1990) *Forests for the Future* **Landscope**, Vol. 6, No. 1, pp.35-43.

Sheldrake, R. (1988) *The Presence of the Past* Fontana, London, 391p.

Sheldrake, R. (1990) *The Rebirth of Nature* Rider, London, 215p.

Sheldrake, R. (2003) *Minds Beyond Brains: New Experimental Evidence* **Scientific and Medical Network Review**, No. 82, pp.10-12.

Singer, E. (2004) *Life Force* **New Scientist**, 4 December, pp.46-49.

Siu, R. G. H. (1957) *The Tao of Science* MIT Press, Cambridge, 180p.

Small, M. F. (1994) *Ay Up, a Chimp wi' an Accent* **New Scientist**, 4 June, pp.33-37.

Smith, D. (2003) *Bacteria Fooled by a Secret Life of Plants* **Sydney Morning Herald**, 21 August. {http://www.smh.com.au/articles/2003/08/20/1061368355182.html?from=storyrhs&oneclick=true}

Sogyal, Rinpoche (1992) *The Tibetan Book of Living and Dying* Harper, San Francisco, 425p.

Sommerhoff, G. (1981) *The Abstract Characteristics of Living Systems* In Emery (Ed), Vol. 2, p. 145-188.

Stein, B. (1996) *It Takes Two to Tangle* **New Scientist**, 28 September, pp. 26-31.

Sternberg, R.J. (Ed) (1990) *Wisdom: Its Nature, Origins, and Development* Cambridge University Press, Cambridge, 339p.

Stewart, I. (2003) *Never Ending Story* **New Scientist**, 27 September, pp.28-33.

Stewart, I. (2003a) *How the Species Became* **New Scientist**, 11 October, pp.32-35.

Susskind, L. (2003) *A Universe Like No Other* **New Scientist**, 1 November, pp.34-41.

Symonds, N. (1991) *A Fitter Theory of Evolution?* **New Scientist**, 21 September, pp. 20-24.

Tart, C. (2001) *Enlightenment and Endarkenment* In Lorimer (Ed), pp.125-149.

Thich Nhat Hanh (1987) *Being Peace* Rider, London, 115p.

Thompson, D. W. (1917) *On Growth and Form* Cambridge University Press, Cambridge.

Thorne, K. S. (1994) *Black Holes and Time Warps: Einstein's Outrageous Legacy* W. W. Norton, New York, 619p.

Tompkins, P. and Bird, C. (1973) *The Secret Life of Plants* Harper Collins, New York, 402p.

Toulmin, S. (1982) *The Return to Cosmology: Postmodern Science and the Theology of Nature* University of California Press, Berkeley, 283p.

Tritton, D. (1986) *Chaos in the Swing of a Pendulum* **New Scientist**, 24 July, pp.37-40.

Tudge, C. (1994) *Going Bats over Conservation* **New Scientist**, 12 March, pp.27-31.

Vines, G. (1994 a) *The Emotional Chicken* **New Scientist**, 22 Jan., pp. 28-31.

von Bertalanffy, L. (1968) *General System Theory: Foundations, Development, Applications* George Braziller, New York, 289p.

Waddington, C. (1977) *Tools for Thought* Jonathon Cape, London, 250p.

Walker, A. (1994) *The Kingdom Within: A Guide to the Spiritual Work of the Findhorn Community* Findhorn Press, Findhorn, 415p.

Wallace, B.A. (2001) *The Potential of Emptiness* **Scientific and Medical Network Review**, No.77, pp.21-24.

Wallace, B.A. (2002) *The Scientific Frontier of the Inner Spirit* **Scientific and Medical Network Review**, No.80, pp.18-19.

Wallace, T. M. and Vogel, S. (1994) *El Nino and Climate Prediction* University Consortium for Atmospheric Research, Boulder, Reports to the Nation on our Changing Planet, No. 3, pp.6-8.

Walsh, R.N. and Vaughan, F. (eds) (1980) *Beyond Ego: Transpersonal Dimensions in Psychology* Jeremy P. Tarcher, Los Angeles, 272p.

Weatherall, D. (2003) *Evolving with the Enemy* **New Scientist**, 22

November, pp.44-47.

Wertheim, M. (1995) *The Way of Logic* **New Scientist**, 2 December, pp. 38-41.

Whitmore, D. (1990) *The Joy of Learning* Crucible Books, UK.

Wilber, K. (1979) *No Boundary: Eastern and Western Approaches to Personal Growth* Shambala Publications, Boston, 164p.

Wilber, K. (1991) *Grace and Grit* Shambala, Boston, 422p.

Wilber, K. (1993) *The Spectrum of Consciousness* 20th Anniversary Ed., Quest Books, Wheaton, 362p.

Wilber, K. (1995) *Sex, Ecology, Spirituality: The Spirit of Evolution* Shambala, Boston, 831p.

Wilber, K. (2000) *A Brief History of Everything* 2nd Edition, Gateway, Dublin, 330p.

Wills, C. (2003) *The Trouble with Sex* **New Scientist**, 6 December, pp.44-47.

Wise, N. (1990) *Rock Body Tree Limb* In Badiner (Ed), pp.99-101.

Woodhouse, M. (2001) *Consciousness and Energy Monism* In Lorimer (Ed), pp.231-246.

Woolger, R.J. (1987) *Other Lives, Other Selves* Thorsons, London, 380p.

Ziman, J. (1978) *Reliable Knowledge: An Exploration of the Grounds for Belief in Science* Cambridge University Press, Cambridge, 197p.

Zimmermann, M. (1989) *Human Physiology* Springer-Verlag, Berlin

Zohar, D. and Marshall, I. (2000) *SQ Spiritual Intelligence: The Ultimate Intelligence* Bloomsbury, London, 324p.

Index

O

is a symbol of the world,
of oneness and unity. O Books
explores the many paths of wholeness and spiritual understanding which
different traditions have developed down
the ages. It aims to bring this knowledge in
accessible form, to a general readership, providing practical spirituality to today's seekers.

For the full list of over 200 titles covering:

ACADEMIC/THEOLOGY • ANGELS • ASTROLOGY/
NUMEROLOGY • BIOGRAPHY/AUTOBIOGRAPHY
• BUDDHISM/ENLIGHTENMENT • BUSINESS/LEADERSHIP/
WISDOM • CELTIC/DRUID/PAGAN • CHANNELLING
• CHRISTIANITY; EARLY • CHRISTIANITY; TRADITIONAL
• CHRISTIANITY; PROGRESSIVE • CHRISTIANITY;
DEVOTIONAL • CHILDREN'S SPIRITUALITY • CHILDREN'S
BIBLE STORIES • CHILDREN'S BOARD/NOVELTY • CREATIVE
SPIRITUALITY • CURRENT AFFAIRS/RELIGIOUS • ECONOMY/
POLITICS/SUSTAINABILITY • ENVIRONMENT/EARTH
• FICTION • GODDESS/FEMININE • HEALTH/FITNESS
• HEALING/REIKI • HINDUISM/ADVAITA/VEDANTA
• HISTORY/ARCHAEOLOGY • HOLISTIC SPIRITUALITY
• INTERFAITH/ECUMENICAL • ISLAM/SUFISM
• JUDAISM/CHRISTIANITY • MEDITATION/PRAYER
• MYSTERY/PARANORMAL • MYSTICISM • MYTHS
• POETRY • RELATIONSHIPS/LOVE • RELIGION/
PHILOSOPHY • SCHOOL TITLES • SCIENCE/
RELIGION • SELF-HELP/PSYCHOLOGY
• SPIRITUAL SEARCH • WORLD
RELIGIONS/SCRIPTURES • YOGA

**Please visit our website,
www.O-books.net**